THE OTHER DOOLITTLE RAID

THE OTHER DOOLITTLE RAID

THE GENESIS OF A WORLD WAR II BOMBER GROUP

EDWARD CLENDENIN

Copyright © 2017—Edward Clendenin

ALL RIGHTS RESERVED—No part of this book may be reproduced in any form or by any electronic or mechanical means, including information storage and retrieval systems, without permission in writing from the authors, except by a reviewer who may quote brief passages in a review.

Published by Deeds Publishing in Athens, GA
www.deedspublishing.com

Printed in The United States of America

Cover design by Mark Babcock. Text layout by Mark Babcock & Matt King.
Cover photo courtesy of Thomas Shumaker via Al Blue.

Library of Congress Cataloging-in-Publications data is available upon request.

ISBN 978-1-944193-75-1

Books are available in quantity for promotional or premium use. For information, email info@deedspublishing.com.

First Edition, 2017

10 9 8 7 6 5 4 3 2 1

In 1989, my father and I were on a tour of the Air Force Muesum in Dayton Ohio. As we approached the *Strawberry Bitch*, the B-24 that is on display, my father commented that he flew that plane. I made some comment like "Yes, I know. You were a pilot and flew B-24's out of Italy."

"That is not what I meant," he said. "I flew THAT airplane."

Thus began my search to confirm his statement. He rarely talked about his experiences during the war. His flight log did not record any details of the aircraft he flew. It was not until I visited the Air Force Historical Research Association at Maxwell AFB in Montgomery, Alabama that my search began to bear fruit. It was there that I found 15 rolls of microfilm, recording much of the history of the 376th Bomb Group. I was futher fortunate in that the clerk for the 512th Bomb Squadron, the unit that my father was in, began to record aircraft tail numbers on the mission reports filed by the crews in early 1944.

On February 16, 1944, Dad piloted aircraft tail number 42-72843, to the Prato, Italy Marshalling Yards. Then, on April 7, 1944, he flew that same plane to Mestre, Italy. That aircraft, group number 24, was the *Strawberry Bitch*.

This work is dedicated to my Dad and all of the men who served in HALPRO, the 1st Provisional Bomb Group, and the 376th Heavy Bombardment Group.

TABLE OF CONTENTS

INTRODUCTION 1

PRE-WAR PLANNING 3
    The First Proposal to Bomb Japan 5
    The Second Proposal to Bomb Japan 7
    A Change in Strategy 8
    A Third Proposal to Bomb Japan 14
    Defense of the Philippines 14

THE ATTACK ON PEARL HARBOR 23

PLANNING FOR ACTION 29
    December 1941 31
    Januray 1942 73
    February 1942 119
    March 1942 155
    April 1942 183
    May 1942 213

REALITY OVERCOMES THE BEST LAID PLANS 229
    May 1942 231
    June 1942 257
    The Fall of Tobruk 278

WE ARE ALL IN 289
    June 1942 291
    July 1942 305
    August 1942 339
    September 1942 357
    October 1942 369
    November 1942 381

December 1942	397
A SUMMARY OF 1942	405
CONCLUSION	409
APPENDICES	411
APPENDIX A — INTERVIEWS	413
Brigadier General Merian Cooper	413
Brigadier General Eugene Beebe	415
APPENDIX B — INDIVIDUAL STORIES	419
June 12, 1942 Mission to Ploesti	419
The "Hijacking" of *Brooklyn Rambler*	443
June 15, 1942 Mission Against the Italian Fleet	444
June 21–22, 1942 Night Mission to Benghazi	450
APPENDIX C — HALPRO CREWS & PLANES	455
APPENDIX D — ORAL HISTORIES	461
REFERENCES	463
INDEX	475

INTRODUCTION

In 2001, Tombstone Pictures released the movie *Pearl Harbor*. Following the depiction of the Japanese attack on the United States Army and Navy bases on the island of Oahu, there is a scene showing a meeting of President Franklin D. Roosevelt with his military and civilian advisors. After getting a status report, he asks what is being done to respond. When he is told that little can be done, Roosevelt demands that the United States bomb Japan. Following Admiral Ernest King's declaration that such a mission is not possible, Roosevelt dramatically stands without the use of his crutches and says: "Do not tell me it cannot be done."[1]

The audience is left with the impression that:

1. this is the first time Roosevelt discussed the bombing of Japan,
2. that Roosevelt proposed the bombing of Japan,
3. the military considered the mission impossible,
4. the Doolittle raid was the only mission that resulted from that meeting.

All of those conclusions are incorrect.

Roosevelt's not so innocent request would cause ripple effects throughout the military planning of the United States and Great Britain. Two hundred and twelve men and 23 B-24D heavy bombers would form the nucleus of an Army Air Corps unit called the Halverson Provisional Group, HALPRO for short. They would be assembled at the Fort Myers, Florida Air Base by May 1, 1942. By the end of the month, they had departed the continental United States. They were about to enter into the political and military vortex of 1942. And not all of these ripples were caused by the enemy.

A few comments by the author are warranted. Decisions are made in real-time. What one person considers important, another person might consider irrelevant. In the traditional historical analysis, the author often makes the editorial decision about what facts to include and what facts to exclude. This author does not claim insight into a decision maker's thought process. When actions seem to be in conflict with the "known facts," then one needs to reflect on what else was occurring.

Immediately after the Japanese attack on Pearl Harbor and for the first few months of the war, Roosevelt and his military and political staffs seemed to be in a constant state of flux. Decisions made were often inexplicably changed before they could be implemented. Assuming the decision makers were not irrational, then the explanation for these changes must be found elsewhere. It is the intent of this work to offer other possible explanations, especially as those decisions that would impact the lives of the men of HALPRO.

Thereore, events are presented in chronological order. An attempt is made to assign each event to an overall strategic objective. Since the American public needed a lesson in geography, maps often accompanied newspaper articles about the battles and strategic objectives. Newspapers used their front-page headlines to convey important events. The use of banner headlines[2] indicated importance.

ENDNOTES

1. Touchstone Pictures, "*Pearl Harbor*", 2001.
2. banner headlines: a large title of a story in a newspaper that stretches across the top of the front page, Cambridge Dictionary Online, http://dictionary.cambridge.org/us/dictionary/english/banner-headline, accessed August 29, 2015.

PRE WAR PLANNING

THE FIRST PROPOSAL TO BOMB JAPAN

Dealing with Japan was not the primary concern of most of the American and British leadership. But there were those in the Roosevelt Administration who opined that China was worth saving and Japan's aggression needed curtailing. The United States had nearly four decades of experience planning for a war with Japan. The plan for a war in Europe was less than a year old. American planning would follow a torturous path putting military objectives at cross-purposes with political hawks.

1940 NOVEMBER 30

In 1937, Claire Chennault went to China to assist the Chinese in the development of an effective air force to fight the Japanese. In July of that year, the Second Sino-Japanese War broke out. Although the conflict had reached a stalemate by 1939, the Chinese Air Force was essentially destroyed. The Chinese needed to be re-armed and its pilots trained.

In late November 1940, Chennault proposed that if China were given a 500-plane force piloted, supplied, and maintained by the United States, the Chinese could virtually annihilate the Japanese forces within China and neutralize Japan's naval striking ability. This "Special Air Unit," operating from bases in China only 650 miles from Tokyo, "could operate independently in attacking Japan proper."[1] The memo pointed out that there were 136 airfields in China; over half would provide an excellent base for operations. T.V. Soong, Generalissimo Chiang Kai-shek's personal representative in Washington D.C. (and his brother in law), presented the proposal to Treasury Secretary Henry Morgenthau on November 30.[2]

Since the supply of any planes to China would entail a diversion of planes currently slated to go to Britain, Morgenthau discussed the Chinese proposal with the British Ambassador to the US, Philip Henry Kerr (Lord Lothian) on December 3. Morgenthau told Lothian that he was going to get four-engine bombers to the Chinese, "with the understanding that these bombers are to be used to bomb Tokyo and other big cities."[3]

Lothian readily agreed.

The next day, the State Department told the Chinese embassy that it was changing its policy regarding aid to China; in particular, military aid. Plus, the United States would no longer oppose Americans from volunteering for service to the Chinese forces as mercenaries.[4]

1940 DECEMBER 8

On December 8, Morgenthau discussed the idea with Roosevelt, who said ". . . it would be nice if the Chinese would bomb Japan." Later that day, he had lunch with Soong. He told Soong that the grant of anything approaching 500 planes was impossible, though a substantial number might become available sometime in 1942. Meanwhile, ". . . what did he [Soong] think of the idea of some long range bombers with the understanding that they were to be used to bomb Tokyo and other Japanese cities?"

Soong was beside himself with joy.[5] Morgenthau asked for additional information regarding possible airfields. The next day, Soong sent him a map of Chinese airfields currently under control of Chiang's forces.[6]

1940 DECEMBER 12

On December 12, Morgenthau discussed the idea with Secretary of State Cordell Hull. Hull surprised Morgenthau by declaring enthusiastic support for sending 500 American planes. It would be even better if they could "start from the Aleutians and fly over Japan just once."

An added plus would be "if we could only find some way to have them drop some bombs on Tokyo."[7]

On the same day, Chiang cabled Roosevelt asking for a quick approval of the plan. Chiang repeated his request four days later in a cable to Morgenthau.[8] Morgenthau told Roosevelt that he had received a communiqué from Chiang that he wanted to attack Japan. At a meeting on December 18, Roosevelt asked if Chiang wanted to fight. When Morgenthau said that subject was what the message was about, Roosevelt said, "That's what I have been talking about for four years."[9]

Roosevelt asked Morgenthau to put finishing details to the plan.[10]

At a Cabinet meeting the next day, the plan was discussed and approved. Following the Cabinet meeting, a second meeting was held with Roosevelt, Morgenthau, Hull, Secretary of War Henry L. Stimson, and Secretary of the Navy William Franklin "Frank" Knox. All agreed it was a great plan. After the meeting Morgenthau told Soong that he needed to find someone who could fly four-engine bombers. Discussions continued the next day with all the parties, with Chennault brought in for additional input. Chennault preferred the B-17 to his original requested Lockheed Hudson, since it had a greater range. He believed, that with fighter escort, the bombers could attack Tokyo, Nagasaki, Kobe, and Osaka.[11]

Somehow, Chief of Staff General George C. Marshall got wind that plans were being made to divert planes from Britain to China. At a meeting on December 22 with Morgenthau, Knox, and Stimson, he voiced vigorous opposition to any such diversion. As an alternative, he did agree to send 100 P-40 fighters.[12] A second meeting was held the next day with Hull and Admiral Harold R. Stark, Chief of Naval Operations (CNO), joining the group. Again, Marshall voiced opposition to diverting the bombers and agreed to send fighters. The group adjourned to communicate their proposal to Roosevelt. Roosevelt accepted the revised recommendation and told Morgenthau to break the news to Soong and Chennault.[13]

1941 JANUARY 1

Ten days later, on January 1, Morgenthau had to explain to Soong and Chennault that the Army had killed the idea of sending long-range bombers to China. As a consolation, the Chinese would receive 100 P-40s. The "Flying Tigers" were about to become operational.[14]

Marshall's opposition was enough to kill the project.

THE SECOND PROPOSAL TO BOMB JAPAN

1941 MAY 10

Five months after Chennault had submitted his request for American bombers, Lauchlin Currie, an economic adviser to Roosevelt, reviewed the proposal and sent it to the Joint Aircraft Committee. They, in turn, forwarded it to Marshall.[15] Currie amended the proposal, reducing the request to 250 planes. This revision was then sent to Knox.[16]

Concerns over a Japanese move into French Indochina spurred Currie to repeat his request on June 11.[17]

The Joint Aircraft Committee approved the plan on July 9. The full Board approved the plan three days later, noting that Americans flying for the Chinese would have an opportunity for the "incendiary bombing of Japan."[18]

1941 JULY 18

Roosevelt was officially notified of the Committee's approval on July 18. He approved it on July 23. Currie explained that sixty-six bombers were to be made available to China before the end of the year. Twenty-four were to be sent immediately and, in addition, the United States would begin training Chinese pilots.[19]

The bombers were never delivered.

A CHANGE IN STRATEGY

1941 APRIL

In early April 1941, Army Air Corps officials decided to begin sending B-17 heavy bombers to the Hawaiian Islands. This decision represented a major deviation from the plans already drawn up, namely American-British-Canadian-1 (ABC-1), ABC-2, and RAINBOW-5. These earlier plans emphasized the Germany First strategy. More importantly, such a transfer represented the first non-stop flight of a military aircraft from the United States mainland to the Islands. Fearing a public outcry if the mission failed, extensive plans were made to gather weather data, establish links with commercial radio stations, and coordinate with the Navy and Pan American Airways. The flight of 21 B-17s occurred without incident.[20]

1941 JUNE 7

While plans were being made to ferry bombers to Hawaii, American, Dutch, and British military commanders were ordered to have joint strategy sessions, similar to the ABC conferences. Their conclusion, known as American-British-Dutch-1 (ABD-1), oozed with confidence in the ability of their combined forces to stop Japanese aggression. They even went so far as to suggest that heavy bombers, based in Luzon, could reach some Japanese cities. Following the theories Giulio Douhet first proposed in his influential book *Command of the Air*, they argued that bombing would cause economic pressure on the average Japanese citizen, who, in turn, would cause the end of hostilities.[21] After reviewing this report, on June 7, Stark and Marshall rejected the study as being unrealistic, writing that:

> *The United States intends to* adhere to the decision not to reinforce the Philippines except in minor particulars.
>
> The principal value of the position and present strength of the forces in the Philippines lies in the fact that to defeat them will require a considerable effort by Japan and may well entail a delay in the development of an attack against Singapore and the Netherlands East Indies. A Japanese attack in the Philippines might thus offer opportunities to the Associated Powers to inflict important loss-

es on Japanese naval forces and to improve their own dispositions for the defense of the Malay Barrier.[22]

Marshall explained his position in his Biennial Report to the Secretary of War for the period of 1941 to 1943. Realizing that this was written nearly two years after the fact, nonetheless, Marshall wrote:

> *Deficiencies in arms and equipment* especially in ammunition and airplanes required for the immediate defense of the Western Hemisphere, the Panama Canal Zone, Alaska, and for the Regular Army and National Guard with supporting troops, were so serious that adequate reinforcements for the Philippines at this time would have left the United Sates in a position of great peril should there be a break in the defenses of Great Britain. It was not until new troops had been trained and equipped and Flying Fortresses, fighter planes, tanks, guns, and small-arms ammunition began to come off assembly lines on a partial quantity production basis in the late summer of 1941 that reinforcements for our most distant outpost could be provided without jeopardy to continental United States.[23]

1941 JULY 17

From this time forward, a change in both personnel and strategy would occur. What exactly transpired in conference rooms at the War Department is not clear. What is known is that on July 17, General Leonard T. Gerow, Chief of the War Plans Division, wrote a memo recommending steps to be taken for the improved defense of the Philippines. One of those steps was to recall General Douglas MacArthur to active duty.[24]

1941 JULY 18

Following the release of his July 17 memo, Gerow released a second memo on July 18. In it, he recommended sending four heavy bomber groups to the Far East, consisting of 272 planes. General Henry 'Hap' Arnold, Chief of the Army Air Corps, forwarded Gerow's study to Marshall the next day.[25] If such a recommendation was enacted, it represented a change in the RAINBOW-5 planning. Now, both the United States Navy and the Army Air Corps would be responsible for thwarting Japanese aggression.

1941 JULY

The United States Army was not the only military branch concerned about Japan. Since the formulation of ABC-1, the Navy knew they would be on point for any action against the Japanese. So when talk circulated in the Roosevelt administration of an oil embargo against Japan, the Navy's planning department, under the direction of Admiral

Richmond K. Turner, conducted its own study, entitled "Study of the Effect of an Embargo of Trade between the United States and Japan." In it, the preparers wrote:

> *It is generally believed that* shutting off the American supply of petroleum will lead promptly to an invasion of the Netherlands East Indies. While probable, this is not necessarily a sure immediate result ... Japan has oil stocks for about eighteen months war operations. Export restrictions of oil by the United States should be accompanied by similar restrictions by the British and Dutch ... An embargo on exports will have an immediate severe psychological reaction in Japan against the United States. It is almost certain to intensify the determination of those now in power to continue their present course. Furthermore, it seems certain that, if Japan should then take military measures against the British and Dutch, she would also include military action against the Philippines, which would immediately involve us in a Pacific war ... An embargo would probably result in a fairly early attack by Japan on Malaya and the Netherlands East Indies, and possibly would involve the United States in early war in the Pacific ... Recommendation: That trade with Japan NOT be embargoed at this time.

Stark forwarded his copy of the report to Roosevelt, writing on the report "I concur in general..."[26]

1941 JULY 26

MacArthur had gone to the Philippines in 1935 as United States Military advisor to the Philippine government. At the time, Philippine independence was planned for 1946. In February 1941, MacArthur sent a personal letter to Marshall, outlining his plans for the defense of the Philippines. While War Plan ORANGE envisioned the defense of Manila Harbor, MacArthur proposed a much more expansive island defense. One should remember that, at this time, MacArthur was not the commander of United States Army forces in the Philippines. On July 26, MacArthur was named Commanding General, United States Army Forces in the Far East.[27]

On the same day, two other actions took place:

1. Roosevelt issued an Executive order freezing Japanese assets in the United States,[28]
2. The British added insult to injury by cutting off oil supplies to Japan. And on July 27, so did the Dutch.[29]

1941 JULY 28

Meanwhile, back in Manila, MacArthur was informed that there would be no reinforcements for the Philippine Forces.[30] So far, the policy of the United States as it pertained to the Philippines was consistent. But this is also where the timeline seemed fragmented. Since February, MacArthur had been advising Marshall that the United States needed to rethink its Philippine strategy. And Gerow had released a report ten days earlier, seemingly in support of MacArthur's recommendation. Yet Marshall told MacArthur to forget about a new plan. What he had on hand was all he was going to get.

1941 JULY 31

But then, the inexplicable happened. Three days later, on July 31, two separate decisions were made by the United States. First, at a Cabinet meeting, Roosevelt agreed to limit oil exports to Japan to 1935/1936 levels.[31] Second, Marshall approved a War Plans Department (WPD) plan to reinforce the Philippine Islands' defense "in view of the possibility of attack" and telegrammed MacArthur of this change. Marshall informed his immediate staff, "It was the policy of the United States to defend the Philippines." Gerow was so surprised by this statement that he entered it in his office diary. MacArthur was told that he would initially receive a squadron of B-17s, with more to follow. MacArthur would not receive his complete bomber force until the spring of 1942.

Why this change in strategy occurred has never been fully explained. One can understand sending scarce heavy bombers to Hawaii as the Islands served as the homeport of the Pacific Fleet. But the sending of bombers to the Philippines was inexcusable. One rationale was that both United States politicians and military planners had bet the future on the ability of the heavy bomber. Apparently, it was believed that heavy bombers could intercept any invasion force while that force was still at sea. The only weakness in the argument was that there were no heavy bombers in the Philippines.

1941 AUGUST 1

On August 1, 1941, the United States joined the British and Dutch oil embargo.[32]

1941 AUGUST 14

Seeing a need for increased defenses in Hawaii, no doubt reflecting the realization that the odds of war with Japan had greatly increased, the Army Air Corps was ordered to study the "air situation in Hawaii." The plan was released on August 14. Entitled "A Plan for the Employment of Bombardment Aviation in the Defense of Oahu," it assumed that the most likely attack would be in the early morning from carrier-based aircraft. It called for 72 B-17s to conduct radial search patterns around the Islands, stretching out 833 miles. Since 72 B-17s represented more bombers than were in the entire Air Corps inventory, the plan was more theoretical than practical.[33]

The Air Corps now had two competing plans with insufficient resources to perform either one of them. While much has been written about the massive war production accomplishments of the United States, such a statement was not true in the summer of 1941, especially as it pertained to heavy bomber production. By the end of 1940, only 134 B-17s had been made, and less than ten B-24s had been manufactured.

Why the Air Corps seemed to place its reliance on an unproven weapon system seems a logical question in hindsight. At the time however, Washington was receiving glowing reports on the success of the B-17s that had been transferred to the Royal Air Force (RAF). They were operating against German targets.[34] Stimson was reading of the "sudden and startling success" of the B-17. The reality, as it turned out, was something else entirely.[35] Why Stimson was receiving reports contradictory to those being filed by the RAF is another mystery surrounding pre-war activities. According to reports Arnold was receiving, the RAF was disappointed in the performance of the B-17. The RAF would withdraw the B-17 from service at the end of the month. To rationalize the RAF decision, the Air Corps blamed the poor showing on the users rather than the system.[36]

1941 SEPTEMBER 5

Apparently the defense of the Philippines outranked the defense of the Pacific fleet. On September 5, the first unit of B-17s promised to MacArthur, the 14th Squadron of the 19th Bombardment Group, departed Hawaii. Just three months earlier, B-17s had flown the 2400-mile distance from California to Hawaii, a record mission at that time. Now, B-17s were embarking on a 10,000-mile journey to Clark Field in the Philippines.[37]

1941 SEPTEMBER 12

On September 12, the nine B-17s that had left Hawaii on September 5 arrived at Clark Field. When this was reported to Stimson, he wrote

> *The four-engined bomber completely changed* the strategy of the Pacific, ... demonstrating our power by air ... and again [the United States would] be in a position to exert that power in the South West Pacific.[38]

1941 OCTOBER 1

After a couple of months of command, MacArthur gave Washington his assessment of the situation in the Philippines. He estimated that, ultimately, he would need 200,000 men trained and organized into divisions. He would also need a strengthened air force. He admitted that it would be impossible to stop the Japanese from capturing Manila if they were able to secure a beachhead on any of the southern islands. Given the multitude of landing options, the "citadel type defense" concept provided for in the ORANGE and RAINBOW plans would not work. The defense of all of the islands was required: "The strength and composition of the defense forces projected here are believed to be sufficient to accomplish such a mission."[39]

In essence, MacArthur had agreed that the Philippines were not defensible without the future commitment of additional forces.

1941 OCTOBER 3

Responding to MacArthur's communiqué, Marshall continued to feed MacArthur's desire for more reinforcements. On October 3, he cabled MacArthur that MacArthur could expect 136 first line bombers with 34 in reserve.[40] However, he would not receive them in the immediate future.

1941 OCTOBER 8

American war planners continued to believe that the United States was pursuing a sound military strategy in the Far East. The Philippines lent itself to an ideal air campaign. Regardless of which direction Japan decided to attack, Air Corps units in the Philippines would be a thorn in the side of those plans. If the Japanese continued to drive through Indochina, Clark Field, the Philippines, was a perfect base from which to harass both Japanese ground and naval units. If the Japanese went east of the Philippines, there was the United States Pacific Fleet. And if Japan struck north into Russia, these units could strike Japanese bases from the rear.

But what about a direct assault on the Philippines itself? The answer was equally as simple. First, the terrain was rugged and offered few invasion points. Second, since Japanese land-based aircraft could not reach the Philippines, any air cover for an invasion had to come from carrier-based planes. And since carrier based planes were inferior to land-based planes, the advantage was with the Air Corps. Plus, the B-17s could harass the carriers and the troop transport ships long before they were within range of the Islands.[41] This was the strategy of the WPD on October 8.

With the increasing emphasis on the Philippine Island strategy, the United States needed a qualified person in charge of MacArthur's air force. Marshall told Major General Lewis H. Brereton on October 3 that he would be sent to Philippines to become the Commanding General. On the same day that the WPD issued the above study, Brereton made an interesting notation in his diary. He repeated Hull's October 2 negotiating points and then wrote the following:

> *The announcement of these points* as a basis for amicable settlement of the Japanese-American differences was considered in some quarters as a virtual ultimatum. The State Department, however, did not consider it as such, and still hoped to avoid war with Japan.
>
> In any event it was the opinion of the War Department that hostilities, if and when they came, would not begin before 1 April 1942. Our plans for the defense of the Philippines contemplated that by April 1942 the complete air reinforcements would be in place, and in addition that very considerable land reinforcements requested by General MacArthur would be in the Philippines.[42]

A THIRD PROPOSAL TO BOMB JAPAN!

1941 OCTOBER 21

On October 21, 1941, Stimson responded to Roosevelt's query about "the proper strategic distribution" of the new B-17 and B-24 bombers.

In his response, Stimson noted the buildup of an air force in the southwest Pacific, commenting that this was another opportunity to contain Japan:

> *That locality [Vladivostok] can possibly* form the base of a northern pincer movement of American influence and power, this time not only to protect against aggression of Japan, but to preserve the defensive power of Russia in Europe. Its operation would fit into and supplement the operation from the south by permitting a circular sweep of these [four-engine] bombers which would greatly increase their safety by permitting those in the south, after passing over Japan and stopping at Vladivostok to proceed to safety in the north in a way similar to the sweeps which Germany is now employing through the North Atlantic from Norway to France. The power of such a completed north and south operation can hardly be over-estimated.[43] (see Figure 1)

DEFENSE OF THE PHILIPPINES

1941 OCTOBER

Wilbur Mayhew wrote about the events in October:

In October 1941 The 7th Bombardment Group and the 88th Reconnaissance Squadron were alerted to prepare for two year's service overseas. The 7th and 88th were scheduled to go to Del Monte Airfield on the island of Mindanao, although at the time all we knew was we were sailing to "PLUM". At this time the 7th and 19th Bombardment Groups were the only combat ready heavy bombardment groups in the Army Air Force. (I hate to think what the others were like.). These two groups had the majority of B-17s in the Air Force (only 87 in the AAF then). The 19th Group had preceded the 7th to the Philippines.[44]

1941 NOVEMBER 4

By November 4, the 30th and 93rd Squadrons of the 19th Bombardment Group had transitioned to Clark Field. Plans called for 165 B-17s to be at Clark by March 1942. For comparison, the entire United States production of B-17s and B-24s was scheduled to be 200 planes.[45]

1941 NOVEMBER 5

The Joint Board reminded Roosevelt that the air defenses of the Philippines would not be at a significant level until the spring of 1942.[46]

On November 10, at least one squadron of the 19th Bombardment Group was ordered to be on alert at all times. They needed to be able to conduct either reconnaissance or bombing missions. Four days later, the Air Staff concluded that current plans for the build-up of MacArthur's Far East Air Force were still lacking. They proposed shipping all "modernized" B-17s to MacArthur. The next day, they broadened the proposal to include every B-24 now coming off the line. If this plan had been implemented, only 17 B-17s would have been left stateside. The Air Staff even considered sending the 12 B-17s currently in Hawaii to the Philippines. Hawaii and the Pacific fleet were now considered of secondary importance.[47]

1941 NOVEMBER 10

After assuming command, MacArthur had written a letter to Washington on October 1 stating the planning assumptions of ORANGE and RAINBOW were obsolete. Those plans needed to be altered. On November 10, the Joint Board revised RAINBOW-5. In the original version, RAINBOW-5 envisioned a defense posture in the Philippines. With the current planned influx of aircraft, RAINBOW-5 now needed to be rewritten to include offensive operations. In the event of war, the USFEAF was to conduct "air raids against Japanese forces and installations within tactical operating radius of available bases."

This change was forwarded to General Brereton on November 21.[48]

1941 NOVEMBER 16

In light of the change to RAINBOW-5, a reorganization of the air forces in the Philippines was ordered. First, the United States Far East Air Force, USFEAF, was established under the command of General Brereton. Second, the bombers were placed into the V Bomber Command, under Lt. Col. Eugene L. Eubank. Two additional commands were also established: an Interceptor Command and a Service Command.[49]

1941 NOVEMBER 21

On November 21, the WPD, in a memo to Hull, reiterated that the United States was not militarily ready to stop any Japanese aggression in the Far East. They emphasized the "grave importance . . . that we reach a *modus vivendi*"[50] with Japan.[51] MacArthur was not sitting idly by. He had asked one of his aides, Major Laurence S. Kuter, to evaluate the air plans for his Far East Air Force. Kuter had his doubts, which he expressed in a report filed on the same day. In particular, he concluded that:

1. Japanese targets were not as flammable as thought, and
2. if the WPD thought it would take 3000 bombers to destroy 154 German targets, what could 200 or so bombers hope to accomplish?

Little did he know that, four years later, 1,000 B-29s, a more advanced bomber than either the B-17 or the B-24, could not bring Japan to her knees.[52] And while the combined bombers of the 8th and 15th Air Forces did destroy much of the German infrastructure, they could not stop German production of fighter planes nor force a German surrender.

Also, Admiral Thomas C. Hart was ordered to work with his British counterparts to devise a plan for point naval operations. The Dutch were also to be consulted with the desire to have a three-power plan. Similar orders were given to General Brereton.[53]

1941 NOVEMBER 22

A sense of desperation can be seen in a recommendation sent by General Carl A. Spaatz to Marshall on November 22. Spaatz wanted every bomber, destined for the Philippines, to be in transit by December 6.[54]

1941 NOVEMBER 26

Marshall convened a meeting to discuss possible Japanese options. Marshall thought that Japan would not attack the Philippines, as he, Marshall, thought it to dangerous a venture. Plus, the Japanese were too afraid of the B-17.[55]

1941 NOVEMBER 28

Both MacArthur and his naval counterpart, Hart, had long-range aircraft that could perform search activities. On November 28, following discussions between the respective air staffs, MacArthur cabled Marshall that the two services had reached an agreement to cooperate rather than overlap search patterns. MacArthur's B-17s with their longer range and higher ceiling was best suited to search towards Formosa. Hart's PBYs would search the southern approaches. Once implemented, these flights revealed large concentrations of troop transports and cargo ships both at anchor in harbors and at sea.[56]

1941 NOVEMBER 29

Meanwhile, General Brereton and his staff were scrambling to find dispersal sites for the B-17s. Clark Field, the Philippines, was the only field capable of handling the bombers. They sought permission to build an air base at Mindanao, southernmost of the large islands in the Philippines. Unfortunately, there were no plans to provide for ground troops to guard any such base. Despite this objection, permission was granted on November 29.[57] MacArthur cabled Marshall of his decision, explaining that keeping the bombers at Clark Field was asking for trouble and stating that

> *The location of potential enemy* fields and types of aircraft indicate that heavy bombers should be located south of Luzon where they would be reasonably safe from attack.[58]

1941 NOVEMBER 30

Between February 1 and November 30, 1941, the Army Air Corps accepted delivery of 210 B-17s and B-24s. Of these, 113 were sold to foreign countries, leaving 97 for use by the entire Army Air Corps. On November 30, the Hawaiian Air Force had 12 B-17s; MacArthur, 35.[59]

1941 DECEMBER 1

Still, there were not enough bombers in the Philippines. Arnold cabled General Frederick L. Martin, commander of the Hawaiian Air Force, that he, Martin, needed to get his bombers to MacArthur quickly.[60]

Roosevelt was also active on December 1. Reports of the build-up of Japanese forces in French Indochina must have reached Washington for Roosevelt met with the British Ambassador Edward Frederick Lindley Wood (Lord Halifax).[61] Roosevelt promised American support when the Japanese struck Thailand or Malaya.[62] Apparently, an attack on the Philippines was not even considered.

The next day, a Japanese reconnaissance plane was spotted over Clark Field, the Philippines, in the early morning hours. The Japanese had been conducting similar flights since the last days of November. But this was the first one spotted by the Americans. Pursuit pilots were ordered to intercept any future flights.[63]

1941 DECEMBER 2

FAR EAST

Meanwhile the British warships, *HMS Prince of Wales* and *HMS Repulse*, reached Singapore, on the tip of the Malaya Peninsula. (see Figure 2) While finding over 100,000 men ready to defend the garrison, there was virtually no means of servicing ships of this size. The British Admiralty ordered both ships to leave due to reports of Japanese aircraft and submarines operating in the area. It was the December 3. The next day, Admiral Thomas "Tom Thumb" Phillips, commander of the two British ships, flew to Manila to confer with MacArthur. MacArthur told Phillips that reconnaissance flights reported that Japanese troop transports were sailing south towards Malaya.[64]

THE PHILIPPINES

While MacArthur and Philips were discussing Japanese troop movements, General Brereton was ordering the B-17s moved to Del Monte, despite the incompleteness of the facility to operate the bombers. The bombers left the next day.

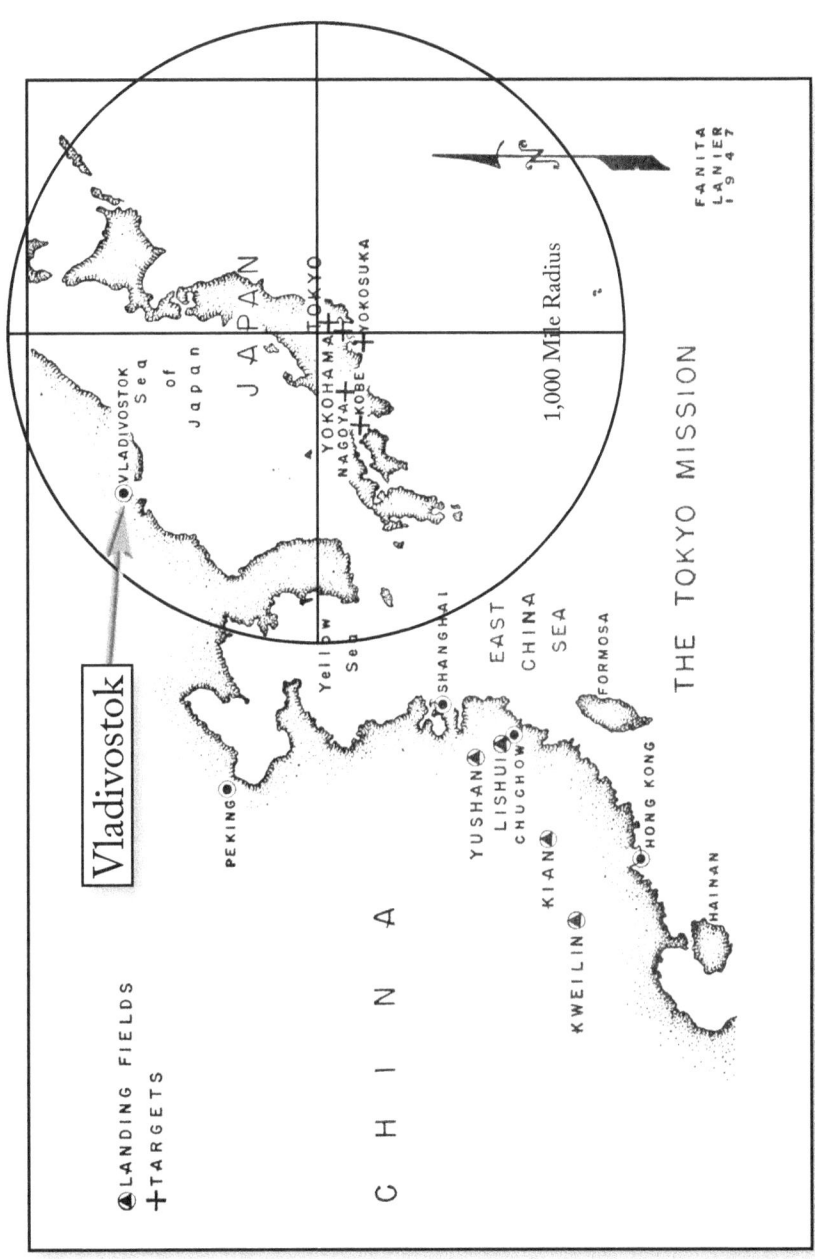

Figure 1 – 1,000 Miles radius from Tokyo

ENDNOTES

1. Schaller, *American Air Strategy in China*, p. 8.
2. Baker, *Human Smoke*, p. 259.
3. Schaller, *American Air Strategy in China*, pp. 8-9; Baker, *Human Smoke*, p. 259.
4. Schaller, *American Air Strategy in China*, p. 8.
5. Schaller, *American Air Strategy in China*, p. 9.
6. Baker, *Human Smoke*, p. 261.
7. Schaller, *American Air Strategy in China*, p. 9.
8. Schaller, *American Air Strategy in China*, p. 10.
9. Baker, *Human Smoke*, p. 265.
10. Schaller, *American Air Strategy in China*, p. 10.
11. Schaller, *American Air Strategy in China*, p. 10.
12. Schaller, *American Air Strategy in China*, p. 11.
13. Schaller, *American Air Strategy in China*, pp. 11-12.
14. Schaller, *American Air Strategy in China*, p. 12.
15. Schaller, *American Air Strategy in China*, p. 15.
16. Schaller, *American Air Strategy in China*, pp. 15-16.
17. Memorandum from Arnold to Lovett, June 11, 1941, JB 355 file.
18. Schaller, *American Air Strategy in China*, p. 16.
19. Schaller, *American Air Strategy in China*, p. 17.
20. Craven, *The Army Air Forces in World War II, Vol. 1*, p. 172.
21. Miller, *War Plan Orange*, p. 61.
22. Watson, *Pre-War Plans*, p 397.
23. Marshall, George, *Biennial Reports of the Chief of Staff of the United States Army. July 1, 1941 - June 30, 1943 to the Secretary of War,*
24. Morton, *The Fall of the Philippines*, p. 17.
25. Craven, *The Army Air Forces in World War II, Vol. 1*, p. 178.
26. Herzog, *Influence of the United States Navy in the Embargo of Oil to Japan*, p. 327.
27. Watson, *Pre-War Plans*, pp. 426-437.
28. ----, *Peace and War,* Statement Issued by the White House on July 26, 1941, pp 704-705.
29. Hoyt, *Japan's War*, p. 207.
30. Morton, *The Fall of the Philippines*, p. 31.
31. Sagan, *The Origins of the Pacific War*, p. 905-906. Contrary to popular belief, Roosevelt did not approve a total oil embargo. Roosevelt left on August 3 for his meeting with Churchill. In his absence, hawks in his Cabinet, such as Harold Ickes and Dean Acheson, took the liberty of making the embargo a total embargo. Upon his return to Washington, Roosevelt did not correct these actions.

There is an interesting comment written by Sagan in one of his footnotes. It reads as follows: It is clear that Hull did not believe that an embargo was in place on August 2, 1941. On August 27 and 28, Roosevelt [after his return from the Atlantic conference] told [Ambassador] Nomura that the Japanese could still purchase oil, and Hull admitted that he "had not checked fully into the matter." It is possible that both men were, at that point, trying to reduce the provocative nature of the action. FRUS, [Foreign Relations United States] The Far East 1941, IV, 359; FRUS, Japan 1931-1941, II, 567, 572; Pearl Harbor Attack Hearings, Pt. I, 265-266.

Also see Ickes, *Vol. III,* P. 588. Ickes had been an advocate of a strong stance against Japan. Ickes must have felt that Roosevelt's decision did not go far enough. He wrote: Notwithstanding that Japan was boldly making this hostile move, the President on Thursday was still unwilling to draw the noose tight. He thought that it might be better to slip the noose around Japan's neck and give it a jerk now and then. Naturally, I am in favor of a complete job as quickly as possible.

32. Hosoya, Miscalculations in Deterrent Policy, p. 110.
33. Craven, *The Army Air Forces in World War II, Vol. 1*, p. 173-174.
34. The first British B-17 had arrived in England on May 7. The bombers were then sent on their first combat mission on July 8.
35. Harrington, *A Careless Hope*, pp. 222-223.
36. Huston, American Airpower Comes of Age, p. 242.
37. Craven, *The Army Air Forces in World War II, Vol. 1*, p. 178; Morton, *The Fall of the Philippines*, p. 38.
38. Harrington, *A Careless Hope*, pp. 225-226.
39. Morton, *The Fall of the Philippines*, p. 66.
40. Harrington, *A Careless Hope*, pp. 224.
41. Harrington, *A Careless Hope*, pp. 226-227.
42. Brereton, *The Brereton Diaries*, p. 10.
43. Memo from Stimson to FDR, Stimson Diaries, entry for October 21, 1941.
44. Mayhew, *transcript of interview*
45. Morton, *The Fall of the Philippines*, p. 38.
46. Sagan, *The Origins of the Pacific War*, p. 918.
47. Craven, *The Army Air Forces in World War II, Vol. 1*, p. 185.
48. Craven, *The Army Air Forces in World War II, Vol. 1*, p. 184.
49. Craven, *The Army Air Forces in World War II, Vol. 1*, p. 182.
50. *Modus vivendi* is a term often used to describe negotiations between Japan and the United States in the last days before Pearl Harbor. It is assumed the reader knows what the term means. For those who do not know, it is a Latin phrase and is often used in diplomacy. It basically means to "agree to disagree." Such an agreement is often temporary, and does not require ratification by the respective ruling bodies. In this particular case, Roosevelt could have reached a *modus vivendi* with Japan without requiring the "Advise and Consent" of the U.S. Senate.
51. Fleming, *The New Dealer's War*, p. 21.

52. Harrington, *A Careless Hope*, p. 232.
53. Craven, *The Army Air Forces in World War II, Vol. 1*, p. 367.
54. Craven, *The Army Air Forces in World War II, Vol. 1*, p. 185.
55. Harrington, *A Careless Hope*, p. 228.
56. Craven, *The Army Air Forces in World War II, Vol. 1*, p. 191.
57. Craven, *The Army Air Forces in World War II, Vol. 1*, p. 188.
58. Morton, *The Fall of the Philippines*, p. 88.
59. ----, *Pearl Harbor Attack, Hearings, Report of, Part IV. Responsibilities In Washington*, p. 165. The information contained herein comes from a War Department memo to General Counsel William D. Mitchell, dated December 13, 1945. The memo did not give deliveries by month or by type. Some `s quoted in the Hearings summary documents were deduced from other sources. For example, other sources state the Air Corps had 109 B-17s by September 1st. That means there were some B-17s available on February 1, when the counting begins. There were no B-24s, as production did not begin until the spring. Regardless of the exact number, the important conclusion is that military and political planner's expectations far exceeded production realities.
60. Craven, *The Army Air Forces in World War II, Vol. 1*, p. 193.
61. Lord Halifax had replaced Lord Lothian as Ambassador in January 1941. Lord Lothian had died on December 12, 1940, shortly after agreeing to diverting British planes to China, for the purpose of bombing Japan.
62. Fleming, *The New Dealer's War*, p. 23.
63. Craven, *The Army Air Forces in World War II, Vol. 1*, p. 191.
64. Herman, *To Rule The Waves*, p. 539. These two ships had been ordered to the Far East as a show of force. They were to rendezvous with the aircraft carrier *HMS Indomitable*, but she had struck a reef off the coast of Jamaica. *Indomitable* was towed to Norfolk Virginia for repairs, and the *HMS Prince of Wales* and *HMS Repulse* sailed on with only four destroyers as escort.

THE ATTACK ON PEARL HARBOR

1941 DECEMBER 7

As stated by President Roosevelt, this day would go down in history as "a date which will live in infamy."[1] At 7:48 a.m. local time, Japanese aircraft attacked the United States Pacific Fleet at anchor in Pearl Harbor. They also attacked the Army Air Corps facilities at Hickham and Wheeler Fields.

The planes were launched from six carriers that had steamed from their naval base in northern Japan. The attack was performed in two waves. The first wave consisted of 183 planes; the second, 171 planes. There was a planned third wave, but Japanese Admiral Nagumo, who was in command of the assault force, chose not to launch it.

MIDWAY

Two Japanese ships fired on the United States Marines stationed on the island.[2] (see Figure 10)

RAINBOW-5

RAINBOW 5 was activated against Japan. It would be activated against Germany following her declaration of war on December 11.[3]

REINFORCEMENTS

Wilbur (Bill) Mayhew was at sea when he heard the news about Pearl Harbor.

About 10:00 A.M. (1000) local time on 7 December 1941 General Quarters was sounded aboard the ship. We were informed that the Japanese had attacked Pearl Harbor that morning. Naturally, we felt a bit helpless, but there was nothing to do but go on. Instead of being frightened out of our wits, as we should have been, we were all cheering like mad. We had been told before we left the United States that we should expect to fight the Japanese before we returned home. However, we also were told that they were incapable of fighting effectively, so we should beat them in about six months. Consequently, when the announcement came over the ship's loud speakers, we all felt we would be returning to the States in six months rather than the two years currently scheduled. We did not know that much of the U.S. Navy already was at the bottom of Pearl Harbor.

Nor did we know what had happened to our 88th Squadron crews who, even then, were arriving in Hawaii in our brand-new B-17Es. These crews remained in the Southwest Pacific, so we never saw any of these men again until after we returned to the States.

We learned later that Lieutenant Ted Faulkner's crew (88th Reconnaissance Squadron) had arrived in a B-24A (early model of the B-24) at Hickam Field, Hawaii on 5 December 1941. The crew consisted of Lieutenant Smith, co-pilot; Lieutenant Campbell, bombardier; Lieutenant Moslener, navigator; S/Sgt. Hobbie, flight engineer; S/Sgt. Grinyer, aerial photographer; Pvt. Polowski, radio operator; and Sgt. Rahier and Cpl. Gradle, gunners. This crew was scheduled to photograph Japanese islands while on their way to the Philippines. The morning of 7 December 1941 S/Sgt. Hobbie was at the plane getting it ready for take-off that day. The first bomb that fell was just 30 feet from the plane, which was later set ablaze by machine gun fire from attacking Japanese planes. As a result of the Japanese attack, Lt. Moslener and Pvt. Polowski were killed and Lt. Smith, S/Sgt. Hobbie and Cpl. Gradle were hospitalized. Sgt. Rahier suffered an ear injury as a result of a bomb concussion. Lt. Faulkner, Lt. Campbell and S/Sgt. Grinyer were uninjured.

We also did not know that the B-17 "Flying Fortress" bombers of the 38th and 88th Reconnaissance Squadrons had reached Oahu in the middle of the attack. The guns for these planes were in boxes inside each plane, with no ammunition on board. These planes were very low on gas, so they had to land as soon as possible. Thus, the planes landed anywhere they could find room. For example, Lt. (eventually Colonel) Frank Bostrom later told me he landed at a par-5 hole on a local golf course. Raymond Swenson's B-17 of the 38th Squadron was set on fire while landing (first American plane shot down in World War II). It was set on fire in the radio compartment and broke in two when it crashed. All the crew members aboard, except the pilot and radio operator, were wounded. Flight Surgeon William Schick, who was also on board, was killed.

Almost immediately after the loud speakers were turned off on our ship, there was a flurry of activity on each of the ships in the convoy. For example, the Meigs, Holbrook, Admiral Halstead and Coast Farmer all had portions of the ships painted a brilliant white (all the rest of the ships were dull gray.). Within minutes, we could see sailors over the sides of these ships painting the hulls gray. By nightfall, those ships were as gray as the rest. The title of First South Pacific Task Force was given to all units in the convoy at this time.

Personnel on the Republic also sprang into action. The 147th and 148th Field Artillery Battalions from Texas that were also on board were assigned immediately to man the three 3-inch and one 5-inch guns on the ship. Ten of us in the

88th Squadron were assigned to man one of the .50 caliber water-cooled machine guns on the sun deck. We manned the foreward gun on the starboard side. I was rated as an expert aerial gunner because I had hit an aerial tow target sufficient times to qualify. However, this was based on four trips to the aerial gunnery range over Great Salt Lake. I knew nothing about machine guns. However, I was assigned to this gun because I was rated as an expert aerial gunner. I had never even seen a water-cooled .50 caliber machine gun before this assignment, but I was now expected to shoot Japanese planes out of the sky with it if the occasion arose. Most of the others assigned to this gun were just as inexperienced as I as far as a machine gun was concerned. Shifts were four hours on, 16 off. One advantage to being on a gun crew: we got to sleep on deck beside our gun turrets.[4]

Bill Hannah was a crewmember flying planes to Canada.

I was on a 12-plane flight delivering planes to Canada Dec 7 of 41 (Detroit). Asked Washington what to do – was told no one knew – Do what you think best. I was $100 down in a red dog card game. So my crew stayed.[5]

Al Story was still in high school.

On the evening of Dec 6th, I had worked from 11PM to 7AM the morning of the 7th. Naturally, I returned home exhausted and, after breakfast, went straight to bed. It must have been around noon or a little after that my brother Marion (aged fourteen) came into the room and woke me with some story about the Japanese attacking Pearl Harbor. I thought he was playing some kind of joke on me and angrily threw a pillow at him for waking me. When I got up and dressed, I discovered that he was not joking.[6]

Rowan Thomas and crew were at sea aboard the *President Johnson* when they heard the news.

On the night of December 7, the news came over the radio that the Japanese had attacked Pearl Harbor. The announcer concluded, his voice husky with tension: "War with Japan is but a matter of hours now."

Here we were, three and a half days out at sea. Our navigators reckoned that we were 670 miles from San Francisco. We were not a task force, merely a detachment sent to strengthen American outposts in the Pacific. Suddenly we realized how wise our Government had been in taking these precautionary measures, and how useless these measures had been because of the obstructionists who were declaring we were in no danger of being drawn into war. Nor did these observations alleviate the feeling we had of being practically dead ducks in an undefended convoy.

The *President Johnson* returned to San Francisco.[7]

ARSENAL OF DEMOCRACY

At the outbreak of war there were 913 United States Army aircraft scattered among the numerous overseas bases. This number of aircraft included 61 heavy, 157 medium, and 59 light bombers and 636 fighters. More than half of the total of heavy bombers and one sixth of the fighters were in the Philippines.[8]

ENDNOTES

1. Roosevelt, Franklin D, Address to Congress Requesting a Declaration of War, December 8, 1941, http://millercenter.org/president/fdroosevelt/speeches/speech-3324, accessed August 19, 2016.
2. Williams, *Chronology 1941-1945 - 1941*, p. 3.
3. Craven, *The Army Air Forces in World War II, Vol. 1,* p. 236.
4. Mayhew, *transcript of interview*
5. Hannah, letters to the author.
6. Story, *Private Story*.
7. Thomas, *Born in Battle*, p. 23.
8. Craven, *The Army Air Forces in World War II, Vol. 1,*193; Army Air Forces in the War Against Japan, 1941-1942, (HQ AAF, 1945), pp. 2 if.

PLANNING FOR ACTION

DECEMBER 1941

For the United States, planning in the abstract suddenly became planning in the real world. Whereas anti-Axis nations had been getting the bulk of United States production, United States needs would now trump their needs. Questions of which of our Allies was facing the greater threat and thus had the greater need were now no longer theoretical. And the Arsenal of Democracy was just beginning to function.

Whereas Germany was always considered the greater threat, it was Japan who seemed unstoppable. Allied defensive bastions, such as the Philippines, Hong Kong, Singapore, once thought to be impenetrable, were suddenly embarrassments to Allied military leadership. Even the Royal Navy, "Ruler of the Waves," would suffer an ignominious defeat at the hands of the Japanese.

The only theater that seemed to offer any hope was North Africa. Erwin Rommel would successfully withdraw his Panzers, thus avoiding a defeat. But it was still a withdrawal.

US civilian and military leadership would experience their first taste of the differences in leadership style between Roosevelt and Winston S. Churchill. Churchill was a meddler. This, combined with Roosevelt's initial tendency to defer to Churchill's experience, continued throughout 1942.

The question remained how the United States would meet the one command from which Roosevelt never backed away – Bomb Japan!

1941 DECEMBER 8

The United States declared war on Japan.

Following the attack on Pearl Harbor, the Joint Board met to assess the situation. The Navy members of the board reported that the damage inflicted by the Japanese on the Pacific fleet was so severe that the fleet would be unable to execute its assigned tasks per the current war plan. The report from Hawaii was that only one battleship survived the attack, thus leaving the entire west coast undefended.[1]

The Air Staff recommended deploying all available air strength to the west coast.[2]

Currie sent a memo to Roosevelt suggesting that Chennault's Flying Tigers be converted from a volunteer group into a regular United States Army Air Force (USAAF or AAF) group. Roosevelt returned the note to Currie asking him to discuss the issue with General John Magruder.[3]

Stimson realized that the British were receiving most of the B-24s being produced. That situation needed to change quickly. He called Air Marshall Sir Arthur Harris and asked him to release the future deliveries back to the United States.[4]

SINGAPORE

During the evening of December 7/8, Japanese troops landed at Singora and Patani, Thailand.[5] Meanwhile, advance units of the Japanese 25th Army landed at Kota Bharu, Malaya, beginning their attack on Singapore. While British defenders repulsed the initial landings, later landings gained a foothold.[6] Kota Bharu is about 300 miles north of Singapore. (see Figure 4)

Admiral Phillips found himself in a dilemma. If he stayed in harbor, the Japanese ground forces would capture him, the rest of his force, and, more importantly, his two ships. Or he could leave, without air cover, and attempt to intercept the Japanese invasion forces. He chose the second option. So, he and the British warships *HMS Prince of Wales* and *HMS Repulse* left Singapore.[7]

THE PHILIPPINES

Within hours of the attack on Pearl Harbor, the Japanese began their invasion of the Philippines. Clark Field was attacked at 12:40 p.m. local time, destroying all but four of the 20th Pursuit Squadron P-40s. Twelve of 17 B-17s were destroyed in a second attack on the field. Another group of Japanese bombers attacked the auxiliary field at Iba, where all but four P-40s of the 3rd Pursuit Squadron were destroyed. That evening, General Brereton was left with only 12 operational B-17s, 40 P-40s, and eight obsolete P-35s.

HONG KONG

The Japanese troops moved towards Kowloon, across from Hong Kong.[8] (see Figure 2)

THAILAND

Prior to the beginning of hostilities, the Japanese had been pressuring the Thai government to allow Japanese troops free access across Thailand. The Japanese needed this access to assist the upcoming invasion of British controlled Burma and Malaya. The Thai government favored cooperation with the Japanese rather than conquest. However on December 7, the Thai government had not officially agreed to such an arrangement. So, in the evening of December 7, the Japanese delivered an ultimatum to the Thai government. They gave the Thais two hours to respond.

By the next morning, the Thai government had not responded. So, Japanese troops, based in French Indochina, crossed the Thailand border and drove towards Bangkok.

Two other Japanese forces landed at Singora and Patani to assist the drive into Malaya.[9] (see Figure 2)

NORTH AFRICA

The British had been fighting the Italians and Germans in North Africa in a seesaw battle since 1940. In November 1941, the British had launched Operation Crusader, an attempt to relieve Rommel's siege of Tobruk. (see Figure 5) Since the start of the campaign, the British had success in driving Axis forces backwards.

On December 8, Axis troops began their retreat towards Gazala.[10] (see Figure 5)

REINFORCEMENTS

The next morning found Albert Story (see Figure 9) at the local recruiting center.

That Monday morning I was standing in front of the recruiting office along with ten or twelve other would-be patriots. I suspect the Army was anxious to recruit as many bodies as possible because they never questioned my assertion that I was eighteen years old.

After passing written exams as well as a physical, I was sworn in and shipped next day to Fort McPherson near Atlanta, GA. At that time as private received 21 dollars per month. My first pay, after deductions for toiletry items, was $3.50.[11]

The next day, Bill Hannah received orders.

Next day, I'm even & Washington ordered us to Hamilton Field. From there to March then Fresno & finally Muroc (now Edwards Air Force Base). We flew 6 planes every other day looking for Japs. The Navy alternated with us. Fortunately no Japs. We had one 50 cal gun total for the 6 planes. But we each had 20 100# bombs – 95# sand 5# marking powder.[12]

1941 DECEMBER 9

THAILAND

After only hours of fighting, the Thai government agreed to the Japanese ultimatum. The Japanese then occupied Bangkok.[13] (see Figure 2)

SINGAPORE

RAF abandoned Kuantan Airfield. Their squadrons diverted to either Singapore Island or the Alor Star Airfield.[14] (see Figure 4)

THE PHILIPPINES

Japanese bombers attacked Nichols Field, the United States Army air base on the outskirts of Manila.[15]

GLOBAL STRATEGY

China declared war on Japan.

Stimson reported to President Roosevelt that:

1. The defense of the Pacific coast was based on the Pacific fleet, and
2. That "the present attack has left the West Coast unprotected."[16]

At 11 a.m., Roosevelt met with some of his Cabinet and military leaders to discuss the Supply, Priorities and Allocations Board.[17] That evening, Roosevelt made a radio address to the nation accusing Hitler of urging Japan to attack the United States.

> *We know that Germany and* Japan are conducting their military and naval operations with a joint plan.
>
> Germany and Italy consider themselves at war with the United States without even bothering about a formal declaration.[18]

Roosevelt knew that what he had just said was not true.[19]

While Roosevelt was having meetings and making an address to the nation, the Joint Board was meeting to discuss allocation of scant resources. Prior to December 7, a convoy was sent to the Philippines with badly needed supplies for the 27th Bombardment Group. The convoy was ordered to return to Hawaii. This order was the first admission that the Philippines would be abandoned.[20]

COMMITMENT TO CHINA

Currie, ever watchful of defending China against Japanese aggression, sent another memo to Roosevelt pointing out the importance of the Burma Road to the supply of material to Chiang. He also pointed out that Chennault's American Volunteer Group (AVG) needed supplies, which were sitting at American bases. One B-24 could deliver seven tons of those supplies to Chennault in ten days via the South Atlantic ferry route.[21] Currie was not in attendance at the 11 a.m. meeting.

1941 DECEMBER 10

SINGAPORE

The British warships *HMS Prince of Wales* and *HMS Repulse* were attacked by land-based Japanese planes and sunk.[22] Admiral Phillips was lost. He had been Commander-in-Chief (CinC) British Eastern Fleet. Admiral Sir Geoffrey Layton assumed command.

The RAF abandoned two airfields - Alor Star (see Figure 4) and Sungei Patani.[23]

NORTH AFRICA

The Axis siege of Tobruk was lifted. The British began to establish a supply depot there.[24] (see Figure 5)

THE PHILIPPINES

Two Japanese forces began landing, one at Aparri; the other at Vigan. (see Figure 6) Japanese aircraft attacked Nichols and Nielson Fields, near Manila, and Del Carmen Field, near Clark Field. The Far East Air Force decided to restrict operations to reconnaissance missions.[25]

MacArthur continued to ask for more planes. He believed that he could still be supplied via aircraft carriers. However, instead of asking for B-17s, he shifted his request for pursuit planes[26] and bombs. He proposed attacking Japanese positions on Formosa.[27] It is important to note that it is over 800 miles from Clark Field to Taipei, Formosa. Only B-24s had a chance of flying a mission of that distance. MacArthur did not have any B-24s, and he had not been told yet that the supply convoy had been ordered back to Hawaii.

This information was withheld because Marshall was meeting with Stimson on how to break the news. Marshall admitted that:

> *He did not like to* tell him [MacArthur] in the midst of a very trying situation that his convoy had to be turned back, and he would like to send some news, which would buck General MacArthur up.[28]

BOMBING JAPAN

Another meeting of Roosevelt's War Cabinet occurred on December 10. As with all of Roosevelt's meetings, no minutes were taken. The source of what occurred can only be found in the notes made by the attendees.

One of those attendees was Stimson. In his diary, he noted that the meeting was off the record. He made no mention of the topic of bombing Japan as being discussed at that meeting. Stimson made the following note in his diary prior to the December 10 White House meeting:

> *He [MacArthur] also telegraphed that* ... the present time offered a golden opportunity by cooperation with the Russians to attack them [Japan] from the north. I

went over to the State Department and talked it over with Hull. We both agreed that the chances of getting the Russians to do much is small. I told Hull, however, that I was trying to see whether I could not send some our B-24s overland via the Atlantic, Africa, and the Middle East to China where the Chinese are said to have some airdromes within reach of Japan. In fact, I started Lovett[29] and others at surveying the route and getting information as to whether that could be done and whether bombs can be secured from India and oil from China.[30]

Stimson was involved with the three previous plans for bombing Japan. Therefore, it was no surprise that he implemented one of them. This dovetails nicely with his October 21 memo on how to use the B-24.

1941 DECEMBER 11
Germany and Italy declared war on the United States.[31]

SINGAPORE
The Allies abandoned two more airfields, both on the east coast of Malaya - Gong Kedah and Machang.[32] (see Figure 4)

THE PHILIPPINES
The Japanese Vigan force headed north and captured Laong and its airfield.[33]

BURMA
The Japanese Air Force attacked the Tavoy airfield.[34] (see Figure 3)

BOMBING JAPAN
The first indication that something was decided at the December 10 meeting occurred when Currie sent the following to Roosevelt: "I am assembling for General Arnold all the material in my office bearing upon the feasibility of China as a base against Japan."[35]

The use of aircraft carriers was not mentioned. Rather, Currie was working on Stimson's idea of using China as a base for heavy bomber operations.

1941 DECEMBER 12

SINGAPORE
Within four days of landing on the Thailand coast, Japanese troops had captured the RAF airfield at Alor Star.[36] (see Figure 4)

THE PHILIPPINES

The Japanese made a third landing on Luzon, this time southeast of Manila at Legaspi. Meanwhile, the Aparri force captured the airfield at Tuguegarao.[37] (see Figure 6)

COMMITMENT TO CHINA

T. V. Soong sent the following to Sumner Welles:[38]

I am in receipt of a cable from General Chiang Kai-shek dated Chungking, December 10th, and reporting among other things,

1. The Chief Soviet Military Advisor expresses his personal opinion — that the Soviet declaration of war against Japan is merely a matter of time and of procedure.
2. The Soviet will make an open declaration of war only after a general coordinated war plan has been arranged between the United States, Great Britain, China and the Soviet.[39]

MIDDLE EAST

General Sir Claude Auchinleck, Commanding General of the British forces in Africa, was told that troops originally intended for him would now be sent to the Pacific theater to help stem the tide of Japanese success.[40]

GERMAN OIL

Roosevelt was given a status report on the German oil situation:

1. To ensure an adequate petroleum supply in case of a major war has been a prime objective of Germany policy since 1933.
2. The Russian campaign has altered this situation in the following respects:

   - (Germany) has proved possible to send Rumanian oil directly North and East, to the Axis armies in the field.
   - Russian petroleum supplies, previously available at the rate of about a million tons a year, have been cut off and this has been only partly offset by the acquisition of the Galician fields.

3. On the basis of present evidence, therefore, it cannot be predicted that Germany will suffer a critical petroleum shortage in 1942.[41]

GLOBAL STRATEGY

The Air Staff Planning Division published their assessment of the tactical situation that faces the United States. In their opinion, Axis airplanes could reach any United States coastal installation in the Western Hemisphere.[42] Exactly how they could do this was unclear. Germany had neither aircraft carriers nor bases in Africa. Japan did have carriers and one could argue they could attack any west coast base. But they had no refueling station on the eastern side of the Pacific.

REINFORCEMENTS

Ed (Hawk) Cave got his "wings as 2nd Lt. Dec. 12, 1941, five days after Pearl Harbor."[43]

1941 DECEMBER 13

NORTH AFRICA

The British began their attack on the German position at Gazala. This conflict would become known as the First Battle of Gazala.[44] (see Figure 5)

HONG KONG

British troops withdrew from Kowloon, across into Hong Kong.[45] (see Figure 2)

SINGAPORE

The Allies sent RAF reinforcements to Ipoh Airfield. Rumors of a Japanese convoy steaming from Saigon resulted in numerous reconnaissance missions in the following days.[46] (see Figure 4)

THE PHILIPPINES

MacArthur continued his incessant drumbeat for supplies. In a radiogram to Marshall, he defined his strategic vision:

> *The Philippine theater of operations* is the locus of victory or defeat. ... If the Philippines and the Netherlands East Indies go, so will Singapore and the entire Asiatic continent.[47]

EAST INDIES

The oilfields in East Sarawak and West Brunei (both British territory) were destroyed to deny their use by the Japanese. This small force then departed for Kuching to defend the airfield there.[48] (see Figure 2)

GLOBAL STRATEGY

With the decision to recall the Philippines supply convoy, military planners realized that the only feasible means of getting badly needed supplies to MacArthur and/or Chennault was via air. And the only viable route was the southern Atlantic - Trans-Africa - India route. On December 13, the War Department entered into a contract with Pan-American Airways to open a supply route from the United States to Singapore.[49] (see Figure 2)

COMMITMENT TO CHINA

Now that the United States was at war with Japan, Roosevelt could make plans with Chiang. Roosevelt asked Stimson to draft a communiqué to Chiang suggesting that Chiang convene a conference with all of the Allied representatives. Roosevelt intended to appoint General George H. Brett[50] as his representative.[51] Roosevelt's selection of an Air Corps general, instead of an Army or Navy officer, was another indication of his desire to bomb Japan.

REINFORCEMENTS

Bill Mayhew wrote about his arrival at the Fiji Islands:

> *About one week later at* 5:00 P.M. (1700) on 13 December 1941 we arrived at Suva, the capitol of the Fiji Islands, in order to refuel. Rumors aboard ship had indicated that the Japanese had already captured the Fiji Islands, so we expected to have to fight our way ashore. We had loaded food and our few guns and the little ammunition we had aboard the lifeboats and rafts. The Navy crew had prepared the Republic for scuttling. This shows how naïve we were at that time. We would not have lasted ten minutes in a fight. Fortunately, it was not necessary. We spent a very nervous 25 hours aboard ship while the convoy refueled.
>
> …We hated to leave Suva so soon, but a Japanese fleet was entirely too close for comfort. So, under the cover of darkness and a storm, at 6:00 P.M. (1800) we started our dash (at 10 knots per hour—the fastest one of the freighters could travel) for Australia (14 December 1941).[52]

1941 DECEMBER 14

SINGAPORE

With the withdrawal of Allied aircraft to Singapore, the airfields there were becoming congested. Plans were formulated to stage the planes in the East Indies.[53] (see Figure 2)

EAST INDIES

Maxwell Hamilton, Chief of the Division of Far Eastern Affairs in the State Department, handed to the British Ambassador two communications regarding the situation in Borneo:

> *The Commander-in-Chief of the United* States Asiatic Fleet [suggests to the British] that steps should be taken forthwith to [destroy] the oil fields in Borneo.
> If [British] oil shortages [result], the United States [will] assist ... the British and the Dutch governments in obtaining the delivery of oil or oil products.[54]

Two days earlier, Roosevelt was given a presentation on the German oil situation. The supply of oil, the primary trigger for Japan's decision to go to war, was working its way into the strategic thinking of Allied planners. In 1941, the United States was the OPEC of its day.

THE PHILIPPINES

While there is no record of any official decision to abandon the Philippines, communiqués between Washington and the Philippines reflect the tactical reality. In a message to Hart, Comander-in-Chief Asiatic Fleet (C-in-CAF), Stark told Hart that Singapore and Luzon were expected to fall. The United States needed to build up forces in northwest Australia.[55] In support of this direction, Hart ordered the remaining aircraft of Patrol Wing 10 and the associated tenders to Australia.[56]

The content of this message reflected the conclusion of Dwight Eisenhower, recently assigned to the War Plans Department. Marshall met Eisenhower when he arrived for work. Eisenhower's first task was to outline a basic strategy for defending the Pacific and winning the war. Hours later, Eisenhower presented his conclusion: supplying the Philippines would take longer than MacArthur could hold out. The defensive line must begin in Australia. Given this gloomy outlook for the fate of Philippines, Eisenhower added that any appearance of abandonment would not sit well with the people of Asia. It must appear that the United States was making every effort possible to provide assistance.[57]

Brereton, on the other hand, was still holding out hope. He sent a message to MacArthur that dive-bombers and pursuit planes were expected in Australia by month's end. If the United States Navy could deliver 250 more such planes by carrier, then his staff would have the fields ready to receive them.[58]

BOMBING JAPAN

Continuing to believe that he was tasked with supporting the bombing of Japan, Stimson made the following notation in his diary:

> *Immediately after the attack on* Hawaii, on Monday morning, I began work looking for a counterattack on the Japanese in their island. I sent Lovett and the Air

Corps to work to find a route for our long range bombers, which could not be interrupted by the Japanese, and which could reach taking off places in China.[59]

And Stimson was not the only one. Gerow gave a second confirmation that Arnold got the same marching orders as Stimson. Found in his Memorandum of December 14 was the following statement:

> *The directive of General Arnold* to General Brett of December 10, 1941 should also be remembered. If we are preparing to strike blows from China, the retention of the Burma Road should receive continuous consideration.[60]

There was still no mention of using aircraft carriers to strike Japan.

1941 DECEMBER 15

EASTERN FRONT

Near Moscow, the Russian army continued pushing German Army Group Central backwards, this time retaking the city of Klin.[61] (see Figure 7)

THE PHILIPPINES

Apparently, Stark was a bit premature in his communication to Hart about the fate of the Philippines. Roosevelt ordered that the Philippines be supplied quickly. MacArthur was informed that the United States was sending 65 new heavy bombers plus 15 of the LB-30 bombers repossessed from the British. He should expect them by February 21 of next year.[62] This movement of aircraft was given the code name Project X.

And while MacArthur was being told that he would be getting 80 new bombers, Brereton was given permission to withdraw his remaining B-17s to Darwin, Australia.[63] (see Figure 2)

MIDDLE EAST

Alexander C. Kirk, the American Minister to Egypt, urged the United States not to forget the Middle East Theater, while its attention was concentrated on the Orient. Also, consideration should be given for American units to operate in North and West Africa, against the Axis.

Commitment of American forces in the Middle East seemed unlikely for two reasons:

1. The Middle East was looked upon largely as a responsibility of the British, and,
2. The defense of the Far East was regarded with greater urgency.[64]

REINFORCEMENTS

Rowan Thomas would soon find himself a member of Project X.

> *Then we received the welcome* news. Everything left in the 7th and 19th Groups was to be joined into one echelon -temporarily called Ferry Project X. We were to be equipped with the latest type Fortresses, B-17 E's, the first to have a "tail stinger" installed, and we were to take off singly for the Philippines and Java, each ship leaving alone as soon as it was ready. Unlike those who had gone before us, we had to take the long way around, the Atlantic route instead of the Pacific. This was the first blow struck from continental United States after the declaration of war—an attempt to supply the planes that MacArthur so urgently needed.[65]

1941 DECEMBER 16

NORTH AFRICA

The Axis forces began a retreat from Gazala towards Agedabia.[66] (see Figure 5)

EASTERN FRONT

Near Moscow, the Russian army retook the city of Kalinin.[67] (see Figure 7)

EAST INDIES

A Japanese invasion force landed on the north side near Miri (Sarawak) and Seria (Brunei).[68] (see Figure 2)

ARCADIA CONFERENCE

During their trans-Atlantic voyage to the Arcadia Conference, the British Chiefs of Staff outlined their plans for Anglo-American strategic cooperation:

1. The United States and Britain would carry on softening-up operations and attrition by strategic bombing, internal subversion, and joint military operations for the purpose of conquering all of North Africa and getting Italy out of the war.
2. Only when Germany had been bled white would a number of simultaneous assaults be launched on the European coast: by the British from the west and the Americans from the south.[69]

1941 DECEMBER 17

HONG KONG

The New York Times published an article by V. H. C. Jarrett, titled: *Hong Kong Strong And In 'Good Heart'*. The third paragraph read:

> *"The colony is in good* heart. There is plenty of food, arms, and ammunition, and the garrison is confident of the outcome."[70]

THE PHILIPPINES

At the start of hostilities, Brereton had 35 B-17s on the Philippines. He began the withdrawal of his remaining 14 bombers on December 17.[71] He established a new headquarters in Australia. The transfer would be completed on December 20. All of the pursuit aircraft stationed on the Philippines were either destroyed or captured.

NORTH AFRICA

Rommel received a shipment of 40 panzers. Along with those his maintenance crews had repaired, he now had 70 operational tanks.[72]

MIDWAY

A squadron of Marine aircraft arrived from Hawaii.[73] (see Figure 10)

EASTERN FRONT

The German Army Group South began its campaign against Sevastopol.[74] (see Figure 8)

BOMBING JAPAN

Presidential advisor Harry Hopkins must have realized that the United States had too few assets chasing too many needs. He sent a note to Roosevelt listing the various needs. One of the questions he asked was:

> *Are we going to make* an effort to get big bombing planes into China and if so by what route, how many, what type and under whose command?[75]

REINFORCEMENTS

The B-17, serial number 41-2446, departed the Sacramento Air Depot for Hickam Field in Hawaii. Once in Hawaii, she was assigned to the United States Navy to fly reconnaissance missions around the Islands.

She was not given a nose art name during her operational life. When her wreckage was discovered in 1972, she was then given the name *Swamp Ghost*.[76]

1941 DECEMBER 18

HONG KONG

Japanese troops landed on the Island of Hong Kong.[77] (see Figure 2)

ARCADIA CONFERENCE

Churchill had different opinions from his military advisors. He stated that it was important to execute an "invasion of the continent of Europe as the goal for 1943."

The Allies could not count on the Nazi regime to collapse on its own. To achieve this goal, the Allies must do three things prior to the invasion:

1. A joint Anglo-American operation was necessary in 1942 to assist in conquering the whole of North Africa, and
2. The next phase would be to take Sicily and seize a foothold in Italy.
3. The strategic bombing of Germany should be intensified.[78]

Which was exactly what would happen!

1941 DECEMBER 19

SINGAPORE

Due to continued Japanese ground gains, the RAF was forced to withdraw its aircraft from Ipoh Airfield. The planes relocated to Kuala Lumpur.[79] (see Figure 4)

BOMBING JAPAN

Hopkins continued to barrage Roosevelt with questions. One of them was "Are we going to make an effort to get an air unit into China in addition to the planes already there?"[80]

Other than the AVG, there were no other United States planes in China.

REINFORCEMENTS

Orders were issued on December 19 to begin the overseas deployment of Project X, the shipment of heavy bombers promised to MacArthur three days earlier. The first leg of that movement would be for the planes to fly to MacDill Field outside Tampa, Florida. This movement would be the first major foreign ferrying job of the Ferrying Command.[81] Although 80 heavy bombers were earmarked for the project, something less than that number would actually leave the United States and an even smaller number would reach the Far East.

1941 DECEMBER 20

THE PHILIPPINES

During the evening of December 19/20, two Japanese task forces arrived off the coast of Mindanao, near Davao. That morning, the Japanese troops landed and captured Davao and its airfield.[82] (see Figure 6)

CHINA

The AVG, which would become affectionately known as the Flying Tigers, flew its first mission.[83]

GLOBAL STRATEGY

The Soviet Union had agreed to fight the Japanese only after a plan was in place. So, the Chinese prepared a strategic plan. They had not been involved with previous Allied discussions and, thus, had a separate view of how the war should be conducted. Their plan was relatively simple:

1. The Allies' forces were to destroy the Japanese air force first, to gain air superiority in July 1942.
2. Then the Allies' air forces were to initiate air attacks on the Japanese army and use aerial bombardment to cut its supplies.
3. Finally, the Chinese army would annihilate the Japanese army, which ended the Pacific War in December 1942.[84]

MIDDLE EAST

As planners began to deal with the logistics of getting war material to the Allies, it was becoming increasingly evident that any supply chain across Africa had to transit the Middle East. Stimson sent a strategy memo to Roosevelt on the subject. In part, it stated:

> *The West African theatre is* vital to us to protect our line of communications and supplies ... with the activities in the Far Eastern theatre through Australia and Singapore on the south and through China on the north.
>
> [But] The Egyptian area is of immense importance psychologically to the British Empire ... In my opinion we should not divert armed forces to that area.[85]

The use of American assets to defend British territory would be a reoccurring point of contention within United States political and military circles.

ARCADIA CONFERENCE
Churchill and his military staff had slightly differing views prior to the Arcadia Conference:

1. Churchill's view was that liberation forces should not land on the European Continent while the German army was at its present level of strength.
2. The British Chiefs of Staff's view was that a landing was possible only after the German army and air force's capacity to resist had been broken. In their opinion, this point would be secured by the strategic bombings that are to be the main war effort in 1943.

These two approaches were forwarded to Washington. Churchill's opinion was that there were no significant differences of views between he and the Chiefs of Staff.[86]

1941 DECEMBER 21

THAILAND
Thailand and Japan had signed an agreement on June 12, 1940. This treaty was subsequently registered with the League of Nations on July 26, 1941.[87]

Thailand and Japan expanded this treaty by signing the Treaty of Alliance at Bangkok, essentially giving the Japanese free access to the Thai infrastructure.

BOMBING JAPAN
In a meeting at the White House, Roosevelt demanded that Japan be bombed as soon as possible to boost public morale after the disaster at Pearl Harbor.[88] It had only been eleven days since the decision had been made to proceed with an attack.

GERMANY FIRST
A *New York Times* editorial argued for the Germany First strategy. Titled: *Germany Still Heart Of Axis War Strength,* the author, Edwin James wrote:

> *The sudden Japanese attacks upon* the United States and their initial successes should not blind one to the fact that Germany remains the heart of the Axis war effort. Unless Hitler wins his partners cannot win in the long run. If the Nazis cave in or miss victory in any other way, Japan and Italy will not be victors in the end.[89]

Yet in the same issue of *The New York Times*, another editorial writer, Hanson W. Baldwin, wrote about Japanese success. The first paragraph of his article, *Japanese Gain Quick Successes In Far-Flung War Of The Pacific*, read:

> *The great Pacific war, spread* over unprecedented longitudes of space, raged last week from the foam-white beaches of the Hawaiian Islands to the steaming jungles of Malaya as the Japanese drove relentlessly for dominion or death.[90]

In retrospect, it can be seen how the public might be conflicted. An argument was being made that Germany was the decisive factor in an Axis success. Yet German advances on African and Russian fronts were being thwarted. And Japan seemed unstoppable. So, given this dichotomy, the response of the Roosevelt administration was to restrict freedom of the press. (See *Freedom Of The Press Restricted For The War* by Arthur Krock.[91])

1941 DECEMBER 22

THE PHILIPPINES

During the evening of December 21/22, three Japanese task forces arrived off the coast of Lingayen Gulf, near San Fernando. That morning, the Japanese troops landed and proceeded unopposed by Allied forces.

Operating from its new base at Darwin, Australia, (see Figure 2) nine B-17s attacked the Japanese forces near Davao. The planes landed at Del Monte Field, Philippines. Four of these planes attacked the Japanese transports off San Fernando the next day.[92] (see Figure 6)

ARCADIA CONFERENCE

Churchill and his military staff had arrived in Washington for the Arcadia Conference. Accompanying him were Admiral of the Fleet Sir Dudley Pound, Field Marshal Sir John Dill, and Air Chief Marshal Sir Charles Portal

Once in the United States, Churchill began waging a public relations campaign to explain British strategic objectives. But he was not the only one. Joseph Stalin's ex-Foreign Commissar, Maxim Litvinoff, was the subject of a *Time* article. In it, Litvinoff explained why Russia would not be declaring war on Japan. As Churchill would say the next day, Germany was the primary threat.[93]

Marshall, after reviewing the British transmittals, sent his comments to Roosevelt. It was a five-page memo outlining his view of the situation. Two key points were noteworthy:

1. China is near complete isolation and needs military materiel. She is conducting

limited offensives to assist in defense of Malaysia. The maintenance of adequate air and sea communications and the safety of China demand the defense of the Singapore - Philippine Dutch Indies area. It follows that Malaysia is a theater of present urgent importance,

2. Reinforce Philippines, Dutch East Indies and Australia to further the security of China and Southwest Pacific.[94]

REINFORCEMENTS

Bill Mayhew related more of his trans-Pacific cruise:

> *On 22 December 1941 we* finally anchored about noon in the harbor of Brisbane. The Norwegian tanker "Falkefjell" and the transport "Chaumont" took oil from the "Republic" so that we could proceed up the harbor to the docks. While we were waiting, one of the troops threw one of the kapok life vests we had been wearing most of the trip, into Brisbane Harbor. It sank in about five minutes. We disembarked at 4:15 P.M. (1615) on 23 December 1941, as the first American troops in Australia. We had spent 32 days on the water and traveled 10,281 miles. As the direct distance is 8,000 miles, this indicates that we had done a great deal of zig-zaging. Dry land never looked so good before!![95]

GERMANY FIRST STRATEGY

During the initial stages of the war, the press often wrote editorials that questioned the strategic conduct of the war. This was especially poignant when the Japanese appeared to be unstoppable. One such editorial, written by Hanson W. Baldwin, was titled *Plight of Hong Kong Force.* The first paragraph read:

> *The expected fall of Hong* Kong after only two weeks of war and a few days of sustained assault is not a major military defeat for the Allies, but it does represent a serious moral setback and reveals surprising and dangerous weaknesses in the allied concepts of strategy.[96]

1941 DECEMBER 23

BURMA

The Japanese began her assault on Rangoon with air attacks. (see Figure 3) The battle for air supremacy lasted until late February.[97]

SINGAPORE

The Allies completed the withdrawal of all of the west coast troops to behind the Perak River.[98] (see Figure 4)

THE PHILIPPINES

MacArthur decided to withdraw his forces to Bataan. An enemy force left Mindanao. Its destination would be Jolo Island.[99] (see Figure 6)

MIDWAY

In the Pacific, Japan resumed her assault on Wake Island. The first attack on December 8 had been repelled. Now, a larger force landed on the Island. The American garrison surrendered.[100] The Japanese now had control of one of the landing strips long enough to accept United States heavy bombers, thus severing the air route between the mainland and the Far East.[101]

A supply convoy in route to Wake was ordered to divert to Midway.[102] (see Figure 10)

EAST INDIES

A Japanese invasion force was spotted off the coast of Kuching, Borneo.[103] (see Figure 2)

PLOESTI

William Bullitt worked in the United States State Department and was assigned to the embassy in Cairo. He became one of the strongest advocates for United States intervention in the Middle East. That position would put him at odds with Marshall. However, since he was with the State Department, he had access to Roosevelt. On December 23, he sent a cablegram to Roosevelt:

1. The British air marshall in Cairo [Arthur] Tedder[104] desired me to call to your attention that the German failure to get through to the oil wells of the Caucasus has made the oil wells of the Ploesti region in Rumania vital to Germany.
2. Tedder believes that three squadrons of Liberators based in Cairo could destroy this oil field in an attack to be sustained over a period of two months.
3. I stated that we were extremely short of Liberators and that I felt almost certain that we should have none to spare for this purpose.
4. Tedder was most insistent that I should present this possibility to you as the vital link in the whole strategic picture.[105]

Interestingly, the cable came via the Navy Department.

REINFORCEMENTS

Rowan Thomas's wait for a plane was soon over.

Finally it was the evening of December 23, two days before Christmas. Major Tate had picked me as his copilot and informed me that we were to start again any minute on Ferry Project X. We were scheduled to make our first flight in the new plane at 10:30 the next morning.[106]

1941 DECEMBER 24

EAST INDIES

The Japanese invasion force landed troops near Kuching, Borneo. The Allied troops had already destroyed the airfield. The Dutch aircraft retreated from Singkawang, Borneo, to Palembang, Sumatra.[107] (see Figure 2)

BOMBING JAPAN

After Roosevelt's prompting on December 21, Stark asked Arnold about the status of the plans to bomb Japan. In response, Arnold stated that:

1. No bombing operations should be undertaken against Japan unless they were strong enough to create substantial damage,
2. The minimum number of bombers should be 50,
3. Unsustained attacks would only tend to solidify the Japanese people.[108]

This seemed to place Arnold in opposition to a Doolittle type raid, which had still not even been conceived.

ARCADIA CONFERENCE

December 24 was the first day of meetings between Roosevelt, Churchill and their respective staffs. One of the first items upon which the two leaders agreed on was that Army Air Corps units would be sent to Britain as previously planned. Then Stark made a surprising announcement: American crews, formed into their own units, would man American planes. Portal responded that this new policy was not in accordance with previous agreements. But those agreements had been made when the United States was just the manufacturer and supplier for the war effort, not a participant in the war. Pearl Harbor changed all that.[109]

COMMITMENT TO CHINA

It was also agreed that Roosevelt would ask Chiang if he would become the Supreme

Commander in China. And it was agreed that the Allies would continue sending supplies to China.[110]

THE PHILIPPINES

Later that evening, Roosevelt and Churchill met in the White House.[111] They discussed the situation in Malaya following the sinking of the British ships, *HMS Prince of Wales* and *HMS Repulse*. Roosevelt remarked that the American reinforcements currently enroute to the Philippines would not be able to fight their way to the Philippines in time to relieve MacArthur.

General Sir Leslie Hollis, the secretary of the British Chiefs of Staff, prepared a memo regarding an agreement between Churchill and Roosevelt.

> *FDR's view was that these* reinforcements should be utilized in whatever manner might best serve the joint cause in the Far East, and ... expressed the desire that the United States and British Chiefs of Staff should meet the following day to consider what measures should be taken to give effect to his wishes.[112]

Roosevelt had just acknowledged the loss of the Philippines and agreed to divert reinforcements destined for the Philippines. Yet, Marshall was still telling MacArthur that supplies were on their way.

Meanwhile, during the night of December 23/24, a Japanese amphibious force arrived in Lamon Bay. They made a landing in the early hours that morning.[113] (see Figure 6)

NORTH AFRICA

Advance units of the British Eighth Army entered Benghazi.[114] (see Figure 5)

1941 DECEMBER 25

SINGAPORE

The next morning, Stimson seemed upset when he wrote in his diary:

> *Generals Arnold, Eisenhower, and Marshall* came in to see me and brought me a rather astonishing memorandum which they had received from the White House concerning a meeting between Churchill and the President and recorded by one of Churchill's assistants ... It reported the President as proposing to discuss the turning over to the British of our proposed reinforcements for MacArthur. This astonishing paper made me extremely angry and, as I went home for lunch and thought it over again, my anger grew until I finally called up Hopkins, told him

of the paper and of my anger at it, and I said if that was persisted in, the President would have to take my resignation.

Harry Hopkins called back and said he asked Roosevelt, in the presence of Churchill, about the diversion of American reinforcements to Malaya. Roosevelt denied that any commitment had been made, and Churchill supported the president. Stimson was dubious and told Hopkins so. "I then read to him extracts from the paper ... and he said that they certainly bore out my view."

Nothing more was said about sending the American forces to Singapore. Stimson kept quiet. Later, he made the following entry in his diary.

> *This incident shows the danger* of talking too freely in international matters of such keen importance without the President carefully having his military and naval advisers present. Hopkins told me at the time I talked with him over the telephone that he had told the ... President that he should be more careful about the formality of his discussions with Churchill.[115]

HONG KONG

The British garrison at Hong Kong surrendered after only a 16-day battle.[116] (see Figure 2)

Eight days earlier, a report was filed with the title: *Hong Kong Strong And In 'Good Heart.'* Now, the garrison surrendered. No explanation was offered about what happened.

MIDWAY

The *USS Saratoga* delivered a Marine fighter squadron.[117] (see Figure 10)

COMMITMENT TO CHINA

On December 13, Stimson had made a proposal to Roosevelt that he, Roosevelt, name Chiang the Supreme commander in the Far East. At the Joint Combined Chiefs of Staff (JCCS) meeting on December 25, Marshall proposed the idea to the British.[118]

REINFORCEMENTS

After a quick trip to Las Vegas to marry his girlfriend, Rowan Thomas was ready.

> *Christmas Day, two weeks after* Pearl Harbor, we were ready to take off.
> Our new B-17, curiously impressive and ominous in its dull olive paint, stood back from the runway as the nine of us, Major Tate in command, filed toward it. We stopped to watch another bomber roar down the field and climb 12,000 feet to skim over the snow-capped mountain peak just south of the airdrome.

In a few minutes we sped down the runway toward the same mountain. The weather was not too promising, but we had great faith in Major Tate's ability as a pilot. Slowly we circled the field and began to climb. But before we could reach the safe height to avoid the jagged peak, two of our engines began to leak oil. One looked serious, and Major Tate feathered the prop, glancing at me out of the corner of his eye as he did so. The Major didn't have to tell me that he didn't intend to climb over that mountain on three motors, let alone two, especially on Christmas Day.[119]

1941 DECEMBER 26

THE PHILIPPINES

Manila was declared an "Open City."[120] According to International Law, the declaration of an "Open City" means that the current defenders would not resist any attempt by an opposing force to occupy the city. Such a declaration supposedly would spare the city of any military attack.[121]

MIDWAY

The supply convoy originally destined for Wake arrived at Midway.[122] (see Figure 10)

BOMBING JAPAN

General Brett had been sent to the Far East to explore how Air Corp heavy bombers could operate from Chinese bases. On December 26, he sent a cable to Arnold from Rangoon (see Figure 3) saying:

1. There were two landing fields available in China for B-17 use.
2. It would be difficult to supply China over the Burma Road.
3. Sustained operations of B-17s from China could not be supported at that time.
4. B-17s should not be sent to China.[123]

1941 DECEMBER 27

THE PHILIPPINES

By Christmas, the situation in the Philippines was nearly untenable. MacArthur radioed that:

> *Enemy penetration in the Philippines* resulted from our weakness on the sea and in the air. Surface elements of the Asiatic Fleet were withdrawn and the effect of the submarines has been negligible. Lack of airfields for modern planes pre-

vented defensive dispersion and lack of pursuit planes permitted unhindered day bombardment. The enemy has had utter freedom of naval and air movements.[124]

NORTH AFRICA

Advance units of the British Eighth Army assaulted Rommel's defensive position at Agedabia.[125] (see Figure 5)

1941 DECEMBER 28

THE PHILIPPINES

With MacArthur's gloomy report, Roosevelt decided that he needed to cheer up the Philippine people. He delivered a speech to them, saying in part:

> *The people of the United* States will never forget what the people of the Philippine Islands are doing this day and will do in the days to come. I give to the people of the Philippines my solemn pledge that their freedom will be redeemed and their independence established and protected. The entire resources, in men and in material, of the United States stand behind that pledge.[126]

Four days earlier, Roosevelt implied to Churchill that the Philippines were lost. Now, he used the verb "redeem." His use of the word "redeem" raises the question of whether he was telling the Filipinos, and thus the whole world, that the islands were lost.

TURKISH NEUTRALITY

Throughout the first months of the war, Allied military strategists knew that if Germany could transit Turkey, their path to the Middle East and Caucasus oil fields would be nearly unopposed. Consequently, keeping Turkey neutral was paramount. Author Ray Brock wrote as much in an article titled *Turkey Awaits Nazi Drive To East*.[127] The accompanying map (see Figure 11) showed how the transit might work.

1941 DECEMBER 29

GLOBAL STRATEGY

At the JCCS meeting on December 29, the conferees decided to approve Marshall's plan to appoint one person as supreme commander in the Far East. They chose British General Sir Archibald P. Wavell. His appointment was announced on January 3.[128] On the same day, Roosevelt announced that Chiang would become the Supreme Allied Commander - China.[129]

REINFORCEMENTS

Bill Mayhew related more experiences:

All the Americans remained in Brisbane about a week, waiting for a ship to try to run the blockade to the Philippines. However, it was finally decided by someone that we would never make it, so on 29 December 1941 the 9th and 88th Squadrons were moved to a little town called Ipswich, about 30 miles inland from Brisbane. The 11th and 22nd Squadrons went to Archer Field nearby. (These two squadrons were sent to Java in January 1942.) Our airdrome (Amberly Field) was just outside the town, so we could get into town quite often. It had a population of only about 20,000 people, but the boys really enjoyed themselves there. As we had no bombers with which to fight (our B-17Es were still at Hawaii where they had landed during the Japanese attack on Pearl Harbor), members of the 9t.h. and 88t.h. squadrons were given the job of assembling some P-40 fighter planes and A-24 dive bombers that we had had in crates on board the convoy, which were destined for the Philippines. More than 150 planes were assembled in five weeks while we were in Australia.[130]

1941 DECEMBER 30

NORTH AFRICA

The British Eighth Army continued a futile assault on Rommel's defensive position at Agedabia.[131] (see Figure 5)

EASTERN FRONT

The German Army Group South continued its campaign against Sevastopol. (see Figure 8) However, the Russians attempted a flanking maneuver by landing troops in east Crimea.

Meanwhile, near Moscow, the Red Army continued to have success pushing the Germans back.[132]

1941 DECEMBER 31

NORTH AFRICA

The British Eighth Army attacked the Axis forces at Bardia.[133] (see Figure 5) Bardia lies on the Libya-Egypt border, on the road from Tobruk to Cairo.

THE PHILIPPINES

Brett had been ordered to Australia. He arrived on December 31 to assume command of the United States Army Forces in Australia (USAFIA). Admiral Hart had withdrawn his surviving vessels to the Philippines and was still C-in-C Asiatic Fleet. Brereton had the remnants of the USFEAF.[134]

The evacuation of Allied troops from Manila was completed. Corregidor was now the headquarters of Allied defensive forces.[135] (see Figure 6)

EASTERN FRONT

The Russian's flanking maneuver succeeded in causing the Germans to stop its assault on Sevastopol.[136] (see Figure 8)

BOMBING JAPAN

While Brett may have been considered the best man on the scene to assume command, Arnold was extremely unhappy with his December 26 report.

> *The job given Brett was* to determine the way to bomb Japan from China with heavy bombers. He was not given the job to determine ways and means for not doing it. The attached is a cable full of "not's." I want to find out how to do, not how not to do it.

Arnold directed Air WPD (AWPD) to study Brett's cable and make recommendations to Major E.H. Alexander, who was in China with Magruder. Arnold said the focus must be "How Can We Bomb Japan From China?"[137]

GERMANY FIRST

In a policy statement, the JCCS reaffirmed the Germany First strategy of ABC-1, "in spite of recent events."[138]

REINFORCEMENTS

Regarding Project X, Arnold issued a logistical memo, splitting the 80 planes evenly between the 7t.h. and 19t.h. Bomb Groups.[139] While Arnold was dealing with logistics, crews, destined to fly the planes, began to receive orders to assemble. Paul Eckerly, who had just graduated from flight school on December 12, was flying submarine reconnaissance missions off the east coast out of Westover Field. He was planning to celebrate New Year's Eve with family and friends, when he received orders to get to Langley Field, Virginia. He arrived on New Year's Eve.[140]

Rowan Thomas and crew arrived on the east coast.

> *Early the next morning, after* the weather was pronounced favorable, our flight

plans filed and approved, we took off for Tampa, Florida, our last stopping-off place in the United States before we hopped the Atlantic.

On December 31 at 11:59 P.M. we were roaring down the take-off strip at Tampa bound for war with as enthusiastic a crew as ever took wings. There was Major Robert F. Tate, of Texas, pilot in command; Second Lieutenant Rowan T. Thomas, of Mississippi, copilot; Second Lieutenant Willard A. Hawkins, of Arkansas, bombardier; Second Lieutenant Merrill K. Gordon, of Montana, navigator; Captain J. D. Gottlieb, of New York, flight surgeon; Master Sergeant William A. Covington, of California, flight engineer; Staff Sergeant Harold E. Vasquez, of New Mexico, armorer and gunner; Staff Sergeant Leslie L. Boone, of Texas, gunner; Corporal Jed M. Rucker, of New Mexico, radio operator.

Their journey to the front would not be uneventful. Crossing the Atlantic, they ran into a thunderstorm. Clearing the storm, they found a landing strip. As they touched down, the tail wheel collapsed. As they parked the plane and cut the engine, an eight-foot section of the fuselage broke away from the rest of the plane. The crew was ordered back to the States.[141]

HEAVY BOMBER PRODUCTION

During December, 94 heavy bombers were manufactured.[142] On January 4, 1942 *The New York Times* published a front-page article by C. P. Trussell extolling increases in the output of aircraft.

War Plane Output Leaps By The Day

Accelerated pooling and coordination of existing manufacturing facilities, reduction in the number of types of fighting planes and greater emphasis on mass output of long-range bombers have put a new face on the program for aircraft production.[143]

Production did not seem to be keeping up with Allied tactical needs, let alone strategic needs.

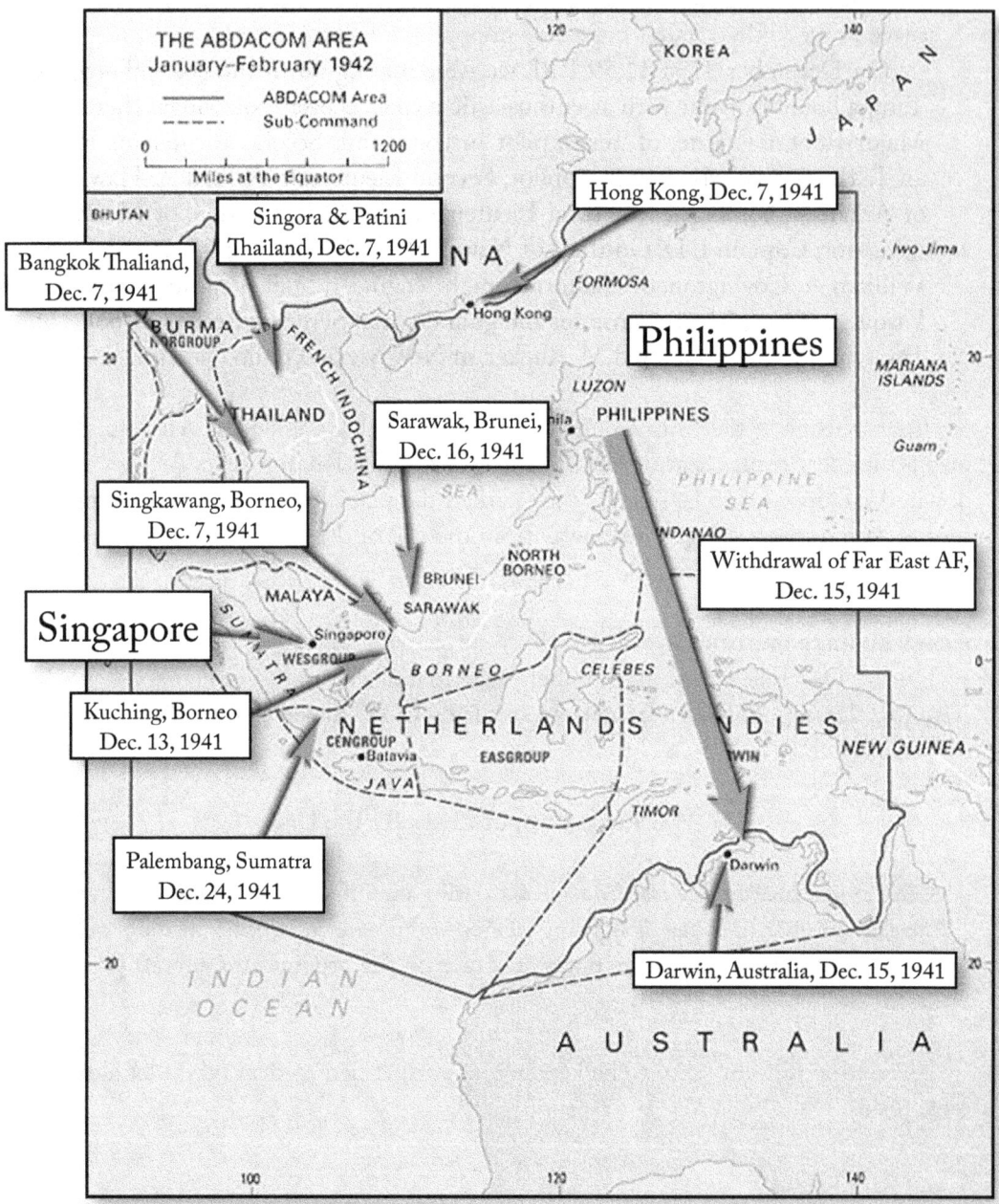

Figure 2 – ABDA – December 1941

Figure 3 – Burma December 1941

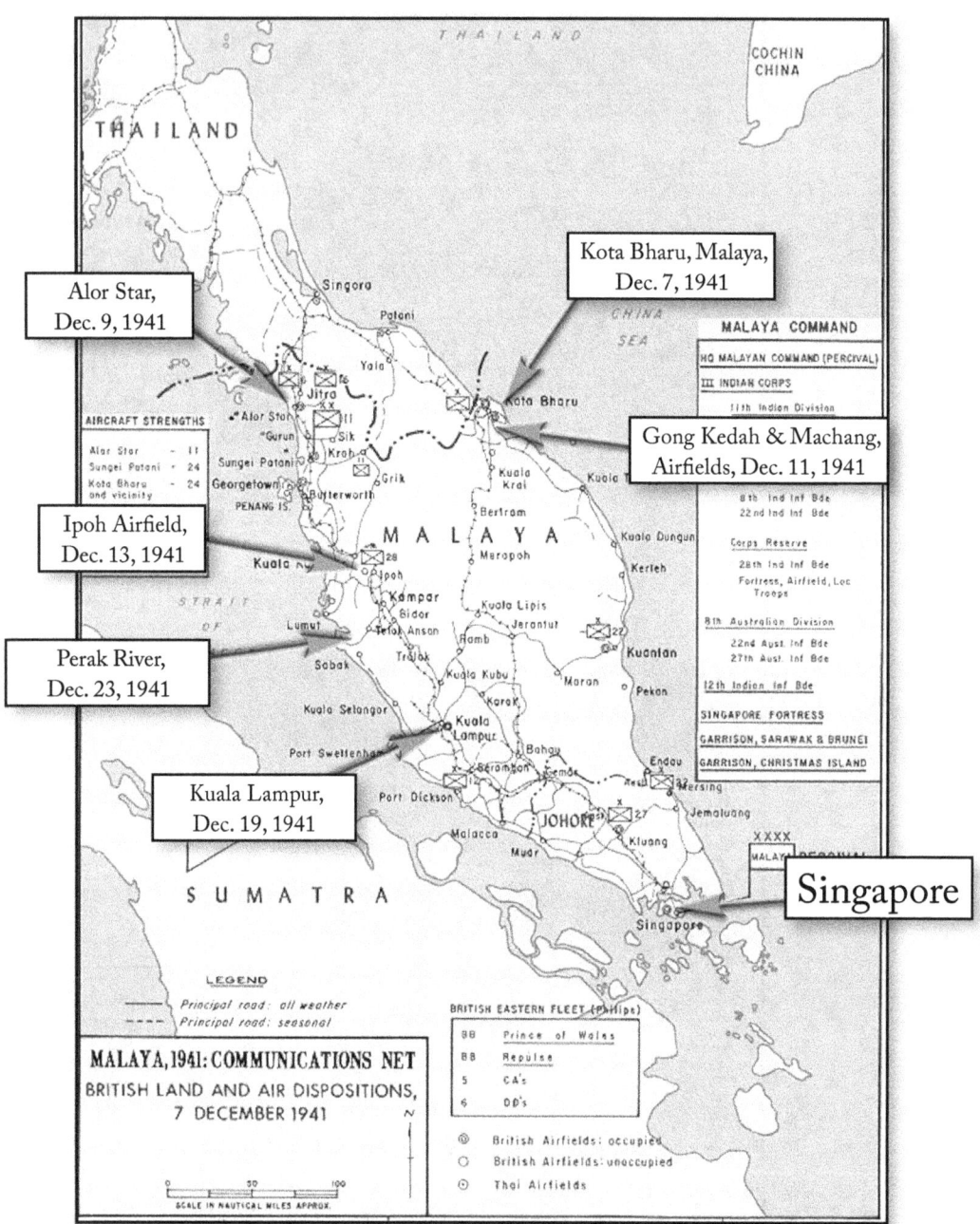

Figure 4 – Malaya – December 1941

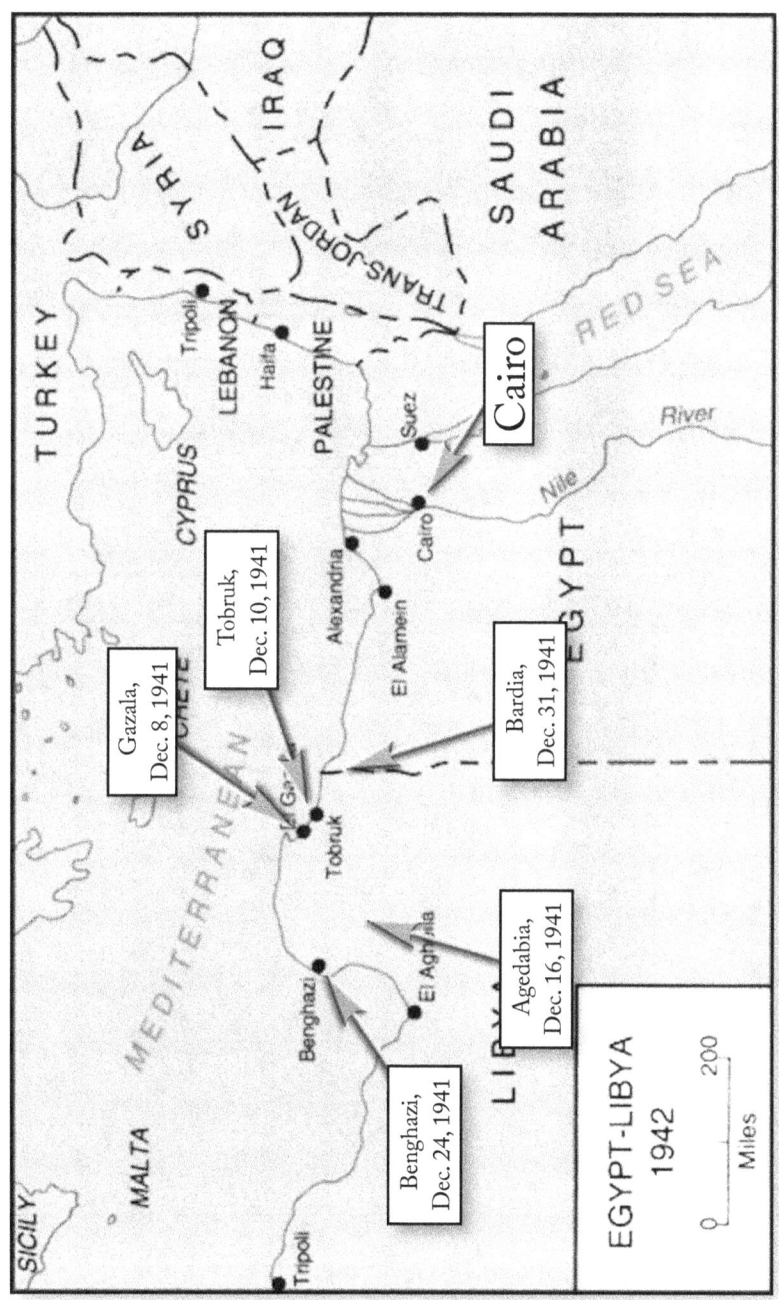

Figure 5 – Mediterranean – December 1941

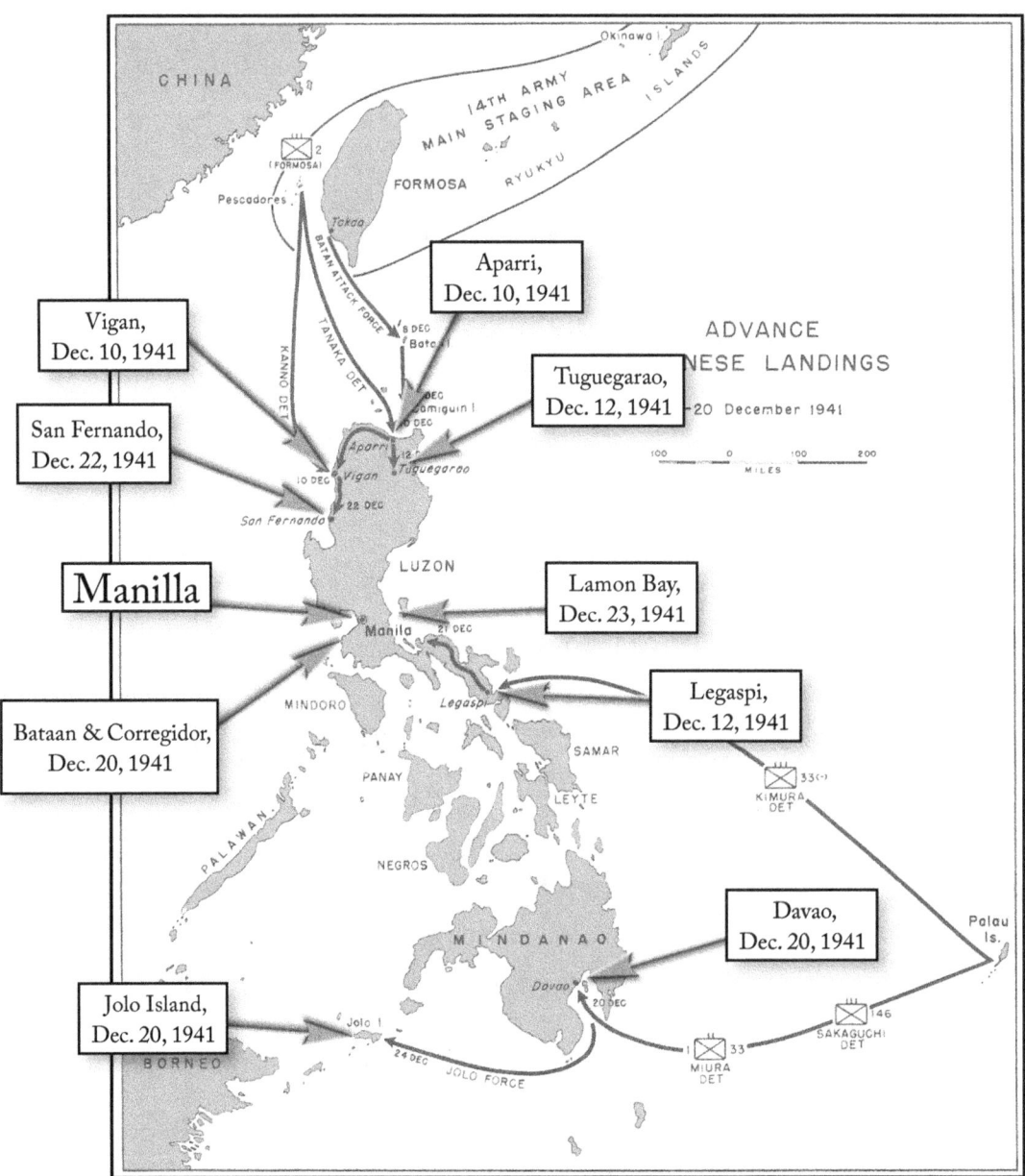

Figure 6 – Philippines – December 1941

Figure 7 – Moscow – December 1941

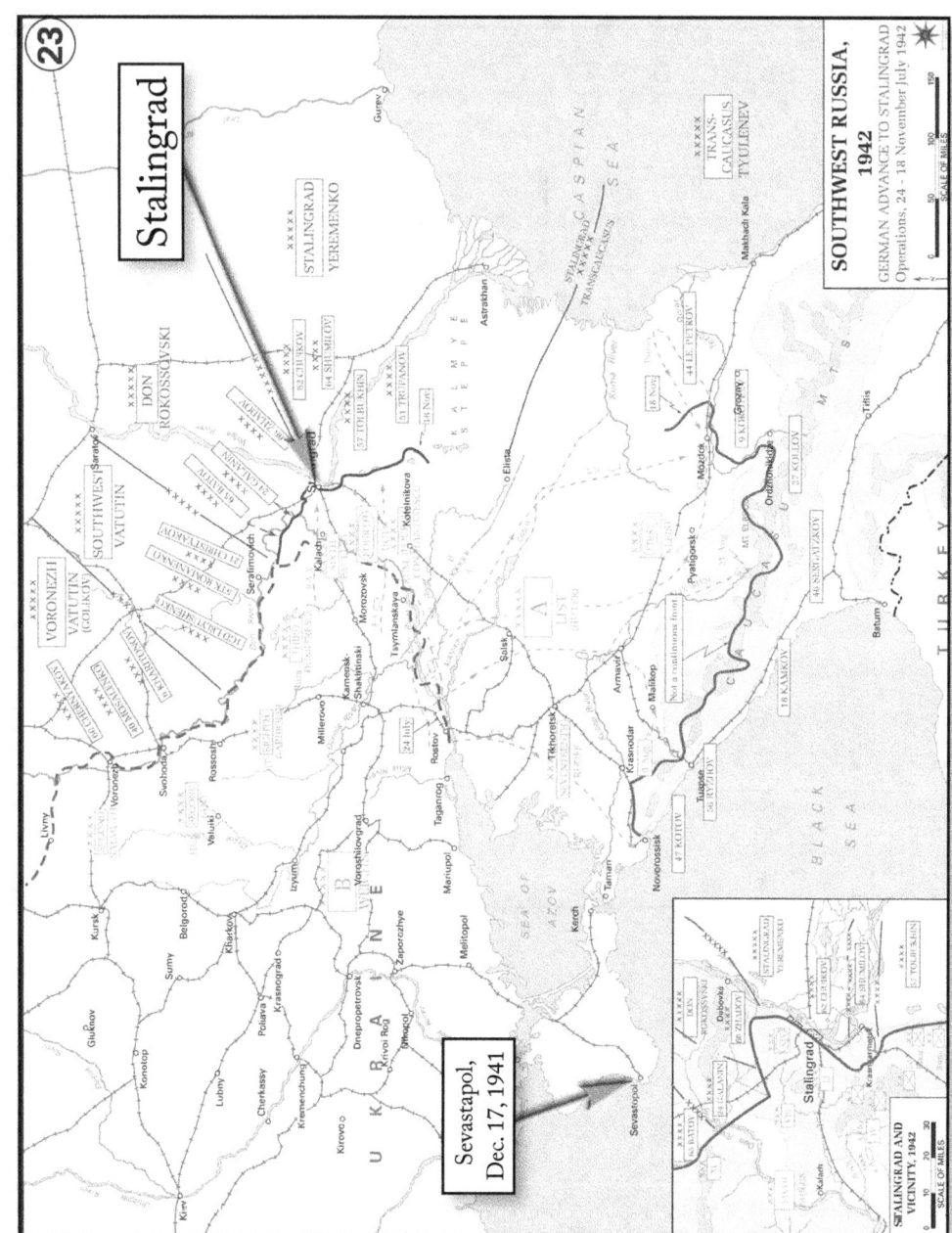

Figure 8 – Stalingrad – December 1941

Figure 9 – Private Albert Story, age 16. December, 1941. (Photo credit: Albert Story)

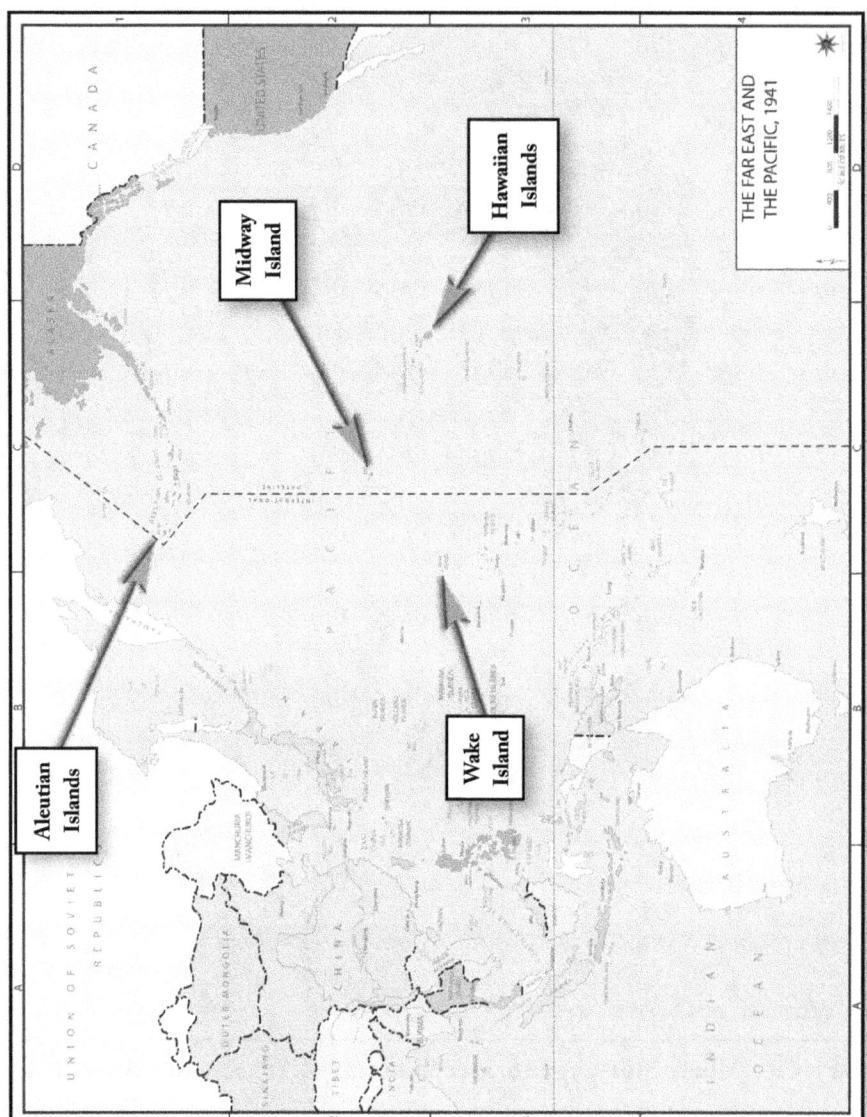

Figure 10 – Wake, Midway, Aleutian, and Hawaiian Islands

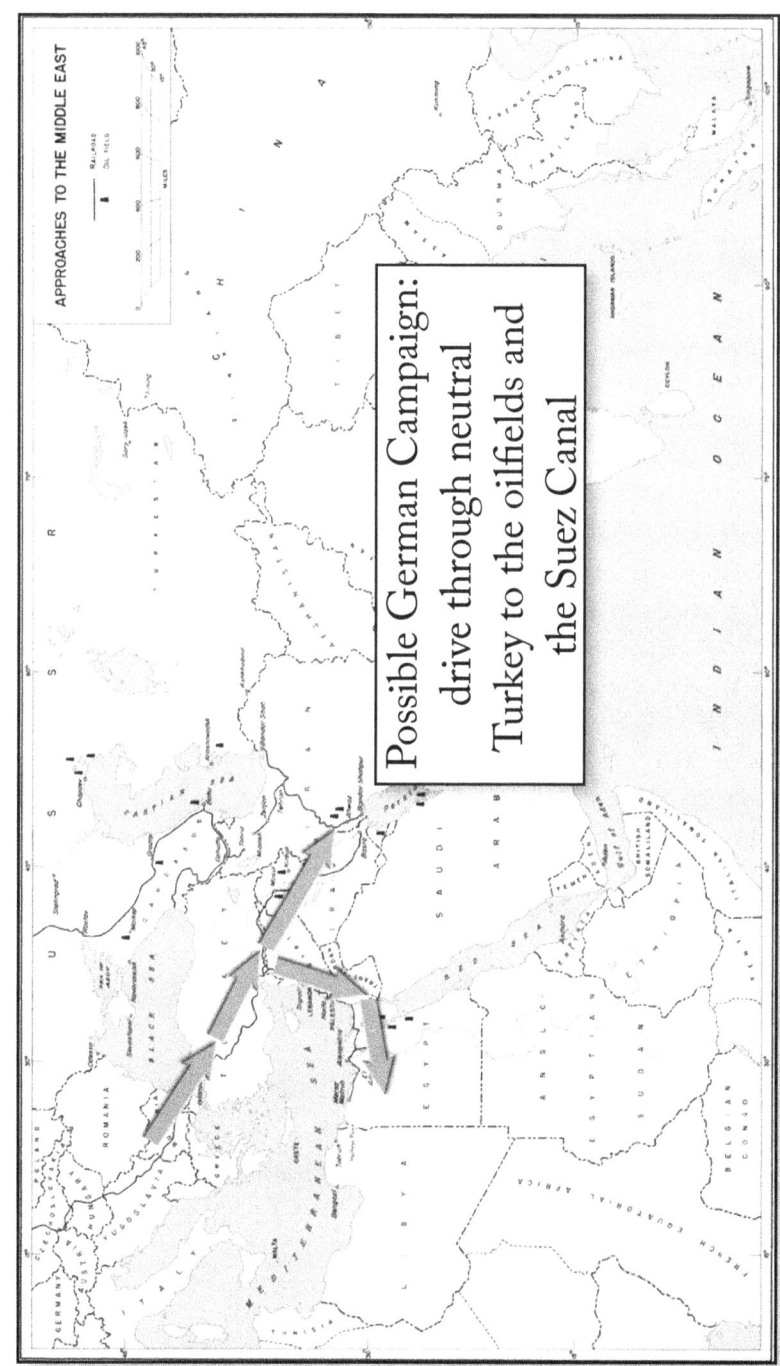

Figure 11 – Turkey awaits Nazi drive to the East

ENDNOTES

1. Craven, *The Army Air Forces in World War II, Vol. 1*, pp. 272-273.
2. Craven, *The Army Air Forces in World War II, Vol. 1*, p. 236.
3. Memo from Currie to FDR, December 8, 1941, China 1941 file, Box 27, PSF, FDRL; Magruder is Chief of the U.S. Military Mission to China.
4. Stimson, *Stimson Diaries*, entry for December 8, 1941.
5. Headrick, *Bicycle Blitzkrieg*, p. 6.
6. Morton, *Strategy And Command: The First Two Years*, p. 138.
7. Herman, *To Rule The Waves*, p. 539.
8. Williams, *Chronology 1941-1945 - 1941*, p. 4.
9. Williams, *Chronology 1941-1945 - 1941*, p. 3.
10. Williams, *Chronology 1941-1945 - 1941*, pp. 3-4.
11. Story, *Private Story*.
12. Hannah, letters to the author.
13. Williams, *Chronology 1941-1945 - 1941*, p. 4.
14. Williams, *Chronology 1941-1945 - 1941*, p. 4.
15. Williams, *Chronology 1941-1945 - 1941*, p. 4.
16. Craven, *The Army Air Forces in World War II, Vol. 1*, p. 272.
17. ----. FDRL, Day by Day, December 9, 1941. http://www.fdrlibrary.marist.edu/daybyday/daylog/december-9th-1941/, accessed July 10, 2012.
18. ----, *Peace and War*, Radio Address Delivered by President Roosevelt From Washington, December 9, 1941, pp. 842-848.
19. Fleming, *The New Dealer's War*, p. 34.
20. Morton, *The Fall of the Philippines*, p. 145.
21. ----, AFHRA, Hap Arnold file.
22. Morton, *Strategy And Command: The First Two Years*, p. 138.
23. Williams, *Chronology 1941-1945 - 1941*, p. 4.
24. Williams, *Chronology 1941-1945 - 1941*, p. 4.
25. Williams, *Chronology 1941-1945 - 1941*, p. 4.
26. The classification of "Pursuit" would be changed later to "fighter".
27. Craven, *The Army Air Forces in World War II, Vol. 1*, p. 222.
28. Morton, *The Fall of the Philippines*, p. 146.
29. Robert A. Lovett, who since December, was a special assistant on air matters.
30. Stimson, *Stimson Diaries*, entry for December 10, 1941.
31. Williams, *Chronology 1941-1945 - 1941*, p. 4.
32. Williams, *Chronology 1941-1945 - 1941*, p. 5.
33. Williams, *Chronology 1941-1945 - 1941*, p. 5.

34. Williams, *Chronology 1941-1945 - 1941*, p. 5.
35. ----, FDRL, PSF, Box 28, Lattimore file, Currie to FDR December 11, 1941.
36. Headrick, *Bicycle Blitzkrieg, p. 7.*
37. Headrick, *Bicycle Blitzkrieg, p. 7.*
38. Sumner Welles was Under Secretary of State and a foreign Policy advisor to Roosevelt.
39. ----, FDRL, PSF, Box 2, China file, Soong to Welles, December 12, 1941.
40. Williams, *Chronology 1941-1945 - 1941*, p. 5.
41. ----, FDRL, PSF, Box 3, Germany file, Status Report, December 12, 1941.
42. Craven, *The Army Air Forces in World War II, Vol. 1*, p. 273.
43. Cave, *transcript of interview.*
44. Williams, *Chronology 1941-1945 - 1941*, p. 6.
45. Williams, *Chronology 1941-1945 - 1941*, p. 5.
46. Williams, *Chronology 1941-1945 - 1941*, p. 5.
47. Morton, *The Fall of the Philippines*, p. 152.
48. Williams, *Chronology 1941-1945 - 1941*, p. 5.
49. Craven, *The Army Air Forces in World War II, Vol. 1*, p. 353.
50. Brett was currently stationed in Cairo, Egypt, attempting to eliminate road blocks to supplying material to the Nritish and Russians.
51. ----, FDRL, PSF, Box 2, China file, Stimson to FDR, FDR to Chiang (draft), December 13, 1941.
52. Mayhew, *transcript of interview*
53. Williams, *Chronology 1941-1945 - 1941*, p. 6.
54. ----, FDRL, PSF, Box 34, Great Britain 1941 file, Note from the British Embassy, December 29, 1941.
55. Morton, *The Fall of the Philippines*, p. 149.
56. Williams, *Chronology 1941-1945 - 1941*, p. 6.
57. Perry, *The Most Dangerous Man in America*, p. 89.
58. Craven, *The Army Air Forces in World War II, Vol. 1*, p. 222.
59. Stimson Diaries. Sun. December 14, 1941.
60. ----, AFHRA, Hap Arnold file, Memorandum For the Chief of Staff, December 14, 1941.
61. Williams, *Chronology 1941-1945 - 1941*, p. 6.
62. Craven, *The Army Air Forces in World War II, Vol. 1*, p. 223.
63. Williams, *Chronology 1941-1945 - 1941*, p. 6.
64. Assistant Chief of Air Staff, *The AAF in the Middle East* , p. 42.
65. Thomas, *Born in Battle*, pp. 30-31.
66. Williams, *Chronology 1941-1945 - 1941*, p. 6.
67. Williams, *Chronology 1941-1945 - 1941*, p. 6.
68. Williams, *Chronology 1941-1945 - 1941*, p. 6.
69. Ben-Moshe, Winston Churchill and the "Second Front, pp. 506-507.
70. Jarrett, V. H. C., *New York Times*, *HONG KONG STRONG AND IN 'GOOD HEART'*, Decem-

ber 17, 1941, http://select.nytimes.com/gst/abstract.html?res=F70913F9345B147B93C5A81789D-95F458485F9, accessed January 28, 2014.
71. Williams, *Chronology 1941-1945 - 1941*, p. 6
72. Mitcham, *Rommel's Desert War*, p 4.
73. Williams, *Chronology 1941-1945 - 1941*, p. 6.
74. Williams, *Chronology 1941-1945 - 1941*, p. 6.
75. ----, FDRL, PSF, Box 3, Hopkins Harry file, Hopkins to FDR. December 17, 1941.
76. ----, *Pacific Wrecks*, http://www.pacificwrecks.com/aircraft/b-17/41-2446.html, accessed February 11, 2014.
77. Williams, *Chronology 1941-1945 - 1941*, p. 7.
78. Ben-Moshe, Tuvia, Winston Churchill and the "Second Front, p. 507.
79. Williams, *Chronology 1941-1945 - 1941*, p. 7.
80. ----, FDRL, PSF, Box 3, Hopkins Harry file, Hopkins to FDR. December 19, 1941.
81. Craven, *The Army Air Forces in World War II, Vol. 1*, p. 332.
82. Williams, *Chronology 1941-1945 - 1941*, p. 7.
83. Xu, *The Issue of US Air Support for China*, p. 460.
84. Xu, *The Issue of US Air Support for China*, p. 477.
85. ----, FDRL, PSF, Box 1, American-British Joint Chiefs of Staff file, Stimson to FDR, 20 December 1941.
86. Ben-Moshe, Tuvia, Winston Churchill and the "Second Front, p. 508.
87. ----, *League of Nations Treaty Series*, http://www.worldlii.org/int/other/LNTSer/1941/37.html, accessed October 21, 2015.
88. Glines, *The Doolittle Raid*, p. 10.
89. James, Edwin, *New York Times*, GERMANY STILL HEART OF AXIS WAR STRENGTH, December 21, 1941, http://select.nytimes.com/gst/abstract.html?res=F40A14FC3C5916738DDDA80A94DA415B-8188F1D3, accessed February 13, 2014.
90. Baldwin, Hanson, *New York Times*, JAPANESE GAIN QUICK SUCCESSES, December 21, 1941, http://select.nytimes.com/gst/abstract.html?res=FA0A14FC3C5916738DDDA80A94DA415B8188F1D3, accessed February 13, 2014.
91. Krock, Arthur, *New York Times*, FREEDOM OF THE PRESS RESTRICTED, December 21, 1941, http://select.nytimes.com/gst/abstract.html?res=F10A14FC3C5916738DDDA80A94DA415B-8188F1D3, accessed February 13, 2014.
92. Williams, *Chronology 1941-1945 - 1941*, p. 7.
93. ----, *Time, The U.S. At War, DIPLOMATICS: Litvinoff's Problem*, December 22, 1941, http://www.time.com/time/magazine/article/0,9171,931955,00.html , accessed July 11, 2012.
94. ----, FDRL, PSF, Box 1, America, Britain, China, and Dutch East Indies (ABCD Powers) I file, Marshall to FDR, 22 December 1941.
95. Mayhew, *transcript of interview*
96. Baldwin, Hanson, *New York Times*, Plight of Hong Kong Force, December 22, 1941, http://select.nytimes.

com/gst/abstract.html?res=F30E1EFC3C5916738DDDAB0A94DA415B8188F1D3, accessed February 13, 2014.

97. Williams, *Chronology 1941-1945 - 1941*, p. 8.
98. Williams, *Chronology 1941-1945 - 1941*, p. 8.
99. Williams, *Chronology 1941-1945 - 1941*, p. 8.
100. Rickard, *Battle of Wake Island.*
101. Craven, *The Army Air Forces in World War II, Vol. 1*, p. 228.
102. Williams, *Chronology 1941-1945 - 1941*, p. 8.
103. Williams, *Chronology 1941-1945 - 1941*, p. 8.
104. Tedder was the British Air Chief in Cairo.
105. ----, FDRL, PSF, Box 2, Bullitt William C file, Bullet to FDR, December 23, 1941.
106. Thomas, *Born in Battle*, p. 32.
107. Williams, *Chronology 1941-1945 - 1941*, p. 8.
108. ----, AFHRA, Hap Arnold file, Stark to Arnold, December 24, 1941.
109. Craven, *The Army Air Forces in World War II, Vol. 1*, pp. 559, 248.
110. King, *Fleet Admiral King*, p. 363.
111. ----. FDRL, Day by Day, December 24, 1941,
112. Brand, H. W., *FDR: Traitor*, p. 243.
113. Williams, *Chronology 1941-1945 - 1941*, p. 8.
114. Williams, *Chronology 1941-1945 - 1941*, p. 9.
115. Brand, H. W., *FDR: Traitor*, p. 243.
116. Williams, *Chronology 1941-1945 - 1941*, p. 9.
117. Williams, *Chronology 1941-1945 - 1941*, p. 9.
118. Craven, *The Army Air Forces in World War II, Vol. 1*, pp. 559, 248.
119. Thomas, *Born in Battle*, p. 36.
120. Williams, *Chronology 1941-1945 - 1941*, p. 9.
121. http://www.merriam-webster.com/dictionary/open%20city , accessed October 19, 2015.
122. Williams, *Chronology 1941-1945 - 1941*, p. 9.
123. ----, AFHRA, Hap Arnold file, Brett to Arnold "Report on B-17 Operations Out of China," 26 December 1941.
124. Morton, *The Fall of the Philippines*, p. 157.
125. Williams, *Chronology 1941-1945 - 1941*, p. 9.
126. ----, The Public Papers of the Presidents of the United States, *The public papers and addresses of Franklin D. Roosevelt. 1941 volume*, p. 604.
127. ----, *New York Times, TURKEY AWAITS NAZI DRIVE TO EAST*, December 28, 1941, http://select.nytimes.com/gst/abstract.html?res=F00D12F83B5F1A7A93CAAB1789D95F458485F9, accessed February 13, 2014.
128. Craven, *The Army Air Forces in World War II, Vol. 1*, p. 243.
129. Miller, *The Chiang-Stilwell Conflict*, p. 59.

130. Mayhew, *transcript of interview*
131. Williams, *Chronology 1941-1945 - 1941*, p. 10.
132. Williams, *Chronology 1941-1945 - 1941*, p. 10.
133. Williams, *Chronology 1941-1945 - 1941*, p. 10.
134. Craven, *The Army Air Forces in World War II, Vol. 1*, p. 370.
135. Williams, *Chronology 1941-1945 - 1941*, p. 10.
136. Williams, *Chronology 1941-1945 - 1941*, p. 10.
137. ----, AFHRA, Hap Arnold file, Arnold to Air War Plans Div, December 31, 1941.
138. Craven, *The Army Air Forces in World War II, Vol. 1*, p. 238.
139. ----, AFHRA, Hap Arnold file, Arnold to A-3, December 31, 1941.
140. Eckerly, *A Pilot's Story*.
141. Thomas, *Born in Battle*, pp. 38-46.
142. ----, Army Air Force Statistical Digest, http://www.ibiblio.org/hyperwar/AAF/StatDigest/aafsd-3.html , accessed January 22, 2014, p. 112.
143. Trussel, C. P., *New York Times*, *WAR PLANE OUTPUT LEAPS BY THE DAY*, January 4, 1942, http://select.nytimes.com/gst/abstract.html?res=F20911FD3D58167B93C6A9178AD85F468485F9, accessed February 14, 2014.

JANUARY 1942

From a planning perspective, the biggest event in January would be the meeting of British and United States leaders in the last half of the month. For the Army Air Corps, the Arnold-Portal Agreement would guide how its resources would support Allied objectives.

But while generals planned, the enemy was not idle. Japan continued her military conquest. Defense of the Philippines would transition to Abandonment of the Philippines. Despite British claims that Singapore could hold out, press reports suggested otherwise.

Lost in the "big picture" were two other activities, destined to collide by year's end. First, the embryo of a United States task force to strike at Japan was taking shape. Second, Rommel was on the move.

1942 JANUARY 1

BOMBING JAPAN

Upon receipt of Arnold's directive to "Bomb Japan," Alexander's assessment was:

1. Everything necessary to sustain operations of a pursuit and bomb group will have to be hauled up the Burma Road.
2. I recommend urgently that no combat units be sent to China. They cannot be supplied, maintained or fought here. Any airplane parked on any airdrome in unoccupied China lies within the operating radius of both Jap bombardment and pursuit.

He recommended that any air effort in the Chinese theatre be based initially in Burma, with subsequent air ground action directed to clearing the Japanese out of Rangoon (see Figure 13) and then rolling the enemy north out of Indochina.[1]

SINGAPORE

Japanese troops attacked Allied positions near Kampar, Malaya, but were unable to overrun the positions. They tried to out flank the Allies by landing amphibious troops on the Bernam River, behind the Allied positions. Meanwhile, Japanese aircraft attacked the Tengah Airfield near Singapore.[2] (see Figure 14)

MIDDLE EAST

The importance of the Middle East increased due to its commanding position on the supply lines to the Soviet Union and to the China-Burma-India theater (CBI). Since the security of the area depended upon the adequacy of its air power, immediate reinforcement of the RAF Middle East Command was imperative. Portal asked Arnold to send two American pursuit groups to this theater at the earliest practicable date.

Arnold favored the idea because:

1. Not only would British military control of the Egyptian district be strengthened, but
2. Our squadrons would acquire valuable training in the coordination of air and ground efforts -- a type of experience that would prove most useful, if in the future the United States wished to employ bombardment units in the locality.[3]

Portal also wanted heavy bombers sent to the Middle East. It is important to recall that on December 23, Tedder had informally sent a message to Roosevelt via William Bullet, asking for three squadrons of B-24s. Now, Portal was making the request through official channels. But, Portal did not want to divert bombers destined for England. So, he asked Arnold if the United States could send a couple heavy bomber groups to the Middle East. Possibly "about March or April" was Arnold's answer.[4]

OPERATION GYMNAST

The new Combined Chiefs of Staff began discussing a quick countermove called GYMNAST, an expedition to North Africa to seize Casablanca. Arnold argued "not very strongly, against GYMNAST."[5]

REINFORCEMENTS

At Langley, pilot Paul Eckerly met the rest of his crew. Lt. James O. Cobb and the flight engineer, Cpl. E. W. Harbaugh had just flown a brand new B-17, serial number 41-2481, nicknamed *Topper*, to Langley from the Boeing Seattle plant.[6]

1942 JANUARY 2

SINGAPORE

Japanese troops continued their attack on Allied positions near Kampar, Malaya.

Their amphibious troops landed at Telok Anson on the Perak River, (see Figure 4) another threat to the Allied positions. The Allies decided to withdraw.[7]

NORTH AFRICA

The Axis forces at Bardia surrendered.[8] (see Figure 5).

MIDDLE EAST

It appeared that the focus on the Middle East was gaining traction. First, in anticipation of the deployment of United States units to the Middle East, tentative plans were outlined for a United States air force. This new Command was directed to establish a small headquarters for interceptor, bomber, and air service commands.[9]

Then, Bullitt sent a message to Roosevelt:

> *I hope that you and* Churchill in your planning are keeping in mind the vital need to retain Egypt as a secure base not only for operations in Libya, Tunisia, Palestine, the Lebanon, Syria, and Turkey, but also for transit of planes to India and the Far East.[10]

OPERATION GYMNAST

A second conference on GYMNAST was held. According to General Joseph W. Stilwell, "every [US advisor] was against it." Marshall explained that the basic objective was to go into Casablanca to protect the Mediterranean Sea lane, thus defending British shipping.[11] This plan seemed to be another example of the perception that the United States was using its resources to save British interests.

The drumbeat for an Allied attack, i.e. a United States Air Corp attack, on German oil supplies slowly increased. Gerow forwarded a memo to Arnold, which included a report from Bonner Fellers, the military attaché in Cairo. Fellers had stated:

1. Germany's oil reserves were 1,700,000 tons (Jan 1942). At the estimated consumption rate of 400,000 tons per month, supplies would be depleted in a little over four months.
2. The Germans might make an effort "toward the Iraq [oil] source some time this winter."
3. He recommended that the WPD take into consideration the bombing of Rumanian oil fields stating:

   - Denial of Rumanian oil will shut off German facility for conducting even static operations.[12]

HALPRO

Subject: Orders
To: Lieutenant Colonel Harry A. Halverson, Fort George Wright; Washington.

1. The Commanding General directs that you proceed on or about January 2, 1942, by commercial aircraft (as authorized by Section II. Circular No. 128. War Department, November, 1940) to Bolling Field, D. C., for temporary duty.[13]

Halverson was Assistant Chief of Staff, G-3, in the Headquarters of the Second Air Force. He had been involved in the first Army Air Corps air-to-air refueling demonstration and was an expert in aircraft mission performance. This was the first assignment of personnel to execute Roosevelt's order to "bomb of Japan".

PROJECT X

In a memo to Office of the Assistant Chief of the Air Staff, Plans (A-3), Arnold outlined plans for delivering heavy bombers to the Philippines.

Directive Memo
Subject: Delivery of HB to "X".

1. Two to three B-17s or B-24 type airplanes per day will be dispatched over Western route to "X" as soon as they can be made available. You will continue dispatching on the basis of three per day over the Eastern route to "X".

By direction of the Chief of the Air staff.[14]

The United States was still trying to send bombers to the Philippines despite the withdrawal of the bombers to Australia the day before.

1942 JANUARY 3

THE PHILIPPINES
The headlines of *The New York Times* said it all:

**MANILA AND CAVITE BASE FALL, ARMY FIGHTS ON
26 NATIONS PLEDGE ALL RESOURCES VICTORY**[15]

(see Figure 16)

1942 JANUARY 4

SINGAPORE

Japanese troops landed on the Selangor River, threatening the Allied position at Rawang, located just northwest of Kuala Lumpur.[16] (see Figure 4)

PLOESTI

Bonner Fellers sent a cable advising that the combined oil output of Rumania and Germany "is now only adequate" to fuel the German war machine for a static war in Russia.

1. He estimated that the frustrations of Hitler's grab for the Caucasus might lead Hitler to turn toward the Iraq to obtain new oil supplies in the future.
2. He recommended that Rumanian oil sources be hit to narrow the scale of German operations.
3. He stated the Rumanian oil supply was within the periphery of Syria-based long-range bombers.[17]

This memo repeated the information he had sent to Gerow two days earlier.

1942 JANUARY 5

THE PHILIPPINES

Meanwhile, Allied forces had completed their withdrawal to the Bataan Peninsula.[18] (see Figure 6)

NORTH AFRICA

The Afrika Korps received badly needed supplies, including 55 new tanks.[19] Recent research indicates that by January 10, Rommel had received 71 tanks.[20]

SINGAPORE

Admiral Layton moved the headquarters of the British Eastern Fleet from Singapore to Tanjung Priok, the port for Batavia, Java.[21] (see Figure 12)

1942 JANUARY 6

NORTH AFRICA

The British began using Derna as a supply depot.[22] Derna is west of Tobruk (see Figure 15) and thus closer to the British front lines.

THE PHILIPPINES

The reality that the defense of the Philippines was a lost cause had finally sunk into the minds of the planners of the United States Air Staff. Heavy bombers were in the air and unless diverted would try to get to the Philippines. Consequently, the flight commanders of the Project X bombers were ordered to proceed as far as Bangalore in southern India, at which point they were to await further orders. Amended orders were issued diverting them to Australia as their final destination.[23]

Project X was the reinforcement of the Philippines. These orders were more confirmation that War Plan ORANGE had been correct — the defense of the Philippines was not possible. Brereton and his Air Corps bombers had already withdrawn from the Philippines.

1942 JANUARY 7

NORTH AFRICA

British reconnaissance units discovered that Rommel had abandoned Agedabia. Meanwhile, a British supply convoy reached Benghazi.[24] (see Figure 5)

EAST INDIES

The Japanese troops that had landed at Sarawak arrived at the Dutch Borneo border.[25] (see Figure 2)

Brereton had withdrawn his bombers a few days earlier. Now *The New York Times* announced that the overall theater commander had also withdrawn to Java:

WAVELL TO DIRECT FORCES FROM JAVA [26]

REINFORCEMENTS

Arnold's AWPD had been working on revisions to their ultimate operational goals for fighting the war. Previous estimates called for an Air Corps with 84 groups. Now, they estimated that the Air Force would need 115 groups.[27]

1942 JANUARY 8

EAST INDIES

The Japanese captured Jesselton, British Borneo.[28] (see Figure 12)

COMMITMENT TO CHINA

Currie had raised the issue of converting the Flying Tigers from a group of mercenaries to legitimate members of the Army Air Corps. He finally got his wish when Arnold issued the following:

Directive Memo For: Air War Plans Division
Subject: Plan for AAF Participation in China Theater.

1. Decision has been made that the Army Air Forces will furnish air support to the Chinese Government in the China Theater. The AVG Group is to be converted into a U.S. Army Air Force pursuit group or utilized as the AVG, considering it in the same category as a U.S.A. Army Air Force pursuit group.
2. It is directed that you prepare a separate plan to…support a U.S.A. offensive air force in China.[29]

AWPD was directed to plan for bomber operations from Chinese bases. It took Arnold a month from Roosevelt's directive to get the AWPD involved. Yet he already had Brett's and Alexander's assessments that China was not a viable option. This would be the third time he asked the same question.

BOMBING JAPAN

Regardless of the reason, Halverson, who had been recalled from Seattle, was identified as the leader of the "USA Offensive Air Force".

Memorandum to: Colonel H. L. George, Air War Plans

1. In reference to Directive No. ___, certain actions have been taken to date as follows:
    a. …..
    b. It is understood that the War Department is establishing a task force for the China Theater…
    c. General Arnold has personally directed Colonel H. A. Halverson to perform a certain mission in the China Theater.[30]

1942 JANUARY 9

THE PHILIPPINES
The Japanese began their assault on Bataan.[31] (see Figure 6)

SINGAPORE
More bad news was coming forth regarding the defense of Singapore. Allied forces

were ordered to withdraw to Johore, located on the tip of the Malaysian mainland, directly across from the island of Singapore.[32] (see Figure 14)

HALPRO

Halverson began to organize his mission. Arnold ordered representatives of his support staff to attend a meeting. And he gave Halverson's mission a label.

Directive Memo For: Air War Plans Division
Subject: Halverson Project.

1. In connection with the Halverson Project,
2. A preliminary meeting will be held…Saturday, January 10, 1942.[33]

This was the first use of the phrase "Halverson Project." It would later be shortened to HALPRO.

1942 JANUARY 10

THE PHILIPPINES

The bad news from the Philippines continued. The Japanese made their first demand for an Allied surrender.[34]

EAST INDIES

Wavell arrived in Java to confer with commanders there. He established his headquarters at Lembang, north of Bandoeng.[35] (see Figure 12)

COMMITMENT TO CHINA

Chiang had already been named supreme commander of the Allied Forces in China. The United States Chiefs of Staff presented a plan to the Combined Chiefs, which was approved on January 10. This entailed:

1. Increasing the security and capacity of the Burma Road,
2. Providing base facilities and technical services to made the Chinese combat operations more effective.
3. Appointing, with the consent of Chiang, a high-ranking military officer to act as the representative of the United States in China.[36]

They also agreed that a high-ranking United States Army officer needed to be sent to

China to act as a liaison between Chiang and the American-British-Dutch-Australian (ABDA) command.[37]

DOOLITTLE RAID

Navy Captain Francis Low, Assistant Chief of Staff for Anti-Submarine Warfare, reported to Admiral King, stating:

1. He thought that twin-engine Army bombers could be successfully launched from an aircraft carrier.
2. He had observed several such bombers operating at the Norfolk naval airfield, where the runway was painted with the outline of a carrier deck for landing practice.[38]

Thus, the seeds of the Doolittle raid were planted.

1942 JANUARY 11

NORTH AFRICA

British units attacked another Axis stronghold at Sollum, Libya, capturing it on January 12. Sollum is just south of Bardia. (see Figure 5) British forces chased Axis forces towards El Agheila.[39] (see Figure 15)

SINGAPORE

The Japanese captured Kuala Lumpur.[40] (see Figure 4)

EAST INDIES

The Japanese made two landings: one at Tarakan, Dutch Borneo, the second in the Celebes at Manado and Kema. Meanwhile, Japanese paratroops were dropped on the airfield near Manado. Allied resistance was minimal and the Japanese placed the airfields at Tarakan and Manado into use.[41] (see Figure 12)

REINFORCEMENTS

Paul Eckerly and his crew departed in their B-17 from MacDill Field in Tampa. This was the first leg of their flight to the Far East. After a flight of almost fourteen hours, they landed at Trinidad.[42]

While this crew was beginning its trip to the Far East, the first three LB-30s were arriving at Java.[43] These planes had used the same South Atlantic route that Eckerly's crew was about to use.

1942 JANUARY 12

PLOESTI

Lt. Col. Fred D. Sharp was the Military Attaché to the United States Embassy in Rumania. He filed a report on the significance of the Rumanian oilfields to the German war effort. His concluding sentence read as follows:

> *The refineries at Ploesti, the* rail centers, the loading and unloading terminals on the Danube and various points on the canals and inland waterways would be among the most vital targets n the whole of German dominated Europe.[44]

OPERATION ACROBAT

General Auchinleck wrote:

> *I am convinced we should* press forward with "Acrobat" for many reasons, not the least in order that Germany may continue to be attacked on two fronts, Russia and Libya.[45]

Operation ACROBAT was the British advance from Cyrenaica to Tripoli in 1941. (see Figure 15)

ARSENAL OF DEMOCRACY

While Roosevelt, Marshall, Arnold and the rest of the military leaders scrambled to allocate scarce resources to the various war fronts, *Time* ran an article in its January 12, 1942 issue, headlined *The Miserable Truth*.

The first three paragraphs were:

The miserable truth was coming out. U.S. production of war materials was still nowhere near enough.

Everybody blamed somebody else. Assistant Attorney General Thurman Arnold blamed "monopoly." The pinko Nation and New Republic blamed capitalism in general. Columnist Westbrook Pegler blamed C.I.O. strikes, A.F. of L. racketeering. Columnist Walter Lippmann put the finger on U.S. industry, for its "disgraceful" commercial boom, chewing up vast quantities of precious strategic metals and rubber. Anti-New Dealers blamed New Dealers, and vice versa. Congressmen blamed the OPM, and vice versa. Little businessmen blamed the Army & Navy.

U.S. citizens — thinking of Marines on Wake Island fighting off the Japs with only four planes, of their soldiers in the Philippines defending bridges with rifles and hand grenades against machine guns — cried a pox on Thurman Arnold et al., a pox on blame-laying, a pox on alibis and confusion and red tape. Bottleneck had become the most hated word in the language, "too little & too late" a phrase too deep for tears.[46]

Roosevelt had used the bully pulpit to argue for his Arsenal of Democracy. Congress had approved huge increases in defense budgets. The public obviously thought the country was ready to go to war. Apparently, the United States was not.

1942 JANUARY 13

EASTERN FRONT
The Russians continued driving the German Army Group Central away from Moscow, retaking Kirov.[47] (see Figure 17)

RAINBOW-5
According to RAINBOW-5, the first United States task force designated for early offensive action was the AAF bombardment force, which was to attack Germany in cooperation with the RAF. This project was officially approved on January 13 by the JCCS.[48]

ARNOLD-PORTAL AGREEMENT
However, other pre-war plans had to be adjusted, in particular, the allocation of United States-made aircraft to American allies. In what became known as the Arnold-Portal agreement, the respective heads of the two Air Forces approved a revised schedule of monthly deliveries of airplanes.[49] The dates and number of aircraft to be delivered to England would depend in large measure upon events in the Far East. Since any offensive against Germany was dependent upon a successful build-up of air power in England, events in the Far East would dictate that schedule.[50]

REINFORCEMENTS
Following the attack on Pearl Harbor, the United States Pacific fleet could no longer contain Japan. That containment was now dependent upon United States air power, which also did not exist. The Germany First strategy of ABC-1 was no longer in effect. Japanese activities (or even potential activities) would now require a response from the western Allies. The two supply routes from the United States had to be kept open. Otherwise the United States, the only major producer of material, would find itself geographically isolated.

One of the first applications of this new thought process was the approval of a report entitled "Defense of Island Bases Between Hawaii and Australia." The plan was written on January 10 and approved three days later.[51] The first paragraph of that report stated the importance of the supply route:

> *There is under development and* approaching completion, an air route suitable for the use of both long and medium range aircraft and extending from Hawaii to Australia. Airdromes are located at Palmyra, Christmas, Canton, American Samoa, Fiji, and New Caledonia. In addition to their use as staging points along the air route, all of these islands are valuable outposts of the defenses of the Hawaiian Islands or of New Zealand and Australia. They will also serve as operating bases for naval and air forces.[52] (see Figure 19)

1942 JANUARY 14

OPERATION SUPER-GYMNAST

Prior to the United States entry into the war, American planners had conceived two different plans to prevent any Axis penetration towards the South Atlantic and South America. One of those plans was an invasion at Casablanca, code name GYMNAST. The British thought an invasion of Tunisia was more appropriate. At Arcadia the joint planners were ordered to merge the two plans into one, code name SUPER-GYMNAST.[53]

ARNOLD-PORTAL AGREEMENT

Portal suggested that one group of heavy bombers be sent to Egypt, even though this reinforcement could be made only by borrowing from Task Force BR [the bomber force for the United Kingdom]. British Air commanders had been clamoring for heavy bombers for the Middle East for some time. However, the previous day, the Joint Chiefs had just approved the buildup of the AAF in Britain as part of the RAINBOW-5 plan. Arnold balked. He then agreed that one group of bombers could be sent to Cairo after the initial Task Force BR increment had been sent to England.[54]

The JCCS process set up by Roosevelt and Churchill seemed to be evolving to one where the British would agree to something, then want to change some aspect of that agreement due to other circumstances. Some American planners recognized what the British were doing and would constantly warn their leaders that the United States should remain focused on the objective – defeat the Axis, not help Britain hold her empire together.

When planners thought about the strategic situation before the outbreak of hostilities, Britain and France had the largest colonial empires. Any expansionist aspirations by any nation would, by its very nature, have to encroach upon the colonies of either or both of these two countries. Japan's aggression into China did not directly threaten any European possession. Therefore, no belligerent action was taken by any other country. But when France fell in June 1940, Japan moved into France's former colonial possessions. With France removed from the picture, only British colonies remained. Japan and Germany needed oil. And to get it, they both needed to trespass across British possessions.

Britain did not own any oil reserves, and Egypt and Singapore stood in the pathway to obtaining it. Therefore, to stop the Axis, the Allies, including the United States, would have to defend British territory.

Another factor that made the appearance of protecting British territory a sensitive issue was the so-called anti-colonial position of the United States. Roosevelt had already shown sympathy for the separatist movement in India by establishing a diplomatic channel to Mahatma Gandhi. Roosevelt also argued that United States aid into the Indian theater should go directly to India, rather than through British supply channels. When Roosevelt asked Churchill to support eventual Indian independence at the Arcadia conferences, Churchill was reported to have "exploded."[55]

ARCADIA CONFERENCE

January 14 would be last official meeting of the Arcadia conference.

1942 JANUARY 15

THE PHILIPPINES

Following the Japanese landings on the Philippines, Admiral Hart had started to withdraw the remnants of the United States Asiatic Fleet. The public was finally told about it.[56] Having withdrawn his surface vessels to the Indian Ocean, he left his submarines in the area of Philippines. Unfortunately, according to United States Navy training at the time, submarines were viewed primarily as scouting vessels, not ships of war. Hence, they were ineffective in attacking the Japanese.

Hart's actions put him at odds with MacArthur. Admiral Conrad E. L. Helfrich of the Dutch Navy would replace Hart.[57]

BOMBING JAPAN

Based upon the recommendations of the JCCS and subsequently approved by Roosevelt and Churchill, the United States Air Staff listed task forces in their order of importance:

1. Task Force X (Project X) - heavy bombers for Australia;
2. Task Force FIVE ISLANDS - for defense of South Pacific ferry bases;
3. Task Force BR - bomber force for United Kingdom;
4. GYMNAST - Northwest Africa;
5. MAGNET - Northern Ireland;
6. Task Force CAIRO - to Egypt.[58]

Notably, the bombing of Japan was not listed. Either the number of planes was too small to be considered a task force, or it was still a secret mission being kept from the view of the British.

REINFORCEMENTS

Paul Eckerly and his B-17 crew departed Trinidad. Six hours later, they landed at Pan-Air field at Natal, Brazil, on the northeast tip of South America. Interestingly, since Brazil was neutral, Germany had the same access as Americans. Lufthansa was flying Ju-52s out of the same field.

We didn't bother them and they didn't bother us.[59]

After being refueled and briefed, Eckerly's crew left for Sierra Leone on the African coast.

1942 JANUARY 16

BURMA

The Japanese surrounded the Allied troops at Myitta, thus threatening to cut-off Tavoy.[60] (see Figure 3)

REINFORCEMENTS

After flying through a series of thunderstorms, Paul Eckerly's crew hit the point of no return. Since the navigator had been unable to get a star fix due to the storms, Eckerly's co-pilot James Cobb, who was flying the plane, elected to return to Natal. When they approached the coast, Cobb asked the navigator "Which way?" The navigator replied "Turn right."

Running low on fuel, Cobb landed at the first airstrip they found. It turned out to be a small village 400 miles northwest of Natal. Being away from civilization, it would take a few days for fuel to reach the crew.[61]

1942 JANUARY 17

NORTH AFRICA

The Axis garrison at Halfaya surrendered. Halfaya is located south of Bardia on the Libya-Egypt border and guards a pass between both countries. With the surrender of Halfaya, the supply lines between Cairo and the British front were now open. This completed the first phase of Operation Crusader, the lifting of Rommel's siege of Tobruk .

British forces continued to probe the Axis position at El Agheila.[62] (see Figure 15)

EAST INDIES

The Japanese landed an invasion force at Sandakan, British Borneo.[63] (see Figure 12)

ARNOLD-PORTAL AGREEMENT

In response to Portal's January 14 request, Task Force CAIRO was set up on paper. Two groups of pursuit planes were committed for June 1942, the 58th and 78th Pursuit Groups. Joining them was one transport group, one air depot group, and essential ground services. When Portal asked if the task force could have a heavy bombardment group, the AAF opposed the idea since the only way to obtain any aircraft was to divert them from Task Force BR.[64]

ARCADIA CONFERENCE

The Arcadia Conferences reached some strategic conclusions:

1. There would be no Anglo-American offensive across the Mediterranean in 1942,
2. There would be no "simultaneous landings in several of the occupied countries" in 1943, one of Churchill's key points,
3. They did not reach an agreement on the Anglo-American landing in French North Africa (code named GYMNAST).

Churchill reported to his Cabinet that he would be able to get the Americans to agree to GYMNAST.[65] Churchill remarked that Americans "were not above learning from us, provided that we did not set out to teach them."[66]

HALPRO

Organizationally, the Halverson project was taking shape. It was to be part of a new Area Command, which was to be comprised of two pursuit and two bombardment groups. It was also to have a transport group.

Memorandum For The Air Adjutant General:
It is directed that you establish an Air Service Area Command in the Calcutta, India area. This Command to have such units of the Air Corps and supporting arms and services as you consider essential to serve a Bombardment Task Force operating in the China area, and to serve and support the American Volunteer Group now operating in the Burma - China area. These combat units will be augmented from time to time and plans must take care of an initial force of two Pursuit Groups, two Heavy Bombardment Groups, and also one Transport Group, all of which may be augmented as the situation develops.

The first echelon of the Bomb-Combat Units is now being organized, and it is further directed that your office maintain close liaison with Lt. Col. H. A. Halverson, commander of this Force.[67]

While the formal organization was taking shape, personnel were being selected. Halverson selected Majors Carl Feldmann and George McGuire to join his staff. McGuire would be his second in command. Both were pilots and had served with Halverson in the 1930s. The third pilot was 1st Lt. Alfred F. Kalberer. Not only did Kalberer have the most experience with four engine aircraft, his previous commercial experience gave him familiarity with the routes that the group would fly on their way to China.[68]

Ulysses S. Nero reported to Headquarters, Mitchell Field and was invited to a meeting with Generals William E. Kepner and Uzal G. Ent, Colonel Halverson, and Lt. Col. McGuire. Nero had already made a name for himself as a crewmember on General Billy Mitchell's Martin-Curtis NBS-1 bomber. On September 5, 1923, Nero was the first bombardier to drop a 1100-pound bomb down the smokestack of a naval vessel.[69]

After being asked if he was willing to volunteer for a dangerous mission, Nero was told that what he was about to hear was top secret. He was ordered to not tell anyone. Then he was told about two projects to bomb Japan.

1. One project is to take four B-24s into Chengtu, China and from there, bomb Tokyo.
2. The other is to attack Japan in another manner. We won't tell you what the other manner is.[70]

This comment, along with the meeting noted below, indicated that the concept of a B-25/carrier based raid had reached the top levels of military.

DOOLITTLE RAID

On January 17, Arnold called Col. James Doolittle in and asked him to take over preparations for a special mission.[71]

COMMITMENT TO CHINA

With efforts to transition the AVG to an official Army Air Corps Group, Chiang did not want to lose his liaison with Chennault. To retain the services of Chennault, Madame Chiang Kai-shek encouraged the United States War Department to appoint him as chief air adviser.[72]

1942 JANUARY 18

EASTERN FRONT

On the southern end of the front, the Russians drove a wedge into the German lines near Izyum, on the Donets River. They also gained ground near Kursk.[73] (see Figure 18)

1942 JANUARY 19

BURMA

The Japanese entered Tavoy. (see Figure 3) As a result of Japanese successes, Allied commanders decided to evacuate Mergui, even though it had yet to be attacked.[74] (see Figure 13)

NORTH AFRICA

With the completion of the first phase of Operation Crusader, the lifting of Rommel's siege of Tobruk, General Auchinleck issued a new set of orders for the British forces. Tripoli was now the objective, with a secondary goal of reinforcing various positions in case a defensive stand was required.[75] (see Figure 15)

A sandstorm hit Libya and effectively grounded the RAF from flying reconnaissance missions over the German positions.[76] Thus, the buildup of Axis forces was unseen.

EAST INDIES

The British garrison at Sandakan surrendered, thereby yielding British Borneo.[77] (see Figure 12)

HALPRO

> Memo to The Commanding General, Air Force Combat Command
> Subject: Personnel for the Halverson Project
>
> 1. It is desired that you select two (2) navigator bombardiers, with extensive experience, ... for assignment to the Halverson Project, Advanced Echelon.
> 2. It is desired that action in this matter be expedited.
>
> By command of Lieutenant General Arnold.[78]

Halverson selected three navigators. The first was 2nd Lt. Francis B. Rang, who had been assigned to the Ferry Command and had been on several around the world flights.

Joining Rang were two other 2nd Lts., Francis H. Smith and Robert B. Kirkaldy. Both were with the 42nd Bombardment Group.[79]

But it takes more than officers to fly a heavy bomber. A flight crew was needed. So on the same day, similar orders were cut for enlisted men.

Subject: Transfer of Enlisted Men

1. It is requested that radio orders be issued transferring…the enlisted men named on enclosure one
2. These men have been selected by Lieut. Colonel Halverson as the advanced echelon of the Halverson Project. They constitute the enlisted personnel of three (3) combat crews to man three (3) of the airplanes being assigned to Colonel Halverson, which will be delivered to Wright Field during the week of January 19, 1942
3. When the Halverson Project is organized, these men will be transferred to it for permanent station overseas.[80]

The Halverson Project now consisted of three planes.

PRESS

Time continued to cover the lack of progress in converting the United States auto industry to munitions manufacturing. On January 19, 1942, it published an article: U.S. AT WAR: OPM FLOPS AGAIN

The first paragraphs read as follows:

> *To its long record of* muddling without muddling through, the Office of Production Management added a finally inglorious chapter last week. OPM sat down with management and labor to bring about the thing the U.S. needs most: immediate conversion of its automobile industry, greatest productive machine in the world, to all-out arms production. Little was accomplished.
>
> The task should have been easy. All 200 men at the meeting — auto executives, labor leaders, OPM and other defense chiefs — had the same goal. Without conversion to munitions the auto industry was finished, because there was no more rubber for its peacetime cars to roll on. Without conversion the union was a union of unemployed. And without conversion the Government would have trouble getting all the materiel that it needs to beat the Axis.
>
> Nevertheless, the job was not done.[81]

It was one thing to talk about being the Arsenal of Democracy, it was quite another to implement it.

1942 JANUARY 20

BURMA

The Japanese crossed out of Thailand, thereby beginning their assault on Moulmein.[82] (see Figure 13)

NEW GUINEA – SOLOMON ISLANDS

Japanese aircraft struck Rabaul, New Britain.[83] (see Figure 19)

NORTH AFRICA

The RAF resumed its reconnaissance flights. They reported that Mersa el Brega, home of Rommel's forward position, was on fire and that ships in the harbor had been destroyed. Rommel's apparent action was confirmation of reports the British had received from their agents in the Italian high command that Rommel was planning a retreat to Tripoli. Therefore, British assumed that Rommel was about to begin another retreat.[84] (see Figure 15)

EASTERN FRONT

The Russians continued to have success near Moscow, this time retaking Mozhaisk, about 60 miles west of Moscow.[85] (see Figure 17)

1942 JANUARY 21

SINGAPORE

Withdrawal of Allied defense forces from the Muar and Segamat fronts continued.[86] (see Figure 14)

NEW GUINEA

The Japanese began their offensive with an aerial attack on Lae and Salamaua.[87] Japanese aircraft also struck Kavieng, New Ireland.[88] (see Figure 19)

NORTH AFRICA

Rommel ordered an attack with the simple directive: "Therefore we shall proceed to attack and destroy the enemy."[89]

Rommel's attack was so successful that British forces were ordered to withdraw to Agedabia, with permission to withdraw further if needed.[90] (see Figure 5)

HALPRO

The HALPRO organization continued to take shape.

> *The primitive conditions expected at* the China bases resulted in close scrutiny of the personal fitness of every candidate. Heading this effort will be Major Edward J. Kendricks, who is ordered to report to the organization from Randolph Field.[91]

1942 JANUARY 22

SINGAPORE

The Japanese completed their conquest of the Muar front.[92] (see Figure 14)

THE PHILIPPINES

MacArthur ordered his troops to withdraw to the final defensive position on the Bataan peninsula. The withdrawal was to be finished by the morning of January 26.[93] (see Figure 6)

NORTH AFRICA

Within a day of beginning his offensive, Rommel's forces occupied Agedabia.[94] (see Figure 5)

NEW GUINEA – SOLOMON ISLANDS

Japanese aircraft struck Rabaul, New Britain.[95] (see Figure 19)

DOOLITTLE RAID

Nero reported to Wright-Patterson Field and met with Col. Doolittle. Doolittle wanted Nero to prepare a parts list for 15 B-25s for a period of three to six months.[96]

1942 JANUARY 23

Suddenly, the military news became ominous.

NEW GUINEA - SOLOMON ISLANDS

Japanese forces captured Rabaul.[97]
Another Japanese invasion force invaded Bougainville.[98] (see Figure 19)

EAST INDIES

One Japanese invasion force steamed through the Makassar Strait to Balikpapan. Another force sailed through the Molucca Passage to Kendari (Celebes).[99] (see Figure 12)

NORTH AFRICA

Rommel's forces continued to Antelat and Saunnu.[100] Saunnu defended the road to Tobruk. (see Figure 15)

CHINA

Pursuant to Chiang's request for an American military aide, General Joseph W. Stilwell was assigned to China.[101]

HALPRO

The WPD continued preparations for sending Air Corps units to China. First, the Halverson Project was given a code name HALPRO.

Subject: Assignment of code designation of project

1. Reference is made to memorandum from this office to you, January 23, 1942…pertaining to a project for which the designation "HALPRO" is requested.
2. The project consists of an offensive air mission in China.
3. Location of the project is as follows:
   - Depot – Calcutta, India
   - Operating base – Vicinity of Chengtu, China.[102]

HALPRO became its official name. It would operate from China. However, it was unclear at this point if it would operate against Japan or Burma.

Then, in a lengthy letter to the War Department, Col. George outlined in detail what steps were to be taken. Magruder was put in command of the United States forces in China and would report to Chiang. Under Magruder were the AVG and HALPRO.

Subject: Army Air Force Participation in China Theater

1. …
2. United States Air Force in China Theater.

   a. General Magruder is commander of the United States Forces in China and he in turn is under the command of Generalissimo Chiang Kai-shek who has been designated as supreme commander of the Allied Forces in China.
      (1) The AVG Group.......
      (2) Halpro. An additional Task Force under the command of Lt. Colonel H. A. Halverson is now being organized and will be dispatched to the China Theater
3. …
4. The mission of the Air Force

a. The mission of the U.S. Air Force in China should be substantially as follows:

　　　　(1) Offensive Mission
　　　　　(a) Air attack on enemy air forces menacing the security of Rangoon and the Burma Road.
　　　　　(b) Air attack on enemy ground forces menacing Rangoon and the Burma Road by interdicting movement of troops and supplies in the area
　　5.　Present Status and Action taken to date

　　　a. Since this theater is already active and operating, it will be necessary to outline ... matters ... as follows:
　　　　(1) ...
　　　　(2) The AVG group ...
　　　　(3) Halpro (Lt. Colonel H. A. Halverson commanding) is being organized and dispatched to the China Theater. The first units of the air echelon expect to arrive in the theater about February 15. The advance air echelon of this force will consist of 3 B-24 D's and 5 C-39 transports. It is contemplated that the eventual strength of this force shall be 50 B-24 D's and 50 C-39's.[103]

This memo was significant for several reasons. First, there was now evidence that Chiang was informed about HALPRO. Second, there was the mention of the bombing of Japan under the mission objectives. Third, Halverson and the initial three planes were expected to arrive by mid-February. And last, HALPRO would eventually grow to 50 planes.

1942 JANUARY 24

EASTERN FRONT
　　On the southern end of the front, the Russians continued their drive into the German lines. They recaptured Barvenkova.[104] (see Figure 18)

EAST INDIES
　　The first group of Allied P-40 fighter planes arrived at Java from their bases in Australia.
　　An Allied naval task force attacked a Japanese naval force off Balikpapan in the Makassar Strait. This engagement was considered the first "big" naval battle of the war. (see Figure 12)
　　Darwin, Australia was merged into ABDA command structure, by order of the Combined Chief of Staffs.[105] (see Figure 2)

NEW GUINEA

Allied forces evacuated Lae and Salamaua.[106] (see Figure 19)

HALPRO

Arnold ordered that HALPRO be increased to six crews.
Subject: Crews for B-24D Airplanes, Halpro Project
To: The Commanding General, Air Force Combat Command, Bolling Field, D.C.

1. It is desired that six (6) crews for accelerated service tests of three (3) B-24D airplanes for the Halpro Project be sent to Wright Field, each crew to consist of:

- one (1) pilot
- one (1) co-pilot
- one (1) aerial gunner
- one (1) crew chief
- one (1) radio operator
- In addition to the above include:
- three (3) bombardiers
- three (3) armorers
- three (3) navigators.[107]

FAR EAST

Currie constantly besieged Roosevelt with reasons why the United States needed to send aid to the Chinese. On January 24, he stated

a number of incidents have contributed to worsen Chinese relations with Britain and the United States. They include:
  (a) The confiscation of lend lease cargoes,
  (b) The reverses in the Pacific,
  (c) The refusal to accept substantial aid from the Chinese in the defense of Burma,
  (d) The small amount of ordinance and aircraft shipped to China,
  (e) The present closing of the Burma Railroad to lend-lease shipments (because of troop movements),
  (f) The niggardly offer of financial assistance by the British, and the misunderstanding of the nature of our offer.

He continued his argument with a prediction that Rangoon (see Figure 3) would shortly fall to the Japanese. If and when that occurred, there would be lots of blame to go around. He suggested Roosevelt consider three actions:

1. Urge the British to accept Chinese assistance,
2. Add a $500 million appropriation to pending lend lease legislation for the Chinese,
3. Establish an airfreight service from Calcutta to China and north Burma.[108]

1942 JANUARY 25

THAILAND

Thailand declared war on the United States.[109]

NORTH AFRICA

Rommel's forces occupied Msus, which lies on the road to Tobruk.[110] (see Figure 15)

By January 25, the British were in full retreat. In five days of fighting, the British had lost 299 tanks and armored vehicles, 147 guns and 935 men. Rommel reported the loss of three tanks and 14 men.[111]

1942 LATE JANUARY

DOOLITTLE RAID

Meanwhile Ralph S. (Scotty) Royce, who had worked on B-24s before the war, had enlisted and was in Washington.

> *I had spent the previous* year working for the British government in the United Kingdom as a B-24 expert, hired from Consolidated Aircraft for that purpose. When the LB-30s started coming over, in late October and November, with a considerable number of changes, I thought I'd better get back to the plant and catch up on the new configurations, as my expertise was slipping away.
>
> It took me almost all of December to get home by way of British aircraft and an American destroyer, but that's another story. Nevertheless, in early January, back at the plant, it became apparent that the U.S. Air Force, or Army Air Corps as it then was, was going to get the bulk of the B-24 production from then on. I spent about three weeks checking over the "D" - B- 24D and was fairly well up on it, in fact it was more like the "A's" than it was like an LB-30, mainly because of the lack of electrical propellers. Late in January, I decided that the Air Force needed me, so I hopped on an airliner and flew to Washington, went down to the War Department to see what I could do. I'd barely entered the Munitions Build-

ing, and was wandering down the corridor looking for some familiar sign, when I heard my name called. I turned around and there was Jimmy Doolittle, then a Lieutenant Colonel. I had known Jimmy all my life, as he was a friend of my father[112]'s, and I had last seen him in London about three months before. Jimmy asked me what I was doing and I explained to him that I wanted to go on active duty with the Air Corps. He said, "Great, I've got a job for you. I have a task force going out and you'll fit right into it as an engineering officer". He told me to go down to a certain room, the number of which I've forgotten, and get sworn in and report back to his office, which I did.

I should say that I had a reserve commission as a 2nd Lieutenant in the Air Force, I was a product of my college ROTC training.[113]

1942 JANUARY 26

GERMANY FIRST

The Germany First strategy continued to perplex some of the American public. Apparently, the public was not convinced. *Time* ran an article on January 26 expressing the public's dilemma over this policy:

> **World Battlefronts, STRATEGY: Dissention among the Allies**
>
> *...The strategy of concentrating on* Hitler might or might not have been adopted... The British would naturally have been pleased by such a plan. They had, in speeches and editorials, been urging just that. Some believe that Winston Churchill came to Washington to sell just that bill of goods. And yet it began to be realized in London last week that the Churchill Government has mishandled affairs in the Orient. The Prime Minister himself knows little of the subject except what he learned as an enthusiastic poloist in a Punjab regiment in Kipling's India. A Cabinet shake-up was demanded.
>
> To Russia the strategy of European priority would be both good news and bad. Stalin's greatest enemy is certainly Hitler, and anything that hurt Germany would help Russia. But Stalin's greatest fear is that Japan may strike at his rear- and therefore he wants continuing Allied resistance in the Far East.
>
> But to two other Allies, the Chinese and the Dutch, the strategy would mean a grim alternative — surrender or death.
>
> Fury. The Chinese and the Dutch were not too happy anyhow. They had been slighted in the formation of the Allied Supreme Far Eastern Command. Generalissimo Chiang Kai-shek had not been taken into the joint counsels beforehand, had been tossed an unlikely bone —operations in Indo-China and Thailand. The

Dutch had been left out altogether. And yet the Allied Supreme Command demanded Chinese troops, airmen and goods in Burma; then proceeded to Java and began to tell the Dutch what to do.[114]

SINGAPORE

The Japanese executed an amphibious landing at Endau on the east coast of Malaya.[115] (see Figure 14)

COMMITMENT TO CHINA

Following the July 1941 approval by Roosevelt of the plan to send bombers to China, Currie had sent a message to Chennault that he, Currie, was positive Chennault would begin receiving the planes by year's end. By January 26, there were still no planes in Chennault's hands and no indication that any were enroute. Chennault fired off a note to Currie asking that he send bombers so that he could "begin attacks on Japan's industry."[116] Chennault wanted to know the location of the planes.

PLOESTI

The British had been peppering Roosevelt with the importance of the Ploesti oilfields for some time. Hopkins must have become convinced of their importance since he posed the following question to Arnold: "Is the bombing of Rumanian oil fields feasible and what would be two or three theatrical approaches to it?"[117]

HALPRO

Arnold may have begun to suspect that HALPRO might be diverted for another purpose. The record is unclear if he sent the next two memos before or after he got the above request from Hopkins. First, he sent a note to A-3 detailing the specific composition of HALPRO.

Subject: Task Force HALPRO

1. A task force, known as HALPRO, is being organized for the purpose of carrying on general Air Force offensive operations from bases located in China...his force will be equipped with B-24-D type heavy bombers, and the advance echelon, consisting of three B-24-D's and five C-39 transport airplanes, is expected to move out for the theater of operations within the next thirty days. The initial planning, insofar as this project is concerned will be based on a total of fifty B-24-D and fifty transport airplanes.

By Command of Lieutenant General Arnold

Second, Arnold sent the following to Office of the Assistant Chief of the Air Staff, A-1 Personnel (A-1) to have HALPRO activated as a group.

Subject: Activation of Unit.

1. It is directed that the following Air Force units be activated:

    • Hq & Hq Squadron of Air Task Force HALPRO

2. This unit will correspond in strength to a wing Hq & Hq Squadron subject to such modifications as recommended by Lt. Colonel H. A. Halverson, commander of this unit.[118]

In a White House briefing paper, Roosevelt received a report on HALPRO's progress: at Eglin Field, Florida, for the HALPRO Project (Aviation Assistance to China):

- 8,700 rds. Cal. .50 armor Piercing and Tracer (4 to 1)
- 8,000 rds. Cal. .45 ball.[119]

1942 JANUARY 27

EASTERN FRONT

On the southern end of the front, the Russians retook Lozovaya, a major rail center 40 west of Barvenkova.[120] (see Figure 18)

EAST INDIES

The Allied troops, defending the airfield at Singkawang, Borneo, were ordered to withdraw.[121] (see Figure 2)

TASK FORCE BR

The Combined Chiefs of Staff (CCS) agreed to send the first two available heavy bomber groups to England. They would operate "independently" but in cooperation with Bomber Command, i.e. the British.[122]

1942 JANUARY 28

SINGAPORE

The Japanese troops arrived at Bentu on the west coast of Malaya.[123] (see Figure 14)

NORTH AFRICA

British forces were ordered to withdraw from Benghazi.[124] (see Figure 5)

BOMBING JAPAN

Roosevelt hosted a meeting with the Allied War Council. In attendance were British military advisors.[125] Roosevelt asked Arnold about the progress of plans to bomb Japan. Arnold answered that "at present a man is working on this proposition of bombing from China or Russia; that it will take a few months to get the gasoline and fields available, after which bombing from China could start."

He continued to tell Roosevelt that these plans contemplated that the bombers would fly to advance bases, land, re-gas, fly over Japan, land at the advance bases, and then return to a base in the rear.[126]

Roosevelt responded that, from a psychological standpoint, it was most important to bomb Japan as soon as possible. He directed Arnold to look into the possibilities of establishing bases in the Aleutians. Arnold pointed out that the distances from there to Japan were too great.[127]

In his book *Global Mission*, Arnold wrote that he had reservations about talking about this mission in this meeting, as not every one knew about it. So, he wrote a memo to Roosevelt saying:

> *There is no present means* of operating heavy bombardment aircraft from Outer Mongolia ... without the definite cooperation of the Russian government.
>
> For this reason I feel that the plan, now in progress, for carrying out an attack upon the Japanese enemy's center of gravity ... is the logical and most effective plan.[128]

It is not clear how far the planning for the Doolittle mission had progressed. Arnold's reference to "a man" seemed to indicate HALPRO was still the primary mission.

93RD HBG

The 93rd heavy bomber group was constituted. It was to be one of the first units deployed to England as part of Task Force BR.

1942 JANUARY 29

EAST INDIES

The Japanese seized Pontianak, on the west coast of Dutch Borneo.[129] (see Figure 12)

BOMBING JAPAN

Charles Grandegerald, an executive with Otis Elevator, identified potential Japanese targets on January 28.[130] His list was forwarded to Arnold and the White House on January 29.[131] This would not be the last time that a civilian had to be consulted for military target advice, indicating that the United States lacked knowledge of Japanese industrial targets.

Meanwhile Chennault was sending notes to Currie about his progress and his needs. On January 29, he reported that the AVG had shot down three Japanese fighters. And, he added:

> *Can begin attacks on Japan's* industries at once if you can send regular or volunteer bombardment group equipped with Lockheed-Hudson as specified by me in June 1941 and American key personnel to operate under my command and control of Generalissimo only.[132]

Chennault thought he could attack Japan even without heavy bombers. This presented an interesting dichotomy. United States military planners were plagued with finding bases and supplies to operate the larger planes with longer range, yet Chennault seemed to think he could get the planes with shorter range close enough to attack the Japanese mainland.

FAR EAST

Churchill sent a note to Roosevelt:

> *I am informed that there* is a danger that fighter squadrons of American volunteer group now helping so effectively in the defence of Rangoon may be withdrawn by Chiang Kai-shek to China after January 31st.
>
> Clearly the security of Rangoon is as important to Chiang Kai-shek as to us, and withdrawal of these squadrons before the arrival of Hurricanes, due the 15th to 20th February, might be disastrous.
>
> I understand that General Magruder has instructions to represent this to Generalissimo, but I think the matter is sufficiently serious for you to know about it personally.[133]

Chiang had just been appointed Supreme Commander of the China Theater. Churchill agreed to this. Apparently when Chiang made a decision not in concert with Churchill's views, Churchill was unwilling to intervene directly. Rather, he wanted Roosevelt to intervene. This was another example of Churchill's meddling.

1942 JANUARY 30

BURMA

The Japanese seized the Moulmein airport.[134] (see Figure 13)

NORTH AFRICA

Churchill should have been more worried about the theaters where the British had a direct interest. *The New York Times* announced the fall of Benghazi:

British Concede Fall Of Benghazi[135]

SINGAPORE

All remnants, except a single squadron, of the Malaya Air Force were withdrawn to the Netherlands East Indies.[136]

EAST INDIES

The Japanese attacked Ambon (Amboina), second largest naval base in Netherlands East Indies. Air units had already been withdrawn.[137] (see Figure 12)

COMMITMENT TO CHINA

After the White House meeting where he answered Roosevelt's query about Japan, Arnold wrote a third memo about China.

> Subject: Army Air Force in China.
>
> 1. ...
> 2. United States Air Forces in China Theater.
>
>   - General Magruder is commander of the United States Forces in China
>   (A) ....
>   (B) Air Task Force #10 - HALPRO. An additional Task Force, under the command of Lt. Colonel H.A. Halverson, is now being organized and will be dispatched to the China Theater.
> 3. ...
> 4. ...
> 5. Present Status and Actions Taken to date
>
>   (2) Task Force #10 (Halpro) - The first units of the air echelon expect to arrive in the theater during February. The advance air echelon of this force will consist of 3

B-24D's and 5 C-39 transports. It is contemplated that the eventual strength of this force will be 50 B-24D's and 50 C-39 transports.

By Command of Lieutenant General Arnold.[138]

Arnold's concerns are unclear. There was no mention of the Lockheed Hudsons that had been promised to Chenault. The statement that "3 B-24Ds" constistute an "advance air echelon" is interesting. Halverson had three planes and three crews, creating the suspicion that Arnold believed they would depart soon, and the other 47 planes and crews would depart later. On January 23, a week earlier, Col. George had issued a memo saying the planes and crews would arrive in mid-February.

GERMAN OIL NEEDS

The British made an assessment of the German oil needs and the alternatives for obtaining the necessary sources. They sent it to Washington on January 30, stating, in part:

1. Germany is becoming short of oil and her air force has been reduced to first line strength of approximately 4,000 operational and 1,400 transport aircraft. ...Germany must secure, at earliest opportunity possible substantial sources of oil. She can only obtain fabric oil in Caucasia or Iraq. For reasons given below, Caucasia is the most likely major operation.
2. A thrust through Turkey (towards) Syria and thence to Iraq and/or Egypt. Advance through Turkey into Iraq would put Germany in possession of oil fields potentially capable of meeting her needs. Would deprive us of oil on which our Eastern Mediterranean naval position partly depends. ...
3. Most probable German major operation likely to be made through Ukraine (toward?) Caucasus to obtain oil. Meanwhile, every effort will be made to control the central and eastern Mediterranean and to obtain and (exploit?) the position in North Africa.[139]

1942 JANUARY 31

BURMA

The Allies forces evacuated Moulmein.[140] (see Figure 13)

SINGAPORE

All Allied forces had completed their withdrawal to Singapore Island. The causeway from the mainland to the Island was destroyed.[141] (see Figure 14)

PLOESTI

Arnold issued a memo directing that every effort be made to "push" the bombing of the Rumanian oil fields. He asked that his office be advised of the status of the project at an early date.[142] Hopkins inquired about Ploesti on January 26.

HALPRO

By the end of January, thirteen enlisted men had been added to the HALPRO roster. They were ordered to report to Patterson Field near Dayton, Ohio. Once there, they began training on three B-24As. At the same time, some of the Washington contingent arrived at Patterson to determine the changes needed to make the B-24 combat ready.[143]

HEAVY BOMBER PRODUCTION

During January, 86 heavy bombers were manufactured. Ten were B-24s.[144]

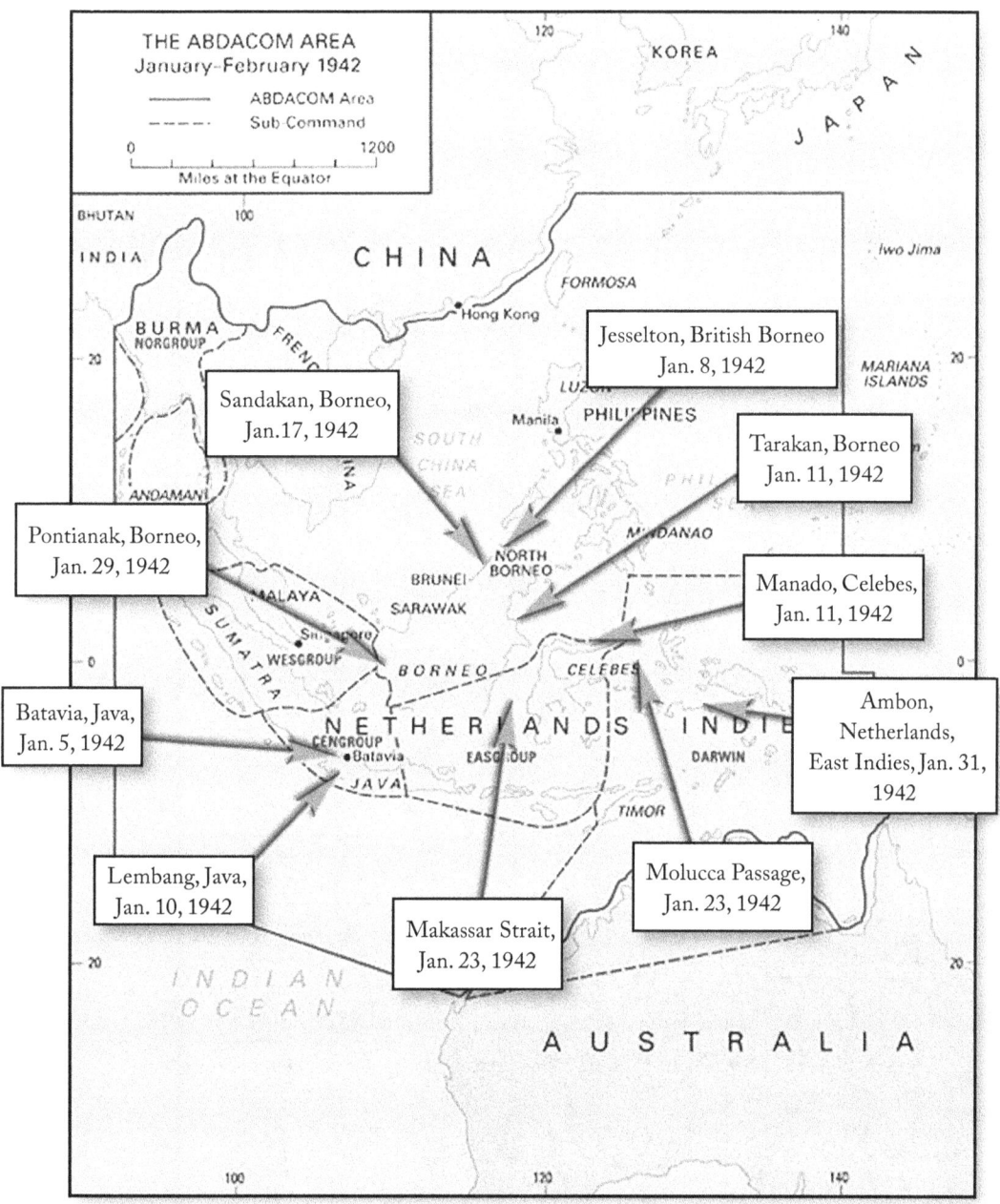

Figure 12 – ABDA – January 1942

Figure 13 – Burma – January 1942

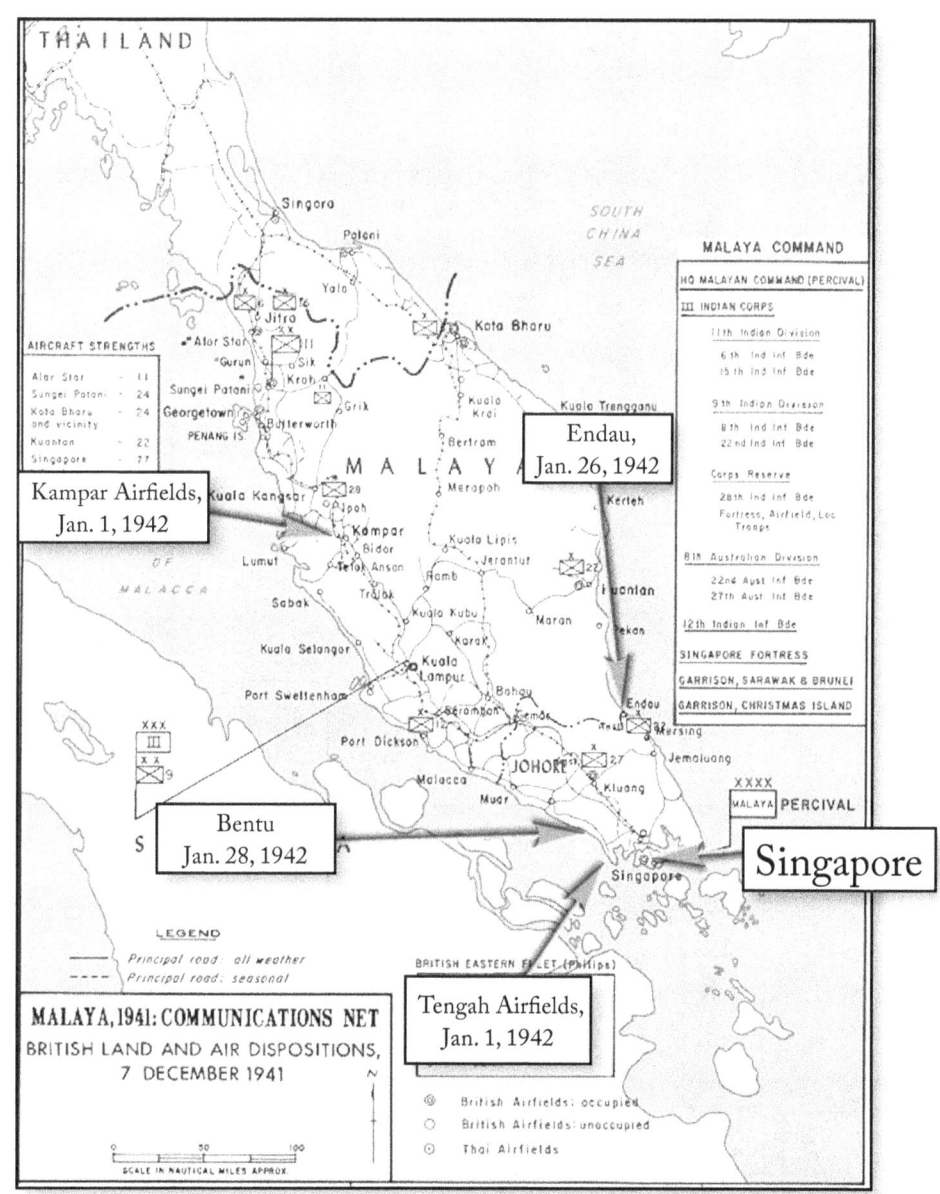

Figure 14 – Malaya – January 1942

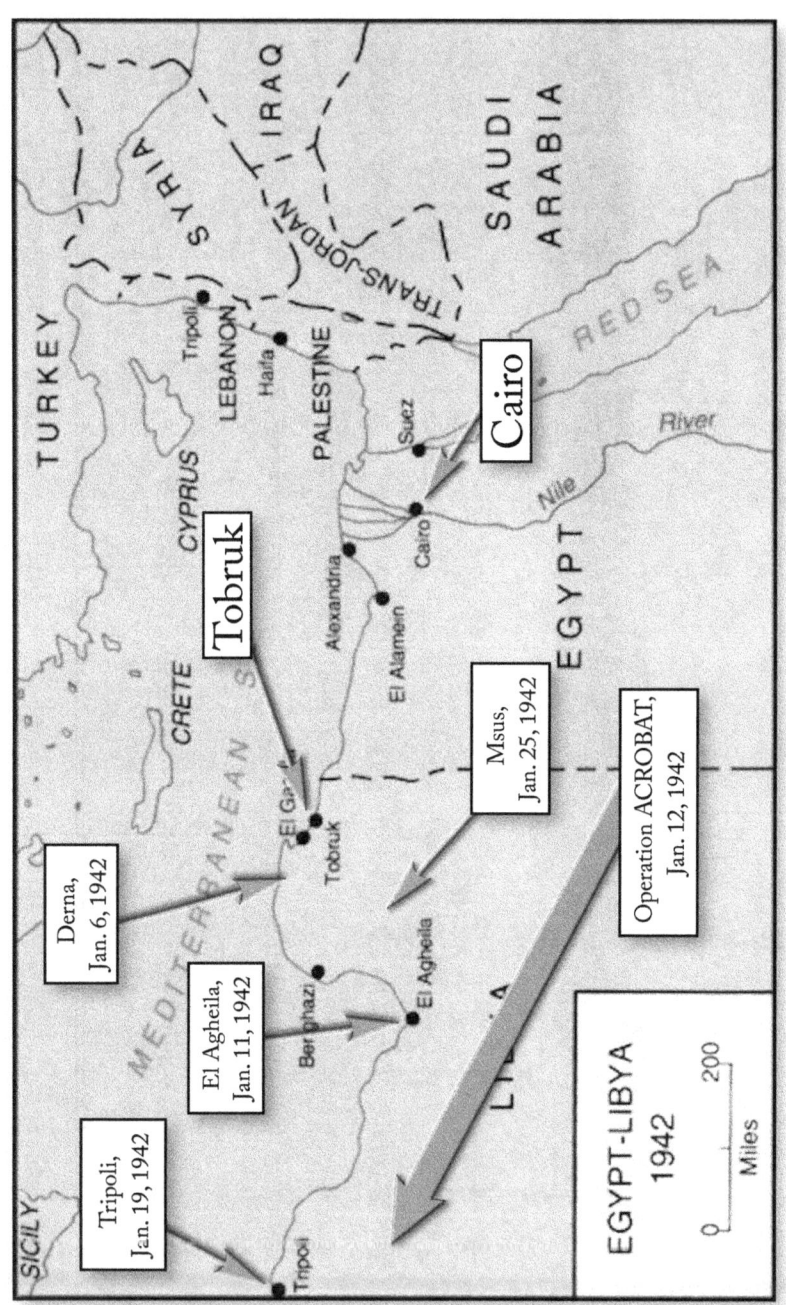

Figure 15 – Mediterranean – January 1942

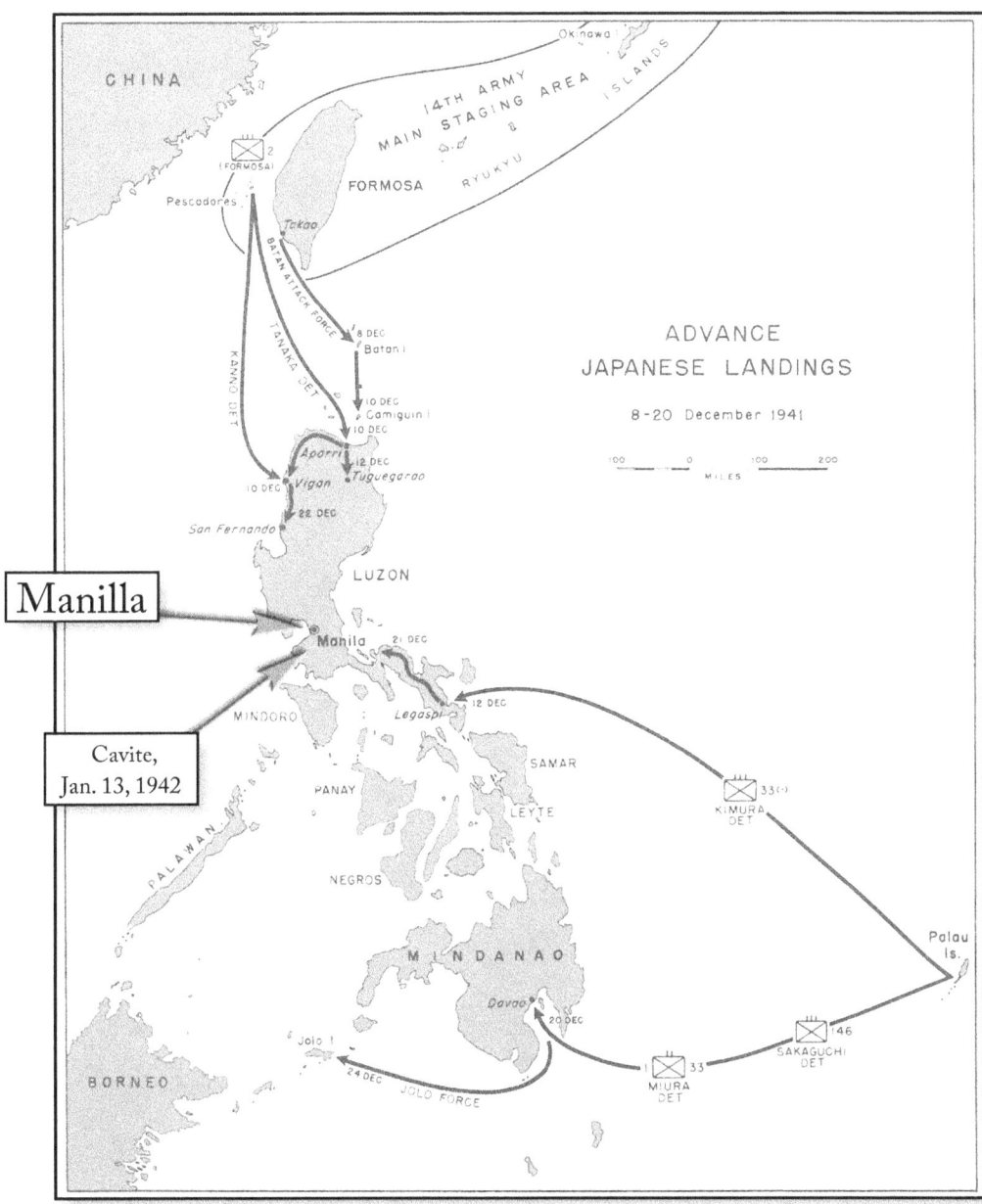

Figure 16 – Philippines – January 1942

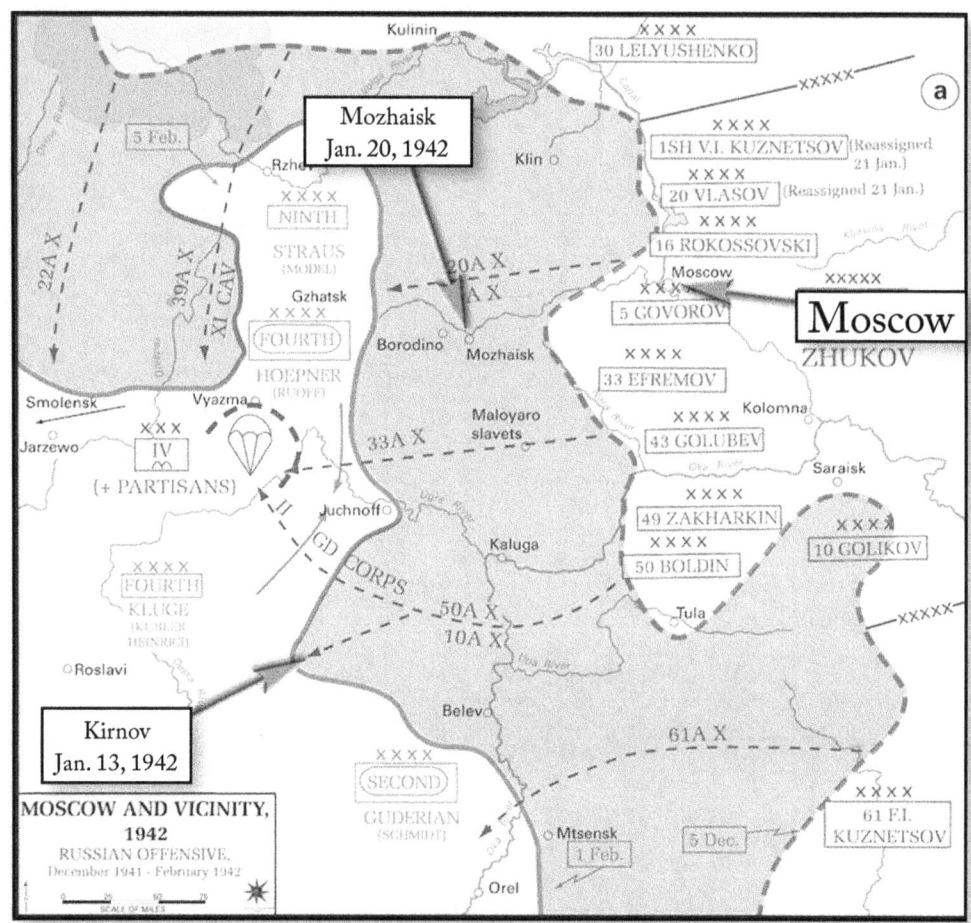

Figure 17 – Moscow – January 1942

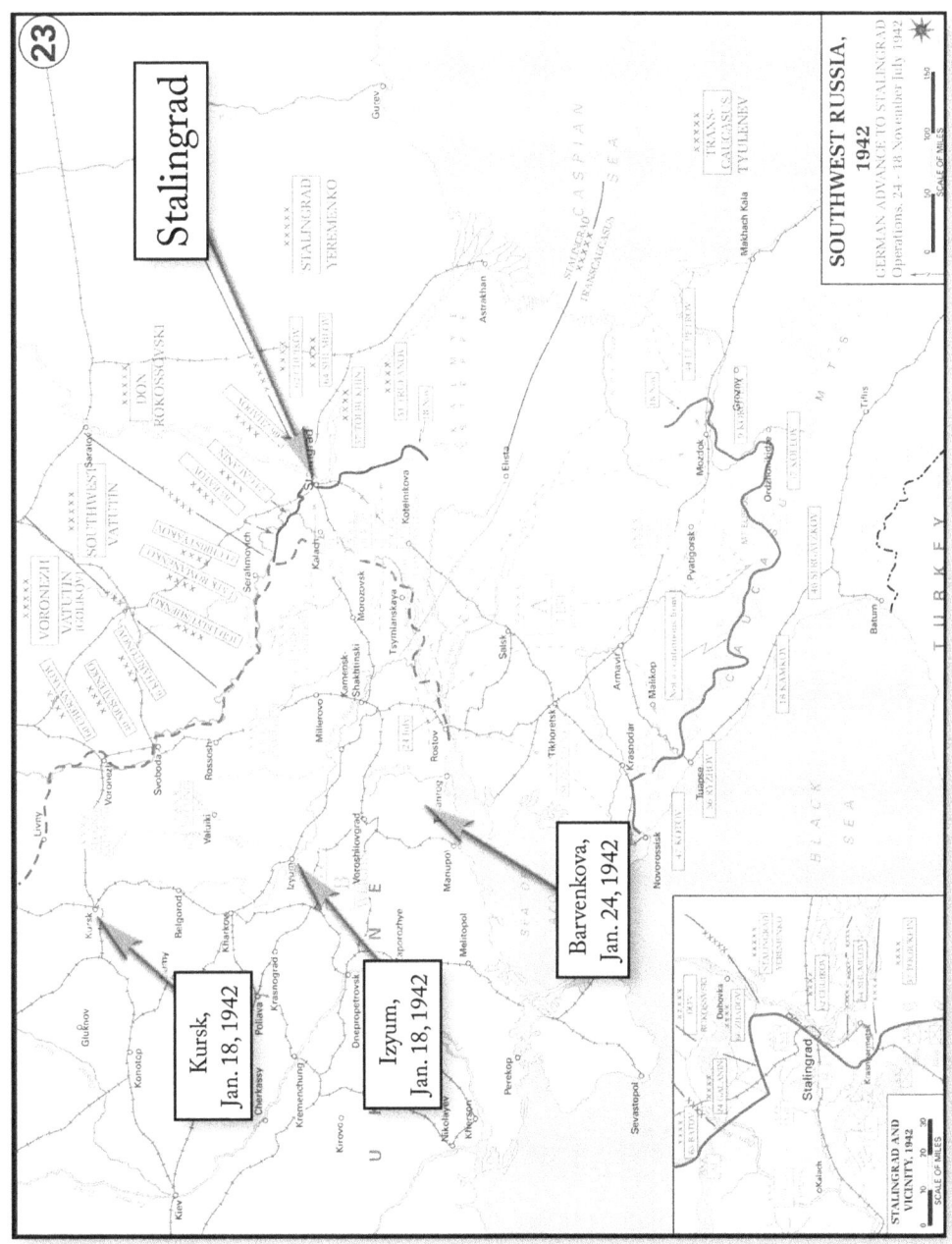

Figure 18 – Stalingrad – January 1942

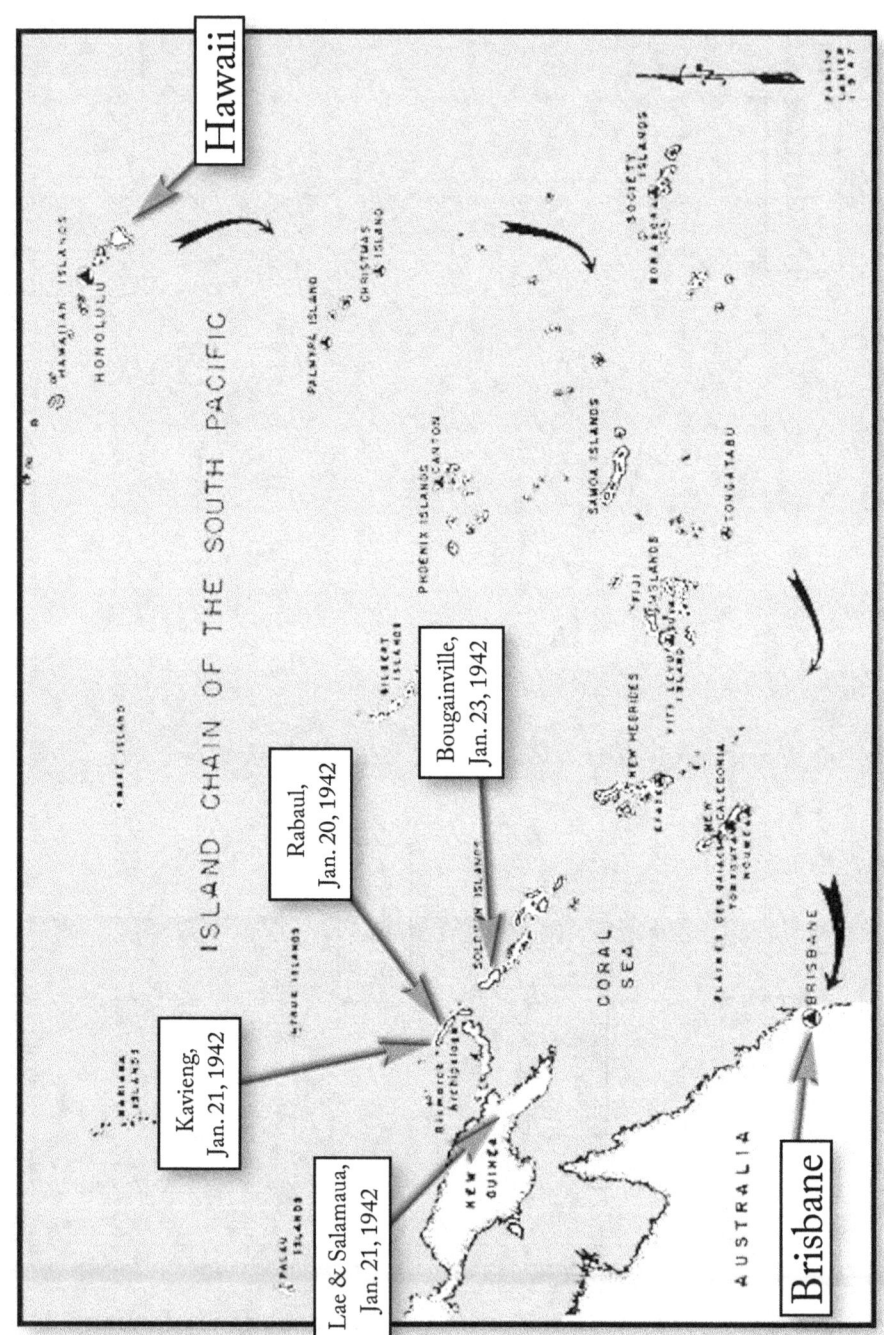

Figure 19 – Supply chain – January 1942

ENDNOTES

1. ----, AFHRA, Hap Arnold file, Letter from Maj. E. H. Alexander, Air Corps, assigned to the Magruder Mission, to Col Robert Olds, January 1, 1942.
2. Williams, *Chronology 1941-1945 - 1942*, p. 11.
3. Assistant Chief of Air Staff, *The AAF in the Middle East*, p. 50.
4. Craven, *The Army Air Forces in World War II, Vol. 1*, p. 560.
5. ----, AFHRA, Hap Arnold file, Stilwell Papers, p. 20.
6. Eckerly, *A Pilot's Story*.
7. Williams, *Chronology 1941-1945 - 1942*, p. 11.
8. Williams, *Chronology 1941-1945 - 1942*, p. 11.
9. Directive memo for AWPD by Lt. Col. C. E. Duncan, on Combat Units and Personnel to the Near East. . ., January 2, 1942.
10. Bullet to FDR, January 2 1942, FDR Library.
11. ----, AFHRA, Hap Arnold file, Stilwell Papers, p. 20.
12. ----, AFHRA, IRIS #1075666, Gerow memo to Gen. Arnold, January 2, 1942.
13. ----, AFHRA, Hap Arnold file, Scherer to Commanding General, Second Air Force. January 2, 1942.
14. ----, AFHRA, Hap Arnold file, Arnold to A-3, January 2, 1942.
15. ----, *New York Times*, MANILA FALLS, January 2, 1942, http://select.nytimes.com/gst/abstract.html?res=F50D14FA3D58167B93C6A9178AD85F468485F9, accessed January 29, 2014.
16. Williams, *Chronology 1941-1945 - 1942*, p. 11.
17. ----, AFHRA, Bonner Fellers file, Fellers' cable to the War Dept, January 3, 1942.
18. Williams, *Chronology 1941-1945 - 1942*, p. 12.
19. ----, wikipedia, http://en.wikipedia.org/wiki/Erwin_Rommel; ---, http://www.wehrmacht-history.com/personnel/r/rommel-erwin-johannes-eugen-heer-personnel-file.htm, accessed February 11, 2014.
20. ----, Operation Crusader, http://rommelsriposte.com/2011/12/06/german-tanks-sent-in-1st-half-of-jan-42/, accessed February 11, 2014.
21. Williams, *Chronology 1941-1945 - 1942*, p. 12.
22. Williams, *Chronology 1941-1945 - 1942*, p. 12.
23. Craven, *The Army Air Forces in World War II, Vol. 1*, p. 333.
24. Williams, *Chronology 1941-1945 - 1942*, p. 12.
25. Williams, *Chronology 1941-1945 - 1942*, p. 13.
26. ----, *New York Times*, WAVELL TO DIRECT FORCES FROM JAVA, January 7, 1942, http://select.nytimes.com/gst/abstract.html?res=F70A12FB3C5D167B93C5A9178AD85F468485F9, accessed February 15, 2014.
27. Craven, *The Army Air Forces in World War II, Vol. 1*, p. 250.
28. Williams, *Chronology 1941-1945 - 1942*, p. 13.
29. ----, AFHRA, Hap Arnold file, Arnold to AWPD, January 8, 1942.

30. ----, AFHRA, Hap Arnold file, Arnold to George, January 8, 1942.
31. Williams, *Chronology 1941-1945 - 1942*, p. 13.
32. Williams, *Chronology 1941-1945 - 1942*, p. 13.
33. ----, AFHRA, Hap Arnold file, Arnold to AWPD, January 9, 1942.
34. Williams, *Chronology 1941-1945 - 1942*, p. 13.
35. Williams, *Chronology 1941-1945 - 1942*, p. 13.
36. Craven, *The Army Air Forces in World War II, Vol. 1*, p. 244.
37. Craven, *The Army Air Forces in World War II, Vol. 1*, p. 493.
38. Glines, Carroll V. The Doolittle Raid: America's Daring First Strike Against Japan. New York: Orion Books, 1988.
39. Williams, *Chronology 1941-1945 - 1942*, p. 14.
40. Headrick, *Bicycle Blitzkrieg*, p. 7.
41. Williams, *Chronology 1941-1945 - 1942*, p. 13.
42. Eckerly, *A Pilot's Story*.
43. Dorr, *7th Bombardment Group*, p. 253.
44. ----, AFHRA, IRIS #1075666, Sharp Report dated 12 January 1942.
45. Jackson, *The Battle For North Africa*, p. 184.
46. ----, *Time, The Miserable Truth*, January 12, 1942, http://www.time.com/time/printout/0,8816,772981,00.html, accessed July 14, 2012.
47. Williams, *Chronology 1941-1945 - 1942*, p. 14.
48. Craven, *The Army Air Forces in World War II, Vol. 1*, p. 240.
49. Craven, *The Army Air Forces in World War II, Vol. 1*, p. 248. The agreed to schedule of deliveries of Heavy Bombers in 1942 was:

Month	*Heavy Bombers*
Jan.	15
Feb.	47
Mar.	20
Apr.	20
May	40
June	45
July	19
Aug.	64
Sep.	43
Oct.	73
Nov.	92
Dec.	111
Total	**446**

50. Craven, *The Army Air Forces in World War II, Vol. 1*, p. 559.
51. Craven, *The Army Air Forces in World War II, Vol. 1*, p. 430.
52. ----, FDRL, PSF, Box 1, Arcadia file, Defense of Island Bases Between Hawaii and Australia.
53. Craven, *The Army Air Forces in World War II, Vol. 1*, p. 240.
54. Craven, *The Army Air Forces in World War II, Vol. 1*, p. 560.
55. Herring, *From Colony to Superpower*, p. 571.
56. ----, *New York Times, U.S. ASIATIC FLEET IS MOVED SAFELY*, January 15, 1942, http://select.nytimes.com/gst/abstract.html?res=F00714F93D58167B93C7A8178AD85F468485F9, accessed February 15, 2014.
57. Helfrich's orders were to defend the Netherlands East Indies at all costs. Hart felt is orders were to preserve his surface vessels to fight another day. Helfrich's mission was more in line with MacArthur. Hart had to leave.
58. Craven, *The Army Air Forces in World War II, Vol. 1*, p. 560.
59. Eckerly, *A Pilot's Story*.
60. Williams, *Chronology 1941-1945 - 1942*, p. 15.
61. Eckerly, *A Pilot's Story*.
62. Williams, *Chronology 1941-1945 - 1942*, p. 16.
63. Williams, *Chronology 1941-1945 - 1942*, p. 15.
64. Assistant Chief of Air Staff, *The AAF in the Middle East*, p. 52.
65. Ben-Moshe, Tuvia, Winston Churchill and the "Second Front, p. 509.
66. Sebrega, *The Anticolonial Policies*, p. 84.
67. ----, AFHRA, Hap Arnold file, George to Adjutant General, January 17, 1942.
68. Walker, *The Liberandos*, pp 2-3.
69. Akin, *Early Air Force Pioneer*, p. 1.
70. ----, Interview of Ulysses S. Nero, May 21, 1974.
71. http://www.historynet.com/countdown-to-the-doolittle-raid.htm
72. Xu, *The Issue of US Air Support for China*, p. 467.
73. Williams, *Chronology 1941-1945 - 1942*, p. 16.
74. Williams, *Chronology 1941-1945 - 1942*, p. 16.
75. Williams, *Chronology 1941-1945 - 1942*, p. 16.
76. Mitcham, *Rommel's Desert War*, p. 29.
77. Williams, *Chronology 1941-1945 - 1942*, p. 16.
78. ----, AFHRA, Hap Arnold file, Hohman to Air Force Combat Command, January 19, 1942.
79. Walker, *The Liberandos*, p 3.
80. ----, AFHRA, Hap Arnold file, Deutsch to Adjutant General, January 19, 1942.
81. ----, *Time, OPM Flops Again*, January 19, 1942, http://www.time.com/time/printout/0,8816,766276,00.html, accessed July 15, 2012.
82. Williams, *Chronology 1941-1945 - 1942*, p. 16.

83. World War II Database, http://ww2db.com/ship_spec.php?ship_id=10
84. Mitcham, *Rommel's Desert War*, p. 29.
85. Williams, *Chronology 1941-1945 - 1942*, p. 16.
86. Williams, *Chronology 1941-1945 - 1942*, pp. 16-17.
87. Williams, *Chronology 1941-1945 - 1942*, p. 17.
88. World War II Database, http://ww2db.com/ship_spec.php?ship_id=10
89. Mitcham, *Rommel's Desert War*, p. 27.
90. Williams, *Chronology 1941-1945 - 1942*, p. 17.
91. Walker, *The Liberandos*, p. 5.
92. Williams, *Chronology 1941-1945 - 1942*, p. 17.
93. Williams, *Chronology 1941-1945 - 1942*, p. 17.
94. Williams, *Chronology 1941-1945 - 1942*, p. 17.
95. World War II Database, http://ww2db.com/ship_spec.php?ship_id=10
96. ----, Interview of Ulysses S. Nero, May 21, 1974.
97. ----, *US Marine Corps in World War II*, http://www.ibiblio.org/hyperwar/USMC/ , *First Offensive: The Marine Campaign for Guadalcanal*, p. 1.
98. Williams, *Chronology 1941-1945 - 1942*, p. 17.
99. Williams, *Chronology 1941-1945 - 1942*, pp. 17-18.
100. Williams, *Chronology 1941-1945 - 1942*, p. 18.
101. Miller, *The Chiang-Stilwell Conflict*, p. 59.
102. ----, AFHRA, Hap Arnold file, Dick to WPD, January 22, 1942.
103. ----, AFHRA, Hap Arnold file, George to War Department, January 22, 1942.
104. Williams, *Chronology 1941-1945 - 1942*, p. 18.
105. Williams, *Chronology 1941-1945 - 1942*, p. 18.
106. Williams, *Chronology 1941-1945 - 1942*, p. 18.
107. ----, AFHRA, Hap Arnold file, Cooley to Air Force Combat Command, January 24, 1942.
108. ----, FDRL, PSF, Box 2, China file, Currie -> FDR, January 24, 1942.
109. Williams, *Chronology 1941-1945 - 1942*, p. 18.
110. Williams, *Chronology 1941-1945 - 1942*, p. 17.
111. Mitcham, *Rommel's Desert War*, p. 32.
112. "Scotty" Royce's father was Major-General Ralph Royce.
113. Royce, Ralph "Scotty," Transcript of an interview.
114. ----, *Time, World Battlefronts, STRATEGY- Dissention among the Allies*, January 26, 1942, http://www.time.com/time/magazine/article/0,9171,932290,00.html, accessed July 17, 2012.
115. Williams, *Chronology 1941-1945 - 1942*, p. 19.
116. Schaller, *American Air Strategy in China*, pp. 17-18.
117. ----, AFHRA, Hap Arnold file, Hopkins to Arnold, January 26, 1942. See also Assistant Chief of Air Staff, *The Ploesti Mission of 1 August 1943*, p. 12.
118. ----, AFHRA, Hap Arnold file, York to A-3, and York to A-1, January 26, 1942.

119. ----, FDRL, Map Room (MR), Box 53, Daily Summary, January 26, 1942.
120. Williams, *Chronology 1941-1945 - 1942*, p. 19.
121. Williams, *Chronology 1941-1945 - 1942*, p. 19.
122. Craven, *The Army Air Forces in World War II, Vol. 1*, p. 560.
123. Williams, *Chronology 1941-1945 - 1942*, p. 20.
124. Williams, *Chronology 1941-1945 - 1942*, p. 20.
125. ----. FDRL, Day by Day, January 28, 1942.
126. ----, AFHRA, Hap Arnold file, discussion of the bombing of Japan.
127. Arnold, *Global Mission*, p.289.
128. ----, FDRL, PSF, Box 2, China file, Arnold to FDR, January 28, 1942.
129. Williams, *Chronology 1941-1945 - 1942*, p. 20.
130. John Carter was a civilian and was appointed by Roosevelt, in 1941, to operate "a small special intelligence and fact finding unit". This information comes from http://educationforum.ipbhost.com/index.php?showtopic=19534 , accessed September 12, 2015.
131. ----, FDRL, PSF, Box 98, Carter file, Carter to FDR, January 29, 1942.
132. ----, FDRL, PSF, Box 28, Currie Out file, Segac to Currie, January 29, 1942.
133. ----, FDRL, PSF, Box 2, China file, Churchill to FDR, 29 January 1942.
134. Williams, *Chronology 1941-1945 - 1942*, p. 20.
135. Levy, Joseph, *New York Times*, BRITISH CONCEDE FALL OF BENGAZI, January 30, 1942, http://select.nytimes.com/gst/abstract.html?res=F40B11FC3C58167B93C3AA178AD85F468485F9, accessed January 28, 2014.
136. Williams, *Chronology 1941-1945 - 1942*, p. 20.
137. Williams, *Chronology 1941-1945 - 1942*, p. 20.
138. ----, AFHRA, Hap Arnold file, George to Arnold, January 30, 1942.
139. ----, FDRL, PSF, Box 1, American-British Joint Chiefs of Staff file, Joint Intel Committee, February 14, 1942.
140. Williams, *Chronology 1941-1945 - 1942*, p. 21.
141. Williams, *Chronology 1941-1945 - 1942*, p. 21.
142. ----, AFHRA, Hap Arnold file, Arnold to file, 31 January 31, 1942.
143. Walker, *The Liberandos,* p 8. The men were:

M/Sgt. Arthur L. Cox, Communications, Headquarters, 4th Interceptor Command, March Field, California

M/Sgt. Jessy C. McConnell, Line Chief, 6th Bombardment Group, Muroc Dry Lake, California

T/Sgt. Vitus Hrubes, Line Chief, 30th Bombardment Group, Muroc Dry Lake, California

T/Sgt. Sidney J. Willis, Bombsight Technician, 30th Bombardment Group, Muroc Dry Lake, California

T/Sgt. George D. McNelly, Mechanic, Bolling Field, Washington, DC

T/Sgt. Gordon H. Hadlow, Mechanic, Bolling Field, Washington, DC

S/Sgt. Paul W. Fitzsimmons, Mechanic, Bolling Field, Washington, DC

S/Sgt. James C. Owen, Mechanic, Bolling Field, Washington, DC

S/Sgt. Alfred C. Colt, Communications, Bolling Field, Washington, DC

S/Sgt. Harry W. Dewald, Communications, Bolling Field, Washington, DC

S/Sgt. Carl W. Edwards, Mechanic, Bolling Field, Washington, DC

Sgt. Andrew Bowan, Mechanic, Bolling Field, Washington, DC

Sgt. Stephan T. Pundzak, Mechanic, Bolling Field, Washington, DC

144. ----, Army Air Force Statistical Digest, http://www.ibiblio.org/hyperwar/AAF/StatDigest/aafsd-3.html , accessed January 22, 2014, p. 112; B-24 production from Author's personal papers.

FEBRUARY 1942

The pattern of events in January continued during February. Leaders planned. The enemy succeeded. Plans were revised.

But a new wrinkle entered the picture. Before, there was no recognition that the enemy's forces would change; no admission that they too would have objectives. And thus there was no discussion by the Allies on how they could thwart those activities.

Now, an overall objective of the Axis moves was being discussed. Rommel was heading east and the Japanese were heading west. The German armies in Russia were idle, but soon the summer campaign would begin. The oilfields of the Middle East provided the perfect location for the intersection of the Axis forces.

Meanwhile, the United States strike force, whose aim was to bomb Japan, continued to grow, adding necessary personnel to carry out its mission. And it had now been given a name: HALPRO. But they were unaware that the seed of a new idea to strike at Japan had been planted.

MIDDLE EAST

More evidence was building that the Middle East was becoming an area of interest for United States military planners. In February, Marshall would write *Comments Of The Use Of U.S. Troops In The Middle East And Africa*. He discussed the potential use of United States troops in Syria, Libya, and Northwest Africa. Of note, he did not include Egypt. Nevertheless, Marshall commented about basing an Air Force in Syria:

> ***Due to the limits imposed*** by the size of the air forces available and our present commitments, it will be impossible to set up a large offensive Air Force in more than one theater during 1942. The pressure of events has forced us to disperse our Air Forces to such an extent that we now, at best, have only a bare minimum of defensive air strength in any theater. This is one of the reasons why the enemy still retains the initiative both in Europe and in Asia . . .
>
> No reserve of Air power exists in the U. S. and the Air Forces set up for this operation must of necessity be allotted from forces available to other theaters.[1]

As noted below, Syria was projected as the only country with an air base close enough to attack the Ploesti oil fields.

REINFORCEMENTS

Patterson Field, near Dayton, Ohio, was becoming a vital center for modifying planes and training crews. Hannah joined one of those crews.

> *Then 4 crews were sent* to Dayton Ohio (I was assigned to Soukup) & given radar equipped planes. Ours was severely damaged in FLA by ground crew – we flew it to Mobile & were given a new plane.
>
> Our orders were verbal and secret – not sure but I think we were to interdict shipping in Japan's home waters. 4 planes. First 2 wrecked – one in Puerto Rico 1 in Belem Brazil. The 3rd got to India & we heard the radar was junked & the plane became a general's personal ship. We were in #4 and when we reached Accra, we were commandeered by a US B.G. & a Brit. Col. It was obvious they were carrying dispatches – each a briefcase chained to his wrist. We were ordered to fly to Khartoum. Took off at night into a hell of an electrical storm. After 5 hours, we had only covered 500 miles.
>
> We did not have enough fuel to get to Khartoum so Soukup set it down on the desert at dawn. We knew there was a Brit base nearby – radio raised them at 9 a.m.. They kept bankers hours – 9 to 5. A US crew in a DC-2 flew 200 gals of gas to us & we eventually reached Khartoum. That was the day Rommel captured Tobruk from the South Africans. We went on to Aden then Karachi & were told to wait for orders.[2]

1942 FEBRUARY 1

NORTH AFRICA

Despite a string of military defeats, a positive spin continued to make the public think that the situation was not that bad.

However, the public was not told that British forces were ordered to withdraw to Gazala.[3] (see Figure 5)

MIDDLE EAST

Kirk filed a report with Hull that there was domestic unrest in Egypt. There were bread riots and public demonstrations in favor of the Axis armies advancing on Egypt.[4]

1942 FEBRUARY 2

NORTH AFRICA

After issuing orders for a withdrawal to Gazala, (see Figure 5) Auchinleck ordered the British forces to hold Tobruk. It was a vital supply port, only 450 miles from Cairo.[5]

GLOBAL STRATEGY

Roosevelt received an assessment by the Joint Intelligence sub-committee on "Germany's Intentions." The very first point was that "Germany is becoming short of oil and her air force has been reduced to first line strength of approximately 4,000 operational and 1,400 transport aircraft."

Later in the report, the authors discussed possible German activities for 1942. The second most likely action identified was

> ... *A thrust in the* south towards Caucasia (while containing Russian army in north and center).
>
> Germany must secure, at the earliest opportunity possible substantial supplies of oil. She can only obtain fabric oil in Caucasia or Iraq. For reasons given below, Caucasia is the most likely major operation.

Another option was

> *A thrust through Turkey [towards]* Syria and thence to Iraq for Egypt. Advance through Turkey into Iraq would put Germany in possession of oil fields potentially capable of meeting her needs. Would deprive us of oil on which our Eastern Mediterranean naval position partly depends. Would bring Germany within striking distance of Abadan and open way to a further advance into Egypt. On the other hand transport of oil from Iraq would be more difficult than from Caucasia, at least until Germany had a secure sea route through Eastern Mediterranean.[6]

The only possibility they did not mention was that Rommel might reach the Middle Eastern oilfields via Egypt.

COMMITMENT TO CHINA

Major E. H. Alexander had sent a report on January 1 explaining his views on sending an air unit to China. His recommendation was that the planes could not be supplied and would be within range of Japanese fighters. Arnold decided that the time was right to forward it to Dwight Eisenhower, who at the time was Chief of War Plans Division.

In the cover letter, Arnold wrote, "I am somewhat fearful that perhaps he [Alexander] has the right idea of what we may expect from the Chinese and the rest of us are doing a lot of wishful thinking."[7]

This was an odd comment by Arnold. His assignment was to figure out a way to bomb Japan and had chastised subordinates for not doing so. He was telling the Chief of the War Plans Division that all of this planning was a pipe dream.

In another peculiar move, Currie sent a note to Chennault asking that he remain in command of:

1. the AVG pilots that wish to remain as volunteers,
2. the regular army pursuit planes,
3. regular army medium bombers, and
4. any Chinese air groups.

Halverson would command the heavy bombers. He also told Chennault that in spite of Chennault's excellent reputation within the Army, General Clayton Bissell would be a better liaison officer than Chennault.[8] So Chiang wanted Chennault as his air advisor and the United States wanted Bissell to be the air advisor.

HALPRO

Arnold added some confusion to the above command structure when he issued the following:

Subject: Project Officer for AVG and HALPRO
To: Lieutenant Colonel H. A. Halverson, Air Corps.

1. ...the HALPRO project is established in this headquarters and will include that project heretofore known as AVG
2. Lieutenant Colonel H. A. Halverson, Air Corps, is designated as project officer.

By Command of Lieutenant General Arnold.[9]

This communication raised the question about who was in command of the AVG - Halverson or Chennault.

BOMBING JAPAN

James Conrad of Republic Steel identified more potential Japanese targets.[10]

DOOLITTLE RAID

Nero completed his task to prepare a B-25 parts list for Doolittle.[11]

At the Norfolk Naval base, two B-25s and their crews were loaded onto the *USS Hornet*. The *Hornet* was the newest United States carrier and had arrived at Norfolk prior to its deployment to the Pacific. After leaving Norfolk, the *Hornet* sailed about 100 miles off the coast. She turned into the wind and the two planes, loaded lightly with fuel, performed a takeoff. The concept had been demonstrated. But these two crews had specifically trained for this demonstration. Now, Doolittle had to find enough crews.[12]

REINFORCEMENTS

And Bill Mayhew was still in Australia.

On 2 February 1942 the ground personnel of the 9th, 88th and Headquarter Squadrons returned to Ascot Race Track in preparation for going to Java. At 2:00 A.M. (0200) on 3 February 1942 we were all awakened because there was a report an aircraft carrier was off the coast. We were formed into several groups, each group to be protected by one individual armed with a .45 caliber automatic pistol. I was one of those with a pistol. We were each issued 12 rounds of ammunition. Although I had been on guard duty my entire stay in Australia and had a .45 pistol on my hip most of that time, I had never fired one. I probably would have been more of a danger to those I was supposed to protect than I would have been to the Japanese. Fortunately, by sunup it was learned that it was an Allied carrier.[13]

PLOESTI

Replying to Hopkins' query about Ploesti, Arnold said that:

- B-17s based within 750 miles could attack the oil fields at Ploesti.

- B-24s based within 1,000 miles could attack the oil fields at Ploesti. (see Figure 26)

- No United States air force had as yet been sent to the Middle East.

- That while the B-24 type aircraft could attack as far as Budapest from bases in the United Kingdom, penetration of German defenses en route would probably result in a prohibitive cost in losses.[14]

1942 FEBRUARY 3

GLOBAL STRATEGY

A Joint Intelligence report stated:

- Germany must secure substantial supplies of oil. She could only obtain fabric oil in Caucasia or Iraq.
- A thrust through Turkey (towards) Syria and thence to Iraq and/or through Egypt were the two most likely scenarios. Any advance into Iraq would put Germany in possession of oil fields potentially capable of meeting her needs, and would deprive us of oil on which our Eastern Mediterranean naval position partly depends.[15] (see Figure 27)

The February 10 scenario regarding the fall of the Middle East was gaining traction.

DOOLITTLE RAID

Four squadrons of the 17th Bombardment Group stationed at Pendleton, Oregon were ordered to Columbia, South Carolina. At Doolittle's request, squadron commanders selected 140 men—enough to make up 24 five-man crews. Only 16 planes were needed to attack Japan, and the eight extra crews and their planes provide "a 50 percent buffer" against losses, as Doolittle put it.[16]

NEW GUINEA

The Japanese bombed Port Moresby.[17] (see Figure 25)

EAST INDIES

Flying from Kendari, Celebes, the Japanese began their invasion of Java with a series of air strikes directed at Soerabaja and Malang.[18] (see Figure 20)

1942 FEBRUARY 4

NORTH AFRICA

British forces completed their withdrawal to Gazala, while the Axis solidified their position on the line between Tmimi and Mechili.[19] Tmimi and Mechili were just west of Gazala and effectively blocked the British from Benghazi. (see Figure 5)

MIDDLE EAST

British troops and tanks surrounded the palace of King Farouk I. The British Ambassador to Egypt, Sir Miles Lampson, presented an ultimatum to Farouk. Farouk needed to appoint a more pro-British government. Farouk had demonstrated some pro-Axis tendencies earlier in the war. Now with Rommel knocking at the door, such sympathies were unacceptable to the British.

EAST INDIES

The Allies surrendered Ambon.

An Allied naval task force, led by Dutch Admiral Karel Doorman, was defeated in the Makassar Strait.[20] (see Figure 12)

REINFORCEMENTS

Bill Mayhew wrote the following:

On 4 February 1942 we (88th, 9th and Headquarter Squadrons) were again loaded on a ship (USAT Willard A. Holbrook). We sailed at about 11:00 P.M. (2300). This time our destination was Java. As most of the islands to the north of Australia had fallen into Japanese hands by this time, we had to go completely around Australia in order to reach our destination. This time I was assigned to a .30 caliber antiaircraft machine gun. Our shifts were 4 hours on and 12 hours off.[21]

COMMITMENT TO CHINA

Roosevelt had asked Arnold to respond to Churchill's query about the withdrawal of the fighters from Rangoon. (see Figure 3) Lord Halifax had repeated Churchill's query on January 30. Arnold prepared a draft response on February 4. The Combined Chiefs had struck a deal with Chiang to delay the withdrawal of the AVG until a group of British Hurricane aircraft arrived. These replacement planes were expected to arrive between February 15 and 20.[22]

1942 FEBRUARY 5

SINGAPORE

An Allied supply convoy reached Singapore. (see Figure 22) Japanese aircraft sank one of the supply vessels, while other enemy aircraft attacked the docks, preventing some of the vessels from off-loading their cargo.[23]

MIDDLE EAST

Britain's repression of the Egyptian demonstrators caused some concern in United States diplomatic circles. Wallace Murray, Chief of the State Department's Division of Near Eastern Affairs (NEA), sent a note to Sumner Welles suggesting that Ambassador

Kirk visit his British counterpart, with the suggestion that Britain focus on fighting the Germans, not the locals. Welles rejected the idea.[24]

HALPRO

Halverson ordered changes to the production B-24s to make them more useable for their mission. These changes were to be performed at Patterson Field near Dayton, Ohio. As originally scheduled, the B-24s were to be flown to Patterson beginning on February 5. Delays prevented their departure.[25]

1942 FEBRUARY 6

EAST INDIES

The Japanese occupied Samarinda, on east coast of Dutch Borneo.[26] (see Figure 20)

BOMBING JAPAN

The third set of potential Japanese targets was identified by Victor Bretandas of Douglas Aircraft.[27]

HALPRO

A "few days later," Arnold invited Nero to dinner where he promised Nero he would tell him to which project, HALPRO or Doolittle, he would be assigned. The next day, Nero got the news that he was assigned to HALPRO.[28]

1942 FEBRUARY 7

MIDDLE EAST

With the British pressing for an air force in Egypt, the WPD addressed the problem of where the planes would come from and sent their opinion to Arnold. Buried in their answer was that:

> *If the strategic situation so* required, the United States was to be prepared to provide a heavy bomber force for the Middle East-- from the first two heavy bomber groups allocated to the United Kingdom.[29]

Rommel's progress in North Africa was causing concern.

1942 FEBRUARY 8

SINGAPORE

Japanese forces landed on the island in the evening hours and created a beachhead.[30] (see Figure 22)

THE PHILIPPINES

Since the beginning of the Japanese invasion of the Philippines, Philippine President Manuel L. Quezón had watched and waited for United States reinforcements. With the Japanese seemingly about to complete her conquest, Quezón cabled Roosevelt that the United States should give the Philippines her independence so that he, Quezón, could negotiate a separate peace with Japan. Francis Sayre, the United States High Commissioner of the Philippines,[31] agreed, assuming that Quezón was correct in his assumption that reinforcements were NOT enroute.[32]

1942 FEBRUARY 9

SINGAPORE

Japanese forces reached Tengah Airfield. Another force landed just west of the now destroyed causeway.[33] (see Figure 14)

COMMITMENT TO CHINA

Rommel's force was not the only army causing the Allies to rethink its strategy. The Japanese were threatening to cut the Burma Road, and thus the supply route to the Chinese. Arnold had been looking at the problem in early February. He thought that if the existing supply line was severed and if the commercial airlines would relinquish their planes, an air ferry program could be set up in a few days. Roosevelt knew that the Chinese had to remain engaged against the Japanese.[34] He thus fired off a message to Chiang:

> *We are rapidly increasing our* ferry service to China via Africa and India. I can now give you assurances that even though there should be a further setback in Rangoon, which now seems improbable, the supply route to China via India can be maintained by air.
>
> The whole plan seems altogether practical and I am sure we can make it a reality in the near future.[35]

Roosevelt's conclusion that the fall of Rangoon was improbable implied that he was reading reports from other sources, or he was confused.

VICTORY PROGRAM

It turned out that Roosevelt was not the only person who was confused. *Time* ran an article in the February 9, 1942 issue titled *The Bombers are Growing*. Buried in the article was the following:

> It has already been established that in building the heavy bombers—with which air power strikes—the U.S. is far & away ahead of the rest of the world. Token proofs are the Flying Fortresses now in Europe; a bigger token, the handful of Flying Fortresses in The Netherlands East Indies, which slashed with grim effectiveness at Japanese naval units, fought off Jap fighter planes, ranged far & wide through the South China Seas—on missions which cannot yet be fully described in print.
>
> There will be more than token proof. U.S. production today is heavily weighted with four-motored bombers—Flying For tresses and swift Consolidated B-24s. In the 185,000-plane goal for 1942-43 set by President Roosevelt there will be a greater proportion of heavy bombers—more heavy bombers than any nation ever built before.[36]

There were plans to put B-17s in Europe, but so far there were no American B-17s in Europe. The British had operated B-17s, but had removed them from front line duty. There was no evidence to support the claim that Brereton's B-17s had "slashed through Japanese naval units." All things considered, the article's title overstated the case that "the bombers are growing."

1942 FEBRUARY 10

SINGAPORE

Even knowing the fate of Singapore, Churchill told General Wavell:

> *There must at this stage* be no thought of saving the troops or sparing the population. The battle must be fought to the bitter end at all costs. Commanders and senior officers should die with their troops. The honour of the British Empire and of the British Army is at stake.

Churchill thus displayed and would continue to display a willingness to sacrifice his troops.[37]

So, Wavell visited Singapore and conveyed Churchill's orders to "Hold." The remainder of the RAF personnel was ordered to Netherlands East Indies.[38]

EAST INDIES

The Japanese landed troops at Makassar, Celebes.[39] (see Figure 20)

BURMA

It was not just the Burma Road that the Japanese were threatening to sever. The air route to Australia from India was also being threatened. George, head of the AWPD, proposed to divert B-17s. The concept was then forwarded to Arnold on February 12. The details of the proposal were:

- A strong probability exists that the ferry route from India to the ABDA area [essentially Australia] was already blocked or might be denied in the near future.
- If that happened, United States heavy bombers, then enroute to ABDA, would become immobilized in India.
- Chennault's ability to support heavy bombers was in doubt. He had been asked to submit an estimate of his capacity to support heavy bomber operations.
- It was recommended that Chennault assume command of any heavy bombers diverted from ABDA until HALPRO was in position to take operational control.
- It was recommended that eight heavy bombers be diverted from planes being shipped to ABDA and assigned to China.[40]

EASTERN FRONT

While the Japanese were active throughout the winter of 1941-42 and Rommel was active in North Africa, the Germans had yet to reveal what their offensive would be once the winter was over. The Joint Military board wrote the following:

- They believe that Germany could not undertake another major offensive until Russia's military power was destroyed. The most probable German major offensive operation during 1942 would be to gain access to the oil of the Caucasus.
- While the German oil supply was limited, it should be sufficient to meet the needs of military operations through 1942.
- The Middle East. A secondary effort would envisage heavy reinforcements to Rommel with a view to overrunning Egypt, seizing the Suez Canal, driving eastward and threatening Basra.[41]

1942 FEBRUARY 11

REINFORCEMENTS

A group of B-17s, which were flying reconnaissance missions around the Hawaiian Islands, was formed into Air Corps units. Under the command of Major John Carmichael, "A" flight of that group began its planned deployment to Australia. The first leg of that deployment was to the Christmas Island. The B-17, *Swamp Ghost*, under the command of Capt. Frederick Eaton, was part of that group.

The next day, the group would fly to Canton Island, followed by another jump to Nandi Airfield, on Viti Levu Island, Fiji. Since Fiji was a colony of the Vichy French, the group was delayed.[42]

1942 FEBRUARY 12

10TH AIR FORCE

At Patterson Field, Ohio, the 10th Air Force was activated on 12 February. It would be assigned to the newly created China-Burma-India (CBI) Theater and was to be based in Karachi, India.[43] (see Figure 21)

FAR EAST

After over a month of trials and tribulations, Paul Eckerly and his crew arrived in Bangalore, India flying his B-17. On February 12, the crew flew to Colombo, Ceylon. (see Figure 32) From there, they would fly the long over water leg to Bandoeng, Java. After a nearly thirteen-hour flight, the crew arrived in Bandoeng, Java the next day.[44] (see Figure 28)

1942 FEBRUARY 13

SINGAPORE

All remaining British naval vessels left the port of Singapore during the night. The commanding Officers of the Army and Navy were evacuated.[45]

EAST INDIES

The Japanese seized Bandjermasin, in southeast Borneo.[46] (see Figure 20)

1942 FEBRUARY 14

SINGAPORE

Japanese aircraft strafed the fleeing British naval vessels. The vessel carrying the two ranking British officers was forced to land on a deserted island. These officers and other men would perish later.[47]

EAST INDIES

An Allied naval task force had departed to engage a Japanese force in the Bangka Strait. However, strong enemy air strikes caused the task force to withdraw.[48]

Japanese paratroops executed a drop at Palembang, Sumatra. (see Figure 20) By the next day, Japanese forces controlled the town and its airfield. Java was now effectively cut off from supplies from the South Atlantic-Africa-India supply route.[49]

REINFORCEMENTS

Paul Eckerly's crew exchanged planes. They and the rest of the group took off for their assigned combat base at Singosari Air Base at Malang in eastern Java. (see Figure 20) Unfortunately, two plus hours into the flight, the flight leader lost one engine and the formation was forced to divert to Djogjakarta.[50]

GLOBAL STRATEGY

British intelligence had sent its assessment of German intentions to Washington on January 30. United States intelligence reviewed this report and sent their comments up the chain of command on February 2. The final assessment was sent to the White House on February 14. Their summary conclusion was:

> *If the Germans can maintain* such a [maximum] schedule, they can gain a major decision against Russia in 1942 and perhaps a decision in the Middle East as well. They would then be in a position to move on India in 1943, if they so elect. The Russian Army might disrupt the schedule. The outcome therefore depends more upon the relative combat value of Russian and German divisions than upon any other factor.[51]

FAR EAST

Marshall sent a report to Roosevelt that 40 heavy bombers had arrived in the ABDA Theater. Twenty-five were enroute.[52]

1942 FEBRUARY 15

SINGAPORE

General Arthur E. Percival surrendered the Singapore garrison, consisting of around 64,000 men.[53] (see Figure 22)

EAST INDIES

Following the Japanese paratroop drop at Palembang, Allied defenders withdrew, having only partially destroyed the oil refineries.[54] (see Figure 20)

SOUTH PACIFIC

Admiral King ordered the *USS Yorktown* carrier Task Force to move toward the Canton Island in the South Pacific.⁵⁵ The island, actually an atoll, is located about halfway between Hawaii and Australia.

REINFORCEMENTS

Paul Eckerly's crew finally made it to the Singosari Air Base at Malang, Java. (see Figure 20) There, they joined up with the remnants of Brereton's B-17s. Eckerly sums up the crew's travels:

> *My trip from Massachusetts to* Malang was about 16,000 miles and took us a month and four days. Our flying time was one hundred and twenty one hours and ten minutes. Project 'X' was now complete for us…⁵⁶

If there was any thought that the Project X B-17s could have assisted Singapore, the distance from Malang to Singapore is slightly less than 900 miles, outside the range of a B-17.

HALPRO

Ed Cave "[w]as sent to Barksdale Field to the 98th Bomb Group, February 15, 1942 to be introduced to the monster B-24. Biggest airplane I had ever seen."⁵⁷

On January 23, George had prepared a plan to send the first three HALPRO planes to China by February 15. This might have been difficult as the planes were at Patterson Field outside Dayton, Ohio. Halverson and his staff were preparing to leave Washington for Patterson Field so the first four B-24s could be modified. The remaining B-24s were being sent to Mobile Air Depot in Alabama for combat modifications.⁵⁸

1942 MID FEBRUARY

HALPRO

Scotty Royce had spent the past few weeks working with Doolittle. Then fate intervened.

> *Jimmy Doolittle shared a suite* of offices with Henry J. F. Miller, also a Lieutenant Colonel and long-time Wright Field technical type.
>
> I was sitting in the secretary's office, waiting to see Jimmy, when Henry J. F. Miller passed through, and again, "Scotty, what are you doing here?" Well I explained the situation to him and he said, "Look, Doolittle has B-25s - if you're a B-24 expert. I need you. I have a task force by the name of Halpro that's going to get the first B-24Ds, and the Air Force really needs someone who understands

these machines because the Air Force experience with B-24s of any kind is very limited. So I'll fix it up with Doolittle for you to go with Halpro".

Not knowing anything about either of these task forces I really had no say in the matter. And being only a 2nd Lieutenant to boot, I could only agree with the Lieutenant Colonel. Furthermore, I really felt that since my expertise was with B-24s, it would be better for me to go with the B-24s. In a short time everything was arranged and I had a ticket to Dayton, Ohio. I joined ... I believe it was the Headquarters, Tenth Air Force, which was being set up in Dayton as a nucleus of what would become the Tenth Air Force, to cover for Halpro.[59]

1942 FEBRUARY 16

EAST INDIES

Wavell sent a rather pessimistic communiqué to Churchill and the Combined Chiefs of Staff (CCS) in Washington. With the limited forces at his disposal, he could not prevent a Japanese landing on Java. His conclusion was that:

> *As an air base, Java* is of value to support naval operations in China seas and recapture Borneo, Celebes, and eventually the Philippines, but ultimate air action against Japan and Japanese line of communications must come through Burma and China. Loss of Java would not directly affect the issue of events in the Philippines, but it deprives us of one line of counteroffensive against Japan. To sum up, Burma and Australia are absolutely vital for war against Japan. Loss of Java, though a severe blow from every point of view, would not be fatal. Efforts should not, therefore, be made to reinforce Java, which might compromise defense of Burma and Australia.[60]

The withdrawal of Allied troops from Palembang was completed. (see Figure 20)
An Allied supply convoy for Timor was recalled to Darwin following attacks by Japanese aircraft.[61] (see Figure 20)

BOMBING JAPAN

J. D. Hitch of Dorr Company recommended that incendiary bombs be used when bombing Japan.[62] The next day, Roosevelt suggested Hitch pass the ideas on to the military staff.[63]

PLOESTI

Arnold's Chief of Staff, Brig. Gen. Martin F. Scanlon, had been reviewing the communiqués from Bonner Fellers regarding an air raid on the Ploesti oil fields. On February 16, Scanlon filed his report on the German oil situation. His opinion was that the German situation was not as dire as Fellers had described. However, he did agree that if Germany was deprived of Rumanian oil, the situation could be as critical as Fellers described.[64]

NORTH AFRICA

Rommel and his operations officer visited Hitler at his East Prussian headquarters, Wolf's Lair. Rommel would spend the next two days trying to convince Hitler and the German High Command of the need for additional supplies. He was unsuccessful as Hitler's attention was focused on Russia and the Eastern Front.[65] While the Allies and the press could see one strategy for the Middle East, the Germans were focused elsewhere.

1942 FEBRUARY 17

GLOBAL STRATEGY

Even the press could look at a map and envision the Axis strategy. The headline of Hanson W. Baldwin's *New York Times* article said it all:

THREAT TO SUEZ RETURNS[66]

1942 FEBRUARY 18

EAST INDIES

Japanese troops landed on the Island of Bali during the night of February 18/19. This effectively isolated Java.[67] (see Figure 20)

NEW GUINEA - SOLOMON ISLANDS

Based upon recent Japanese activities, Admiral King speculated that the Japanese intended to drive southeast into the Solomon Islands, thereby cutting off the Allied ferry route to Australia, the Southern Pacific Supply Route. To prevent that from happening, the United States had to act and occupy islands in the chain. King knew that he needed Army units to help garrison the islands. Thus, he sent a letter to Marshall asking for his approval of the plan.[68] (see Figure 25)

1942 FEBRUARY 19

EAST INDIES

Japanese carrier based planes struck Darwin, Australia.[69] (see Figure 20)

REINFORCEMENTS

Bill Mayhew was still trying to enter the fray.

> *We picked up the rest* of our convoy the morning of 13 February 1942.
> We then proceeded to Fremantle, the principal seaport of Western Australia. We reached Fremantle at 1:00 P.M. (1300) on 18 February 1942. The anti-aircraft machine guns had to be manned constantly, even though we were in port. This was really impressed upon us on 19 February 1942 when 150 Japanese carrier-born aircraft attacked Darwin, Australia. Four carriers from the Pearl Harbor force led the attack. They damaged harbor installations and sank a number of ships. Although this was several hundred miles north of our position, it was entirely too close for comfort.[70]

1942 FEBRUARY 20

EAST INDIES

Japanese troops invaded Timor.[71] (see Figure 20)

NEW GUINEA - SOLOMON ISLANDS

The *USS Lexington* and her escorts steamed towards Rabaul. (see Figure 19) However, heavy Japanese air attacks caused the task force to withdraw. In the process of repelling the task force, the Japanese suffered heavy losses.[72]

SECOND FRONT

Hopkins sent the following to Roosevelt:

> *Stalin enquired as to likelihood* of opening second front in Europe. We replied that we should not be able to in the immediate future but that one of the subjects of the Libyan campaign was to secure a base from which we would attack Italy.[73]

Italy had never been mentioned, except by Churchill who always favored a Mediterranean campaign.

REINFORCEMENTS

Carmichael's "A" flight of B-17s finally arrived Garbutt Field, a Royal Australian Air Force (RAAF) base near Townsville, Australia. They would form the 88th Reconnaissance Squadron of the 7th Bomb Group. They were still under the control of the United States Navy. Since it was believed that Townsville was within reach of Japanese planes, the group was deployed to Cloncurry, another base near Townsville.[74] (see Figure 30)

1942 FEBRUARY 22

THE PHILIPPINES

Roosevelt ordered MacArthur to leave the Philippines.[75]

GLOBAL STRATEGY

By the middle of February, the British situation was indeed grave:

- Burma had been invaded and was fast being overrun
- Singapore had surrendered
- And in North Africa, British forces had fallen back upon a line of defense running south from Tobruk.[76]
- The British chiefs of staff proposed to the CCS a new "Policy for Disposition of United States and British Air Forces," which greatly extended the scope of diversions from earlier agreements. The United States was asked to:
- Conduct bomber operations from China against Japan,
- Assist the British with heavy bombers in the Burma-Indian Ocean theater and, if necessary, in the Middle East.[77]

FAR EAST

The British were not alone in realizing that the situation for the Allies was getting desperate. Brereton had withdrawn the remnant of his Far East heavy bomber force from Manila to Java. He was currently stationed in Bandoeng, (see Figure 28) which is in the western part of Java, closest to the Japanese forces. Paul Eckerly and crew had just arrived in Java from the United States. So, Brereton received the following order:

1. Your own headquarters will be withdrawn in such a manner, at such a time, and to such a place within or without the ABDA area as you may decide, for its timely withdrawal is important. Dutch should be allowed to decide which of their representatives with your headquarters should leave or stay and also destination of any personnel withdrawn.
2. ...
3. When you withdraw, report to whom you have transferred command of Java.[78]

REINFORCEMENTS

The 88th Reconnaissance Squadron of the 7th Bomb Group flew its first bombing mission. Its target was Simpson Harbor at Rabaul. (see Figure 19) Nine bombers departed in the evening. But only five reached the target the following morning. One was the B-17 that would be named *Swamp Ghost* after it was recovered.

Unable to release her bombs on the first pass, she made a second pass. This time, she released her bombs, reportedly hitting a Japanese freighter. During the second pass, a Japanese anti-aircraft shell passed through the starboard wing. It failed to explode but left a gaping hole in the wing.

After leaving the target, she was intercepted by a group of Japanese Zero fighters. Her tail gunner, S/Sgt John Hall, claimed one fighter, while waist gunner T/Sgt Russell Crawford claimed two more. Still, the Japanese fighters inflicted mortal damage to the *Swamp Ghost*.

The Zeros withdrew and *Swamp Ghost* made it to the northern coast of New Guinea. Running out of fuel, Eaton attempted a wheels up landing on what he thought was a landing strip. In reality, the "strip" was a swamp. Fortunately, the entire crew walked away from the accident.[79]

Bill Mayhew's ship arrived at Fremantle to join a convoy to Java.

When we arrived in Fremantle, we tied up beside the "USS Langley," the first American aircraft carrier. The "Langley" had been converted to a seaplane tender. Now its deck contained 32 P-40s that our group (7th Bombardment Group and the 88th Reconnaissance Squadron) had assembled at Amberly and Archer Fields (near Brisbane). We were both to become a part of Convoy MS-5, heading for Java. Convoy MS-5 consisted of five ships, escorted by the cruiser "USS Phoenix". The other ships were the "Langley", "Holbrook", "Duntroon", "Katoomba" (we called it "Smoky Joe"), and " Sea Witch". We left Fremantle on 22 February 1942, but returned to Fremantle before we were out of sight of land. Java was beginning to be evacuated. The next day (23 February 1942) at noon we set sail again, but headed for India rather than Java. Some time after midnight (I was on gun duty at the time) the "Langley" left the convoy, headed for Java. (It was sunk on 27 February 1942 before it reached Java.) The "Sea Witch" left the convoy 25 February 1942 to make a run for Java (it had 18-crated P-40s in its hold. It reached Java, but there was no one to assemble the planes, so the crates were dumped in the harbor of Tjilatjap, Java and the "Sea Witch" successfully escaped.) The "Phoenix" left our convoy on 1 March 1942 when a British light cruiser joined us.

I think the higher authorities did not know exactly what to do with us, since we did not have any equipment with which to fight. From 18 February to 5 March 1942 we did not see land. The hold of the ship was completely filled with explosives of all types that had been loaded aboard in Melbourne, which made all of us on board very uncomfortable. To add to our discomfort, a Japanese submarine was sunk at the entrance of the harbor of Colombo, Ceylon the night before we arrived. We doubled the watch and most of us saw submarine periscopes all night. Fortunately, it was the only one in the region.[80]

1942 FEBRUARY 23

EAST INDIES

Japanese reported the capture of Ambon. (see Figure 12) General Brett left Java for Australia.[81]

10TH AIR FORCE

When viewed from a global perspective, the withdrawal of the remnants of United States Far East AF to Karachi, India, thus joining the 10th Air Force, seemed to put the heavy bombers in a position to strike either the Japanese or the Germans. (see Figure 27)

FAR EAST

General Wavell was also ordered to leave Java for India. Ironically, three days earlier, Wavell had been told that there would be no withdrawal and no reinforcements. But the situation had deteriorated so fast that the withdrawal of leadership was deemed appropriate.[82] Like Churchill's previous order to Singapore, the common soldier was considered expendable and thus left behind.

PEARL HARBOR AND THE PHILIPPINES

Roosevelt apparently decided that he needed to reassure the American public. In his Fireside Chat on February 23, he downplayed the American losses so far in the war:

> *You and I have the* utmost contempt for Americans who, since Pearl Harbor, have whispered or announced "off the record" that there was no longer any Pacific Fleet -- that the Fleet was all sunk or destroyed on December 7th -- that more than a thousand of our planes were destroyed on the ground. ...
>
> Very many of the ships of the Pacific Fleet were not even in Pearl Harbor. Some of those that were there were hit very slightly, and others that were damaged have either rejoined the Fleet by now or are still undergoing repairs. And when those repairs are completed, the ships will be more efficient fighting machines than they were before.
>
> The report that we lost more than a thousand planes at Pearl Harbor is as baseless as the other weird rumors. The Japanese do not know just how many planes they destroyed that day, and I am not going to tell them. But I can say that to date -- and including Pearl Harbor -- we have destroyed considerably more Japanese planes than they have destroyed of ours.[83]

The facts are slightly different. The carriers were the only ships not at Pearl Harbor on December 7. The Japanese sank six battleships, damaged two, and sank three cruisers and two destroyers. As for the airplanes at Pearl Harbor, the Japanese destroyed 180 and damaged 128, at a cost of only 29 Japanese planes. On the Philippines, they had eliminated nearly all of the 277 planes MacArthur started with.[84] The heavy bombers, the only planes with enough range to withdraw, were now on their way to India.

However, Roosevelt did not tell the American public that he had just ordered MacArthur to leave the Philippines for Australia. The loss of the Philippines was imminent, and the Allies could not give the Japanese a public relations coup by capturing MacArthur.

PRESS

Churchill must have begun to feel the pressure of not delivering a victory on any front. On February 23, *Time* published an article titled *Great Britain: Sticks and Stones*, the first sentence of which read:

> *For the first time since* Winston Churchill became Prime Minister, millions of Britons last week began seriously to question his abilities.[85]

Adding to Churchill's discomfort was the fact that the British Navy, once the most powerful naval force in the world, had inexplicably let two German battleships, (the *Scharnhorst* and *Gneisenau*), the heavy cruiser *Prince Eugen*, and their escort vessels to transit the English Channel to their home port in Germany.[86]

Refusing to believe in German technical superiority, Churchill looked for scapegoats. Not willing to accept the superiority of the German leadership, he began to question the skill of his military leaders.[87]

And it was not just Churchill's leadership that *Time* began to question. The politicians of Washington came under the same microscope. In the same issue, *Time* ran an editorial titled *Worst Week*.

After recounting all of the bad news of the preceding week and Roosevelt's press conference proclamation that the war was now a war of attrition, the writer predicted that the bad news would prevail over good news for the foreseeable future. He then asked:

> *Could the people take it?* Washington did not seem to think the people could. Army and Navy communiqués stressed good news, toned down bad. The President, in a testy mood, appeared to feel that the people did not yet understand the war. Congressmen, bogged in gloom and desperately trying to save face on their self-pension bill, potshot at each other (see p. 16).
>
> But the nation seemed to think it could take it. Up & down the country editorial writers, living close to the people of their own communities, worried more

about apathy than the collapse of morale. They wrote with bold strokes: AMERICA CAN LOSE; THE WAR CAN BE LOST; THIS SHOULD AWAKEN US.

The cry was for more bad news, for the truth. Washington had coddled the nation too long; had let them believe that the war could be fought to the finish in the antiseptic, sealed atmosphere of Lend-Lease.[88]

1942 FEBRUARY 24

EASTERN FRONT

On the central portion of the front, the Germans were successful in stopping the Russians from retaking Smolensk. They were also successful in halting the Russian advance in the Ukraine.[89] (see Figure 24)

EAST INDIES

Brereton was ordered to leave Java two days earlier. On February 24, he and a small staff left on two of his remaining heavy bombers.[90] In his diary, he wrote:

> *Everyone realized that it was* a completely hopeless task to defend Java. My desire for some time had been to give the Japs territory and get back where we could reorganize the striking forces, and I didn't care whether it was India or Australia. Brett gave me my choice and I picked India, maybe because I was sick of islands, even one as big as Australia.[91]

The emphasis on India as a strategic center of Allied activity was expressed in a memo from George to Marshall. George pointed out that:

1. The 10th Air Force had just been created and set for establishment in Karachi, India. (see Figure 27)
2. Brereton was withdrawing his meager force of heavy bombers to Karachi and should be put in command of the 10th.
3. The HALPRO project should be cancelled and the force of men and planes should be given a numerical designation.[92]

It seemed that Arnold's premonition about the fate of the HALPRO mission could become a reality.

NEW GUINEA - SOLOMON ISLANDS

Marshall responded to Admiral King's request for Army troops to aid in the proposed

Solomon Island campaign. Basically, Marshall wanted to know why Marines could not perform this task. Marshall concluded his letter by noting that "In general, it would seem to appear that our effort in the Southwest Pacific must for several reasons be limited to the strategic defensive for air and group troops."[93]

GLOBAL STRATEGY

The Americans were not the only people looking at the Asian map and the unstoppable march of the Japanese forces. Chiang had been in India for consultations with other Allied leaders. Upon his return to China, he wrote a communiqué to Soong, conveying his impressions. He could not believe that the British would let the situation reach the state it had. His conclusion was that: "If the solution is postponed until after the Japanese armies enter India, then it will be certainly too late. If the Japanese should know of the real situation and attack India, they would be virtually unopposed."[94]
If that happened, not only would the Burma Road be severed, but also any hope of supplying Chiang by airplane would be virtually impossible.

1942 FEBRUARY 25

GLOBAL STRATEGY

Roosevelt apparently was becoming increasingly concerned about what to do about India. He had already suffered a verbal harangue from Churchill and did not want to cause additional distress. The Allies were withdrawing slowly from Burma and retreating into India. Soong had just forwarded Chiang's assessment of the situation. So, he sent a message to John G. Winant, the United States Ambassador to England.

> *As you may guess, I* am somewhat concerned over the situation in India, especially in view of the possibility of the necessity of a slow retirement through Burma into India itself. From all I can gather the British defense will not have sufficiently enthusiastic support from the people of India themselves.
>
> In the greatest confidence could you or Harriman or both let me have a slant on what the Prime Minister thinks about new relationships between Britain and India? I hesitate to send him a direct message because, in a strict sense, it is not our business. It is, however, of great interest to us from the point of view of the conduct of the war.[95]

Roosevelt's last paragraph encapsulated the issue his military advisors had been raising: What does the United States do when British interests collide with Allied strategy?

1942 FEBRUARY 27

THE PHILIPPINES

A Japanese force landed on the island of Mindoro, Philippines.[96] For all intents and purposes, the Allied forces on Bataan and Corrigador were isolated. (see Figure 6 and Figure 23)

EAST INDIES

The battle of the Java Sea occurred when a composite group of Allied vessels attacked a Japanese task force. The Japanese lost no vessels, while the Allies lost two cruisers and three destroyers. The Allied ships withdrew and, in the process, lost the three remaining cruisers and four destroyers. All that remained of the Allied force were three destroyers.[97] (see Figure 20)

1942 FEBRUARY 28

10TH AIR FORCE

Brereton arrived in Karachi. (see Figure 27) After nearly two months of fighting, he had finally formulated his own plan of attack:

> *I stated that the U.S.* Tenth Air Force, which I was to command, would not be committed piecemeal nor employed until its operational training was completed. I had had enough of a fighting in dribbles. I insisted on building a striking force with a punch to it.[98]

EAST INDIES
It's a good thing Brereton and his weakened force left when they did. The Japanese began landing on the northern coast of Java on the night of February 28.[99]

HEAVY BOMBER PRODUCTION

During February, 134 heavy bombers were manufactured. Sixty were B-24s.[100]

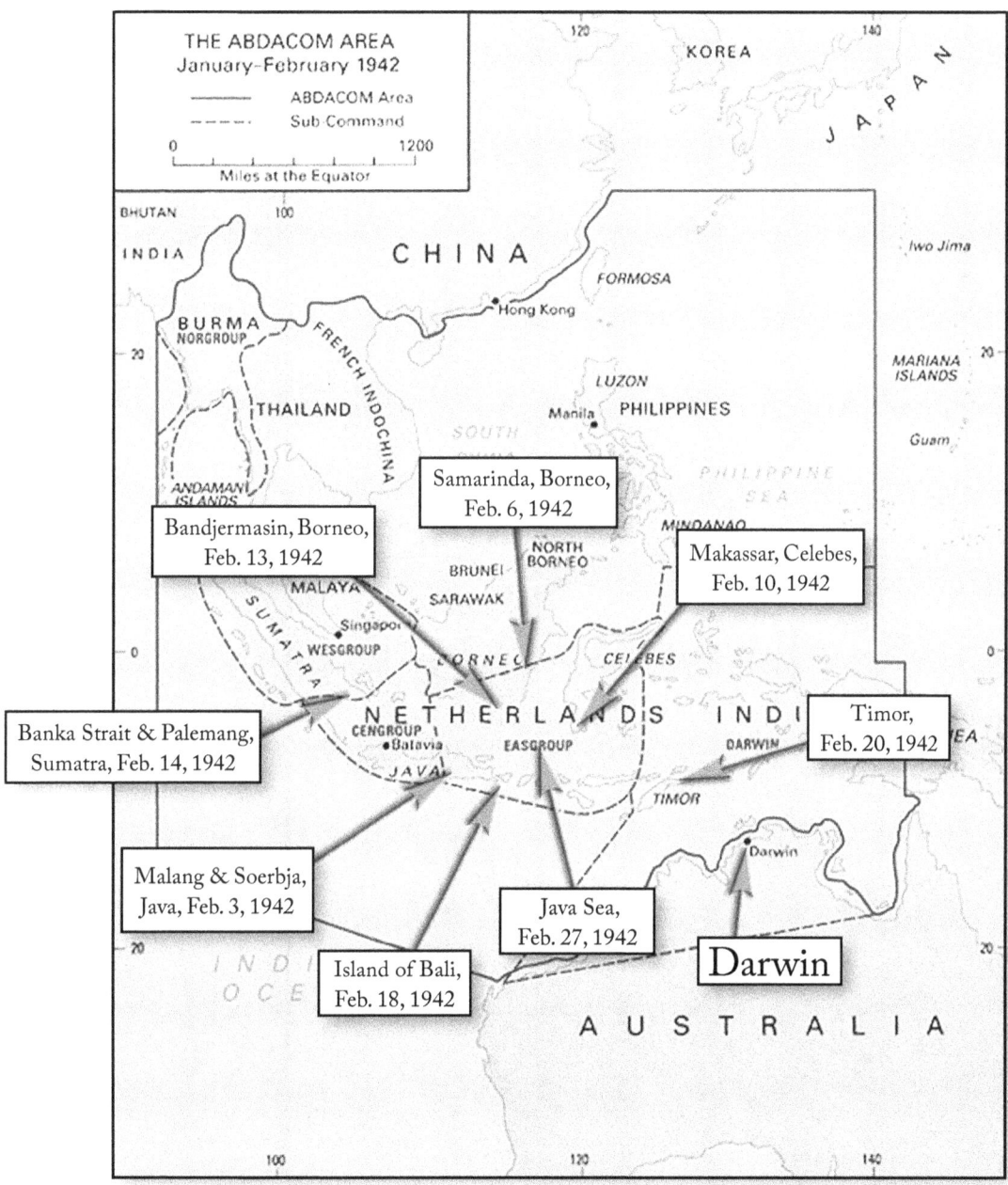

Figure 20 – ABDA – February 1942

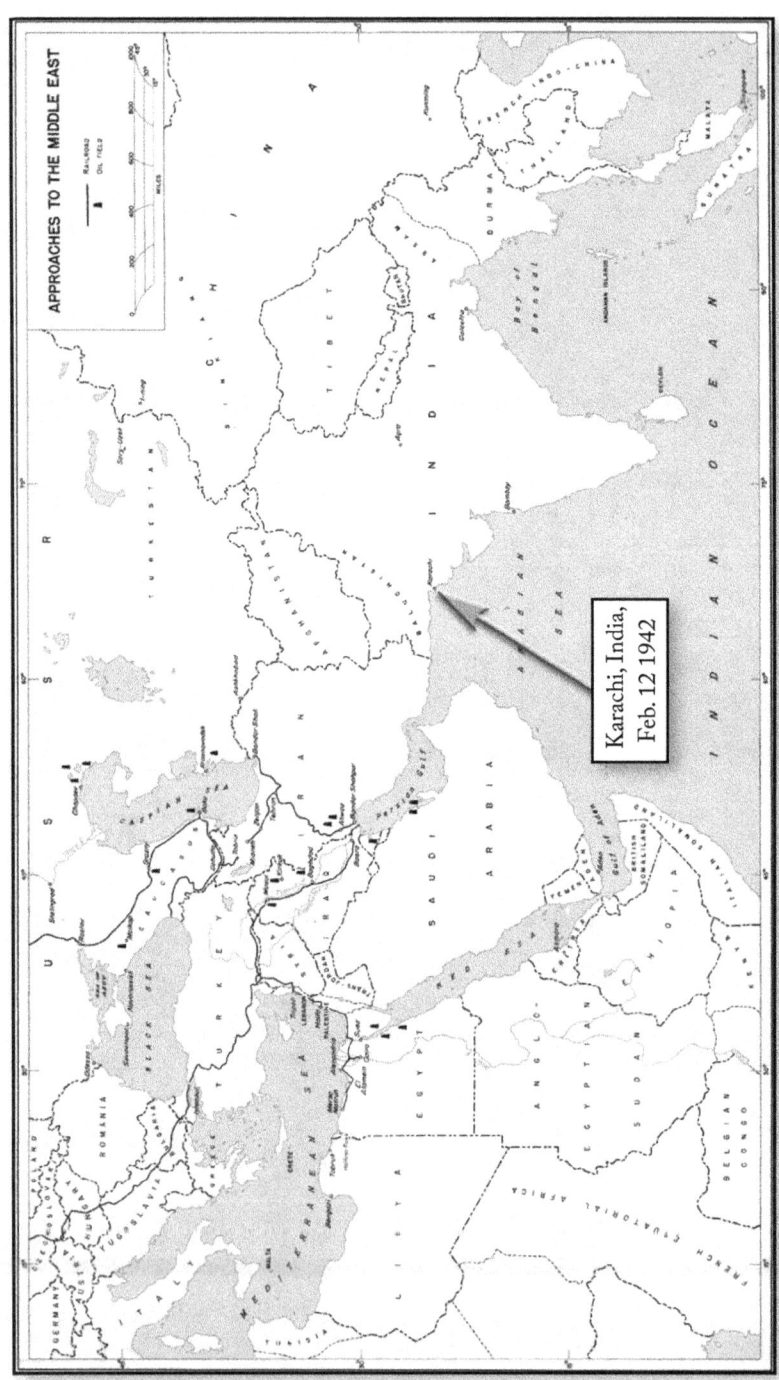

Figure 21 – India – February 1942

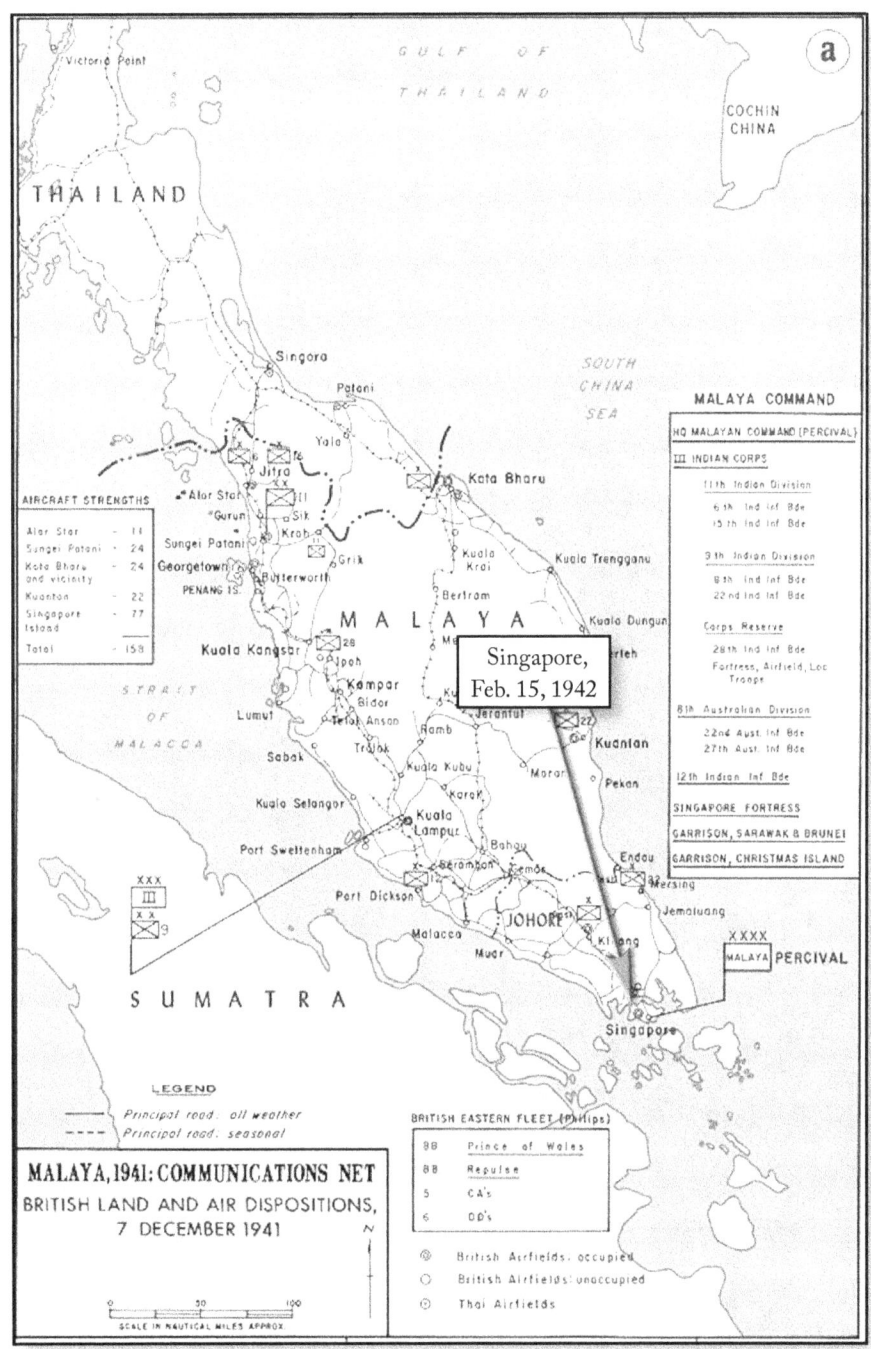

Figure 22 – Malaya – February 1942

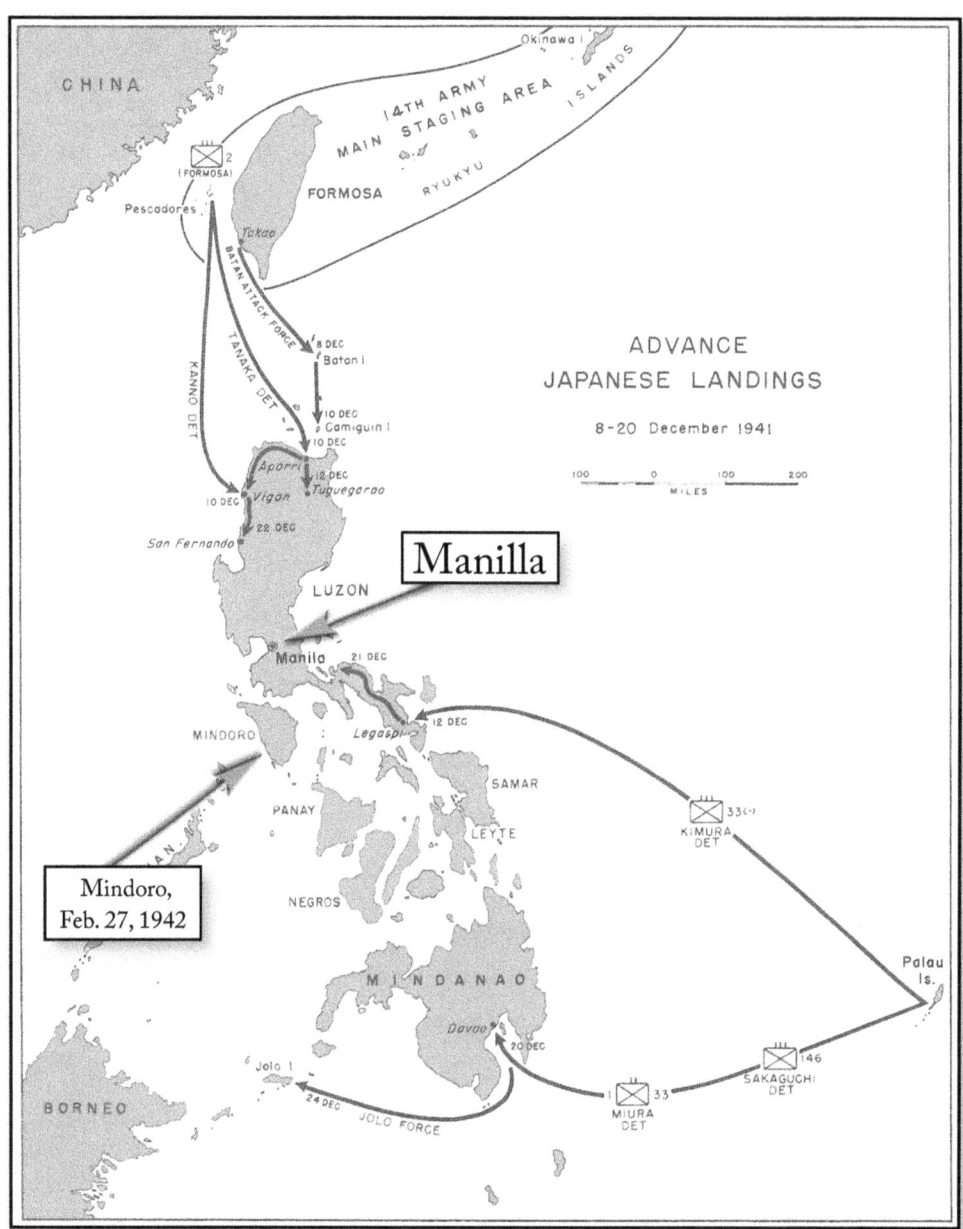

Figure 23 – Philippines – February 1942

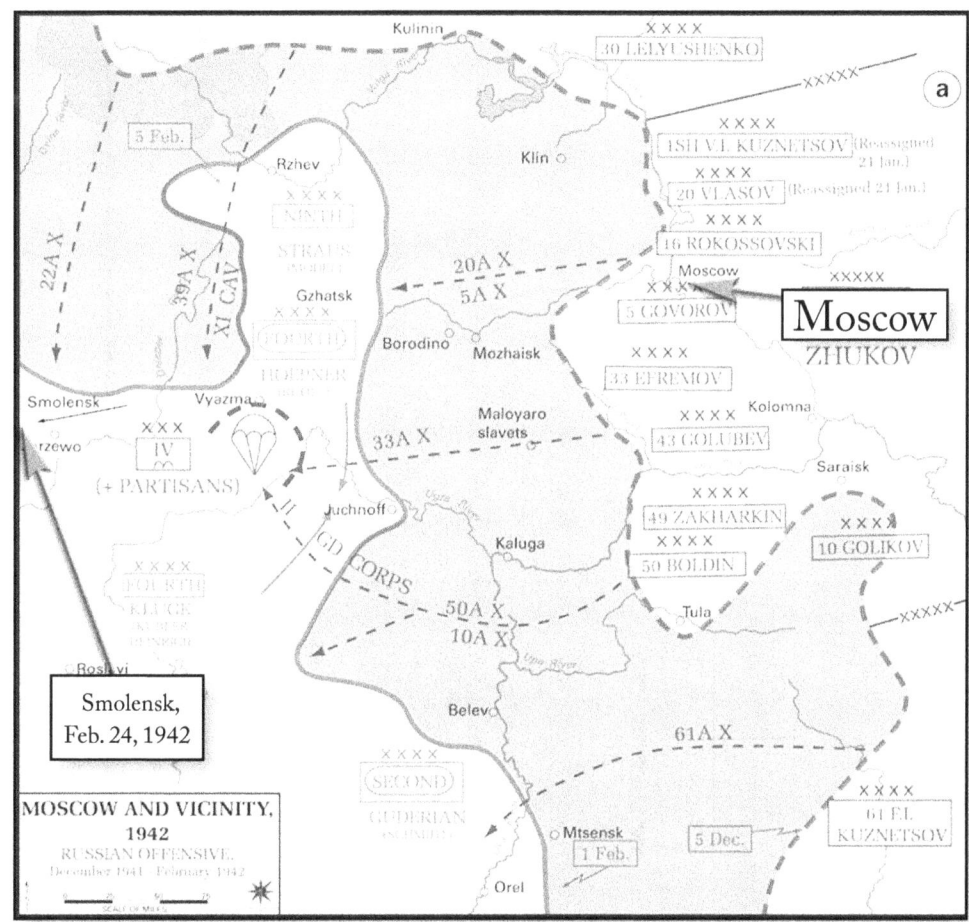

Figure 24 – Moscow – February 1942

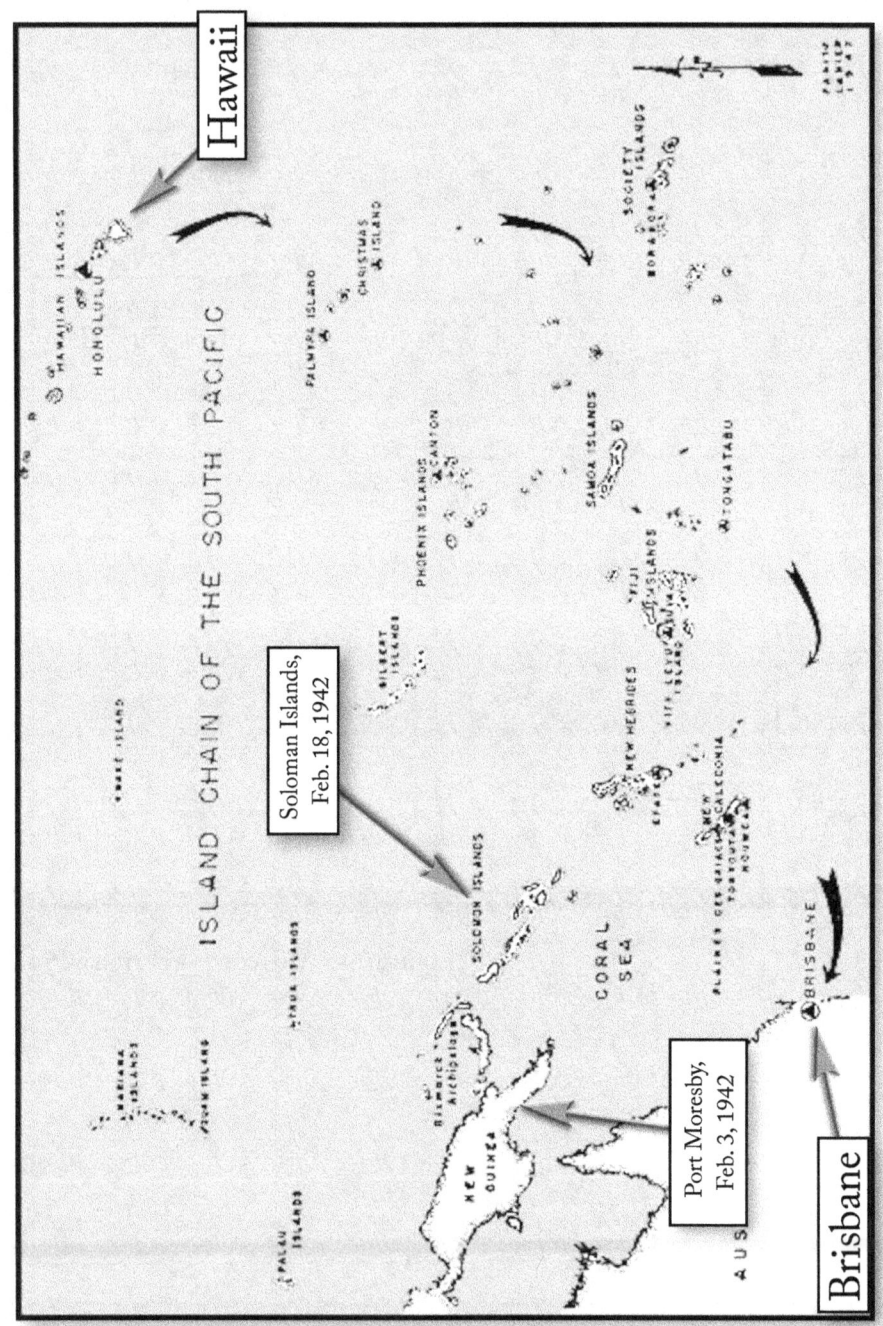

Figure 25 – Supply Chain– February 1942

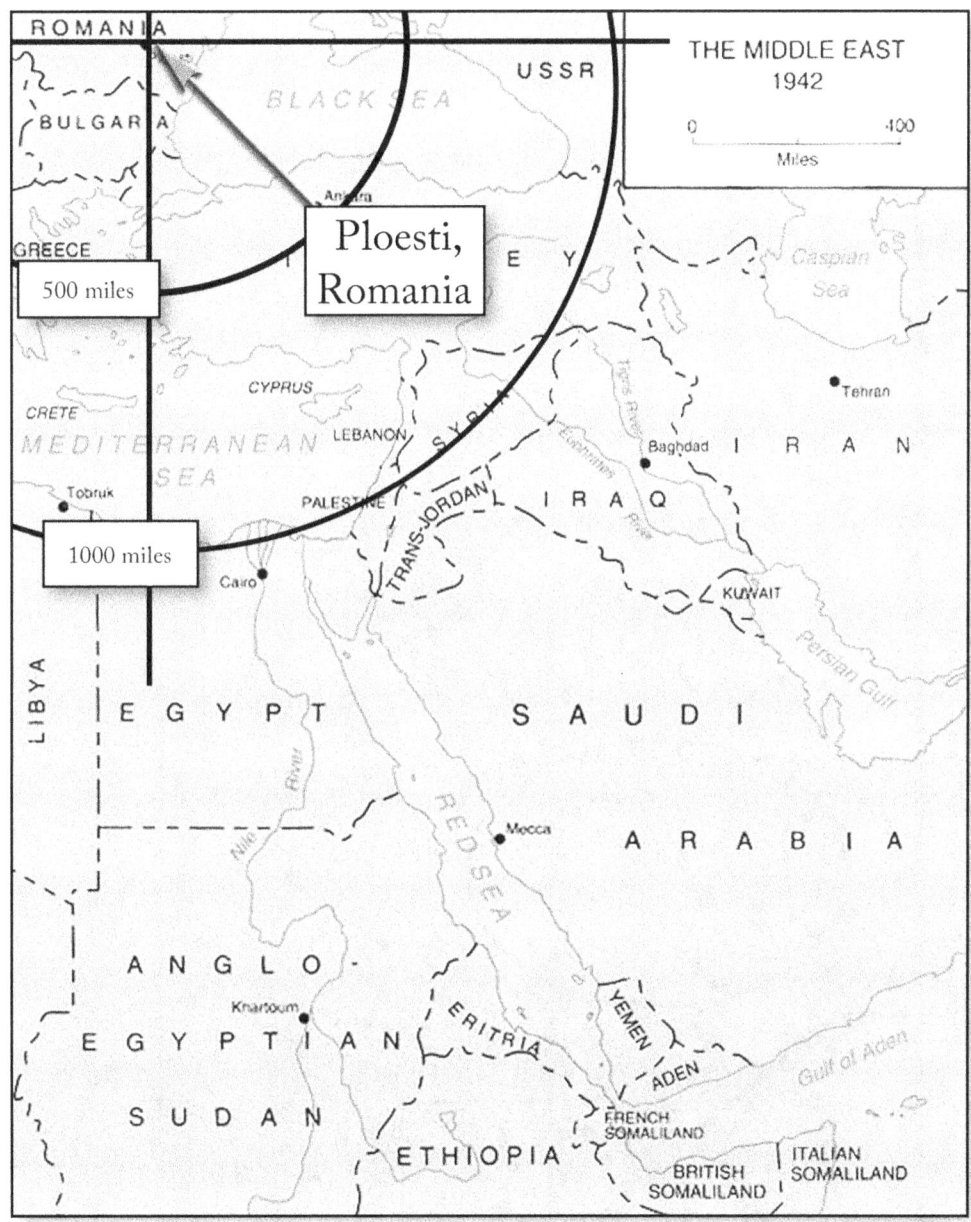

Figure 26 – Distances from Ploesti

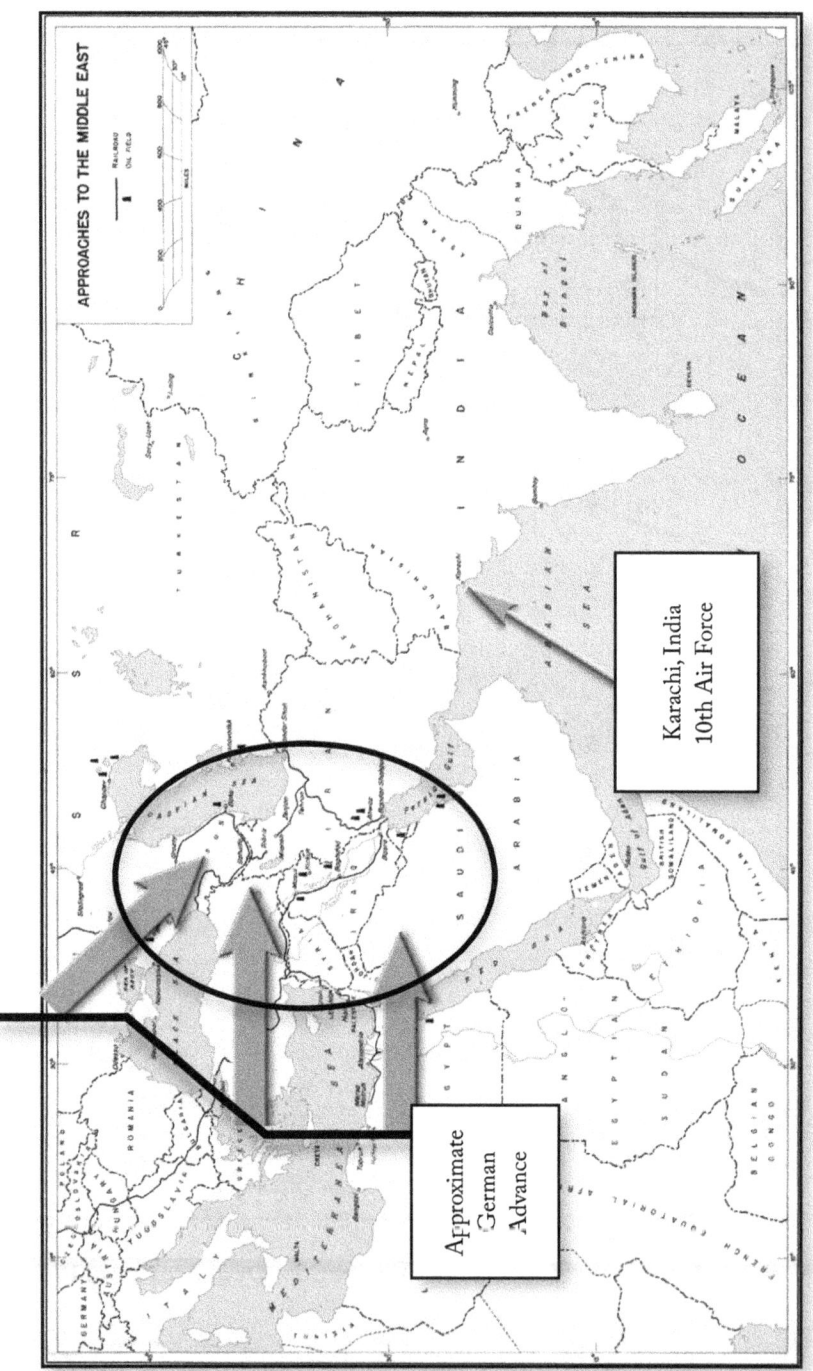

Figure 27 – Possible Axis Strategy

ENDNOTES

1. ----, FDRL, PSF, Box 3, Marshall, George C., 1941 - 4/14/42 file, Comments on the Use of U.S. Troops in the Middle East and Africa.
2. Hannah, letters to the author.
3. Williams, *Chronology 1941-1945 - 1942*, p. 21.
4. Buchanan, *A Friend Indeed?*, p. 282.
5. Williams, *Chronology 1941-1945 - 1942*, p. 21.
6. ----, FDRL, PSF, Box 4, Navy Department 1934 - February 1942 file, McCrea-->FDR, February 2, 1942.
7. ----, AFHRA, Hap Arnold file, Arnold to Eisenhower, February 2, 1942.
8. ----, FDRL, PSF, Box 28, Currie In file, Currie to Segac, February 2, 1942.
9. ----, AFHRA, Hap Arnold file, Cooley to Halverson, February 2, 1942.
10. ----, FDRL, PSF, Box 98, Carter file, Carter to FDR, February 2, 1942.
11. Interview with Ulysses S. Nero, May 21, 1974.
12. http://www.historynet.com/countdown-to-the-doolittle-raid.htm .
13. Mayhew, *transcript of interview*.
14. Assistant Chief of Air Staff, *The Ploesti Mission of 1 August 1943*, p. 12. See also ----, AFHRA, Hap Arnold file.
15. Joint Intel Report, Summary of Recent Joint Intelligence Sub-Committee's Appreciation of Germany's Intentions-2/3/42, February 3, 1942, FDRL, PSF, Box 1, American-British Joint Chiefs of Staff.
16. http://www.historynet.com/countdown-to-the-doolittle-raid.htm .
17. Williams, *Chronology 1941-1945 - 1942*, p. 21.
18. Williams, *Chronology 1941-1945 - 1942*, p. 21.
19. Williams, *Chronology 1941-1945 - 1942*, p. 22.
20. Williams, *Chronology 1941-1945 - 1942*, p. 22.
21. Mayhew, *transcript of interview*.
22. ----, FDRL, PSF, Box 2, China file, FDR to Halifax, February 4, 1942. See also ----, FDRL, PSF, Box 6, War Department file, Arnold to FDR, February 4, 1942.
23. Williams, *Chronology 1941-1945 - 1942*, p. 22.
24. Buchanan, *A Friend Indeed?*, p. 282.
25. Richardson, *Forgotten Force*, p. 13.
26. Williams, *Chronology 1941-1945 - 1942*, p. 22.
27. ----, FDRL, PSF, Box 98, Carter file, Carter to FDR, February 6, 1942.
28. Interview with Ulysses S. Nero, May 21, 1974.
29. Assistant Chief of Air Staff, *The AAF in the Middle East*, p. 145, footnote 106.
30. Williams, *Chronology 1941-1945 - 1942*, pp. 22-23.
31. The High Commissioner was a position created by the Tydings–McDuffie Act of 1934. The High Commissioner was the liaison between the U.S. President and Philippines President. His primary function

was to facilitate the transition of the Philippines to an independent country. Independence was scheduled for July 1, 1946.

32. Perry, *Dangerous Man*, p. 127
33. Williams, *Chronology 1941-1945 - 1942*, p. 23.
34. Xu, *The Issue of US Air Support for China*, p. 466.
35. ----, FDRL, PSF, Box 27, China 1942 file, FDR to CKS February 9, 1942.
36. ----, *Time, U.S. At War: The Bombers are Growing*, February 9, 1942, http://www.time.com/time/magazine/article/0,9171,777569,00.html, accessed July 18, 2012.
37. Ben-Moshe, *Winston Churchill and the "Second Front*, p. 530.
38. Williams, *Chronology 1941-1945 - 1942*, p. 23.
39. Williams, *Chronology 1941-1945 - 1942*, p. 23.
40. ----, AFHRA, Hap Arnold file, George to Harman, February 10, 1942.
41. ----, FDRL, PSF, Box 1, American-British Joint Chiefs of Staff file, Joint Intel Committee, February 14, 1942.
42. ----, *Pacific Wrecks*, http://www.pacificwrecks.com/aircraft/b-17/41-2446.html, accessed February 11, 2014.
43. Assistant Chief of Air Staff, *The Tenth Air Force 1942*, p. 1.
44. Eckerly, *A Pilot's Story*.
45. Williams, *Chronology 1941-1945 - 1942*, p. 23.
46. Williams, *Chronology 1941-1945 - 1942*, p. 23.
47. Williams, *Chronology 1941-1945 - 1942*, p. 24.
48. Williams, *Chronology 1941-1945 - 1942*, p. 24.
49. Craven, *The Army Air Forces in World War II, Vol. 1*, p. 388.
50. Eckerly, *A Pilot's Story*.
51. ----, FDRL, PSF, Box 1, American-British Joint Chiefs of Staff file, Joint Intel Committee, February 14, 1942.
52. ----, FDRL, PSF, Box 3, Marshall, George C., 1941 - 4/14/42 file, Marshall to FDR, February 14, 1942.
53. Williams, *Chronology 1941-1945 - 1942*, p. 24.
54. Williams, *Chronology 1941-1945 - 1942*, p. 24.
55. Frank, *Guadalcanal*, p. 11.
56. Eckerly, *A Pilot's Story*.
57. Cave, *transcript of interview*.
58. Walker, The Liberandos, p. 10.
59. Royce, Transcript of an interview.
60. Brereton, *The Brereton Diaries*, p. 92.
61. Williams, *Chronology 1941-1945 - 1942*, p. 24.
62. ----, FDRL, PSF, Box 98, Carter file, Carter to FDR, February 16, 1942.
63. ----, FDRL, PSF, Box 98, Carter file, FDR to Carter, February 17, 1942.
64. ----, AFHRA, Hap Arnold file, Scanlon to Arnold, February 16, 1942.

65. Mitcham, Rommel's Desert War, p. 35.
66. Baldwin, Hanson, *New York Times*, THREAT TO SUEZ RETURNS, February 17, 1942, http://select.nytimes.com/gst/abstract.html?res=F00912FA3B5F167B93C5A81789D85F468485F9, accessed February 11, 2014.
67. Williams, *Chronology 1941-1945 - 1942*, p. 24.
68. Letter, CominCh to CofS, U.S. Army, 18 February 1942. Naval Records and Library. (Hereinafter NRL.), (footnote 13), ----, *US Marine Corps in World War II, Guadalcanal Campaign*, http://www.ibiblio.org/hyperwar/USMC/Guadalcanal/USMC-M-Guadalcanal-1.html , p. 4.
69. ----, *The Campaigns of the Pacific War*, p. 30.
70. Mayhew, *transcript of interview*.
71. Williams, *Chronology 1941-1945 - 1942*, p. 25.
72. Williams, *Chronology 1941-1945 - 1942*, p. 25.
73. Memo, Hopkins to FDR, February 20, 1942, FDRL, PSF, Box 3, Great Britain.
74. ----, *Pacific Wrecks*, http://www.pacificwrecks.com/aircraft/b-17/41-2446.html, accessed February 11, 2014.
75. Williams, *Chronology 1941-1945 - 1942*, p. 25.
76. Assistant Chief of Air Staff, *The AAF in the Middle East*, p. 53.
77. Craven, *The Army Air Forces in World War II, Vol. 1*, p. 561.
78. Brereton, *The Brereton Diaries*, p. 97.
79. ----, *Pacific Wrecks*, http://www.pacificwrecks.com/aircraft/b-17/41-2446.html, accessed February 11, 2014.
80. Mayhew, *transcript of interview*.
81. Williams, *Chronology 1941-1945 - 1942*, p. 25.
82. Craven, *The Army Air Forces in World War II, Vol. 1*, p. 396.
83. ----, *Fireside Chats of Franklin D. Roosevelt*, February 23, 1942, http://www.mhric.org/fdr/chat20.html, accessed July 26, 2012.
84. Fleming, *The New Dealer's War*, p. 129.
85. ----, *Time, GREAT BRITAIN: Sticks and Stones*, February 23, 1942, http://www.time.com/time/magazine/article/0,9171,884462,00.html, accessed July 26, 2012.
86. ----, *Time, BATTLE OF BRITAIN: Through The Strait*, February 23, 1942, http://www.time.com/time/magazine/article/0,9171,884457,00.html, accessed July 26, 2012.
87. Ben-Moshe, *Winston Churchill and the "Second Front*, p. 535.
88. ----, *Time, Worst Week*, February 23, 1942, http://www.time.com/time/printout/0,8816,884434,00.html , accessed December 3, 2008.
89. Williams, *Chronology 1941-1945 - 1942*, p. 25.
90. Craven, *The Army Air Forces in World War II, Vol. 1*, p. 395.
91. Brereton, *The Brereton Diaries*, p. 98-99.
92. ----, AFHRA, Hap Arnold file, George to Marshall, February 24, 1942.
93. Letter, Marshall to King, 24 February 1942, NRL, (footnote 14), ----, *US Marine Corps in World War*

II, Guadalcanal Campaign, http://www.ibiblio.org/hyperwar/USMC/Guadalcanal/USMC-M-Guadalcanal-1.html , p. 4.

94. ----, FDRL, PSF, Box 2, Chiang Kai Shek file, Chiang to Soong, February 24, 1942.

95. ----, FDRL, PSF, Box 3, India file, FDR -> Winant, February 25, 1942.

96. Williams, *Chronology 1941-1945 - 1942*, p. 26.

97. ----, *The Campaigns of the Pacific War*, p. 31.

98. Brereton, *The Brereton Diaries*, p. 105.

99. Craven, *The Army Air Forces in World War II, Vol. 1*, p. 398.

100. ----, Army Air Force Statistical Digest, http://www.ibiblio.org/hyperwar/AAF/StatDigest/aafsd-3.html , accessed January 22, 2014, p. 112; B-24 production from Author's personal papers.

MARCH 1942

March continued the pattern of January and February. More Allied plans, more Axis successes and more changes occurred. The realization that the Allies were reacting rather than taking the initiative was becoming increasingly apparent.

Roosevelt and Churchill's relationship began to change. The seemingly endless string of British military losses caused Churchill to meddle even more, which began to infuriate United States military leaders. His ever-growing pessimistic view of the British position would cause him to advance military strategies in direct opposition to those to which he had just agreed.

Replacement men and equipment were slowly being fed into the fray. Often, they would depart for the front lines, only to be sent elsewhere due to enemy successes. HALPRO was now acquiring enough men to form crews. They were receiving planes, which had to be modified for combat.

1942 MARCH 1

EAST INDIES

From Allied air bases on Java, the remaining aircraft made a morning raid on the Japanese landing force. Upon their return to the air base to refuel and rearm, Japanese fighters swept over the field, virtually destroying the Allied aircraft as they sat on the tarmac. This ended the operations of the 17th Pursuit Group. The remaining personnel now joined the other troops looking for a way off the island. Five B-17s and three LB-30s were all that remained of the heavy bombers of the Far East Air Force. Now they became transport planes. Each B-17 evacuated 31 men; each LB-30, 35 men. When the last plane left for Broome, Australia, the Japanese were only 18 miles away.[1]

The 19th Bombardment Group dispersed to two airfields in Australia. The 40th Squadron flew to Townsville; the other three squadrons, to Cloncurry.[2] (see Figure 30)

With the collapse of Java and no usable ports, the Allied naval vessels were ordered to withdraw. Unfortunately, of Admiral Doorman's original force, only four destroyers made it to Australia. Included in the losses was Doorman's flagship *HNLMS De Ruyter*, with Doorman aboard.[3] The sinking of the *De Ruyter* was considered the last engagement of the broader Battle of the Java Sea. (see Figure 20)

HALPRO

The HALPRO task force departed for MacDill Field in Florida. For the next month, the newly formed crews flew around-the-clock, becoming familiar with the flight characteristics of the B-24D. The expansion of the organization to 24 crews had already been decided.[4]

93RD HBG

The 93rd HBG was activated and would operate B-24s. Initially, it would be engaged in antisubmarine operations over the Gulf of Mexico and the Caribbean Sea as part of the III Bomber Command. It would serve from May through July 1942.

1942 MARCH 2

EAST INDIES

The Japanese claimed the occupation of Batavia. The government of Netherlands East Indies was forced to relocate to Bandoeng.[5] (see Figure 28)

NEW GUINEA - SOLOMON ISLANDS

The Japanese began an aerial bombardment of New Guinea, in preparation for their invasion of Huon Gulf.[6] (see Figure 30)

Admiral King continued to question the Germany First strategy and pushed for his Solomon Island campaign. He replied to Marshall's rejection of Army support. Army units were needed to garrison the targets taken by the Marines and then vacated by the Marines, who would be needed for the next amphibious operation.[7]

GLOBAL STRATEGY

With the strategic significance of India increasing by the day, the importance of the South America-Africa-India supply route increased proportionately. Brereton pointed out the obvious in a memo to Arnold when he assumed command of the 10th Air Force on the March 2: whether they liked it or not, the Middle East had become a vital link.[8]

1942 MARCH 3

OPERATION SUPER-GYMNAST

The plans for SUPER-GYMNAST, the invasion of North Africa, were shelved. Rommel's progress in Libya brought into question the attitude of the Vichy government, which technically still had colonies there. The Allies believed that they would support the Germans as long as they continued to win.[9]

North Africa was not the only geographic area where planning had been impacted by Axis progress. In January, American planners had realized that the South America-Africa-India supply route to Australia had to traverse the Indian Ocean. Survey teams had been sent to the area to scope out the existing facilities. However, Japanese successes in the Netherlands East Indies had severed the potential link. Surveyors were ordered to abandon their task for that section of the route.[10]

10TH AIR FORCE

Brereton had flown to Calcutta to confer with Stilwell. Upon his return to Karachi, (see Figure 27) Brereton reported that:

> *[P]ending the arrival of air* reinforcements from the United States, I did not expect to be able to undertake full active operations before May. His attitude was very encouraging, his policies clear. He said: "First, get results. Second, ensure American command and control. Third, use your own judgment. Finally, if you need help from me, let me know."
>
> We agreed on one main point at once: Burma must take 'first priority.'[11]

DOOLITTLE RAID

Everyone was assembled at Eglin Field in Florida by March 3 when Doolittle arrived. This base was near the sea so that the navigators could practice over-water navigation. It had an auxiliary field for training in short-field takeoffs and was relatively secluded, free from curious eyes. Doolittle introduced himself in typically blunt fashion—"My name's Doolittle"—and told the men the project "will be the most dangerous thing any of you have ever done. Any man can drop out and nothing will ever be said about it. This entire mission must be kept top secret".[12]

1942 MARCH 4

GLOBAL STRATEGY

American planners must have realized that they needed to relieve the pressure on Allied troops in Southeast Asia. And the only way to do that was to have Japanese attention diverted to another area of their empire. And Russia was the only Allied nation with any sizeable force that could threaten the Japanese. But both Japan and Russia realized that any conflict between them would result in a second front for each nation. Consequently neither party was willing to antagonize the other. Nevertheless, Roosevelt ordered Marshall and Stark to study possible offensive options.[13]

Roosevelt received long cablegrams from Churchill on March 4 and 5. After assessing the world situation, Churchill inquired:

- About the plans for the American Air Force
- If the United States could send an American division to New Zealand and Australia in lieu of Britain recalling one New Zealand and one Australian division currently stationed in the Middle East
- If the United States naval forces could increase protection in the ANZAC area
- He recommended that:
- Planning should be made for an expeditionary force on the United States west coast for attacks upon the Japanese in 1943
- Delaying in the movement of American troops into Northern Ireland (MAGNET), and
- Postponing GYMNAST.[14]

MIDDLE EAST

Despite the growing significance of the Middle East as crucial to the supply of material to both India and Russia, there were no immediate plans to send troops to the Middle East. The Middle East would continue to be designated as a British theater.[15]

EASTERN FRONT

The months of March and April would consist mostly of a stalemate between the opposing forces.[16]

1942 MARCH 5

GLOBAL STRATEGY

Churchill confided to Roosevelt: "When I reflect how I have longed and prayed for the entry of the United States into the war, I find it difficult to realise how gravely our British affairs have deteriorated ... since December 7th."[17]

Roosevelt must have shared Churchill's concerns with his military advisors. King responded directly to Roosevelt:

As to Britain's lines of military effort:

1. It is apparent that we (U.S.) must enable the British to hold the citadel and arsenal of Britain itself by means of the supply of munitions, raw materials and food - and to some extent by troops, when they will release British troops to other British military areas.
2. The Middle East is a line of British military effort which they - and we - cannot afford to let go. This effort should continue to receive our (U.S.) munitions.
3. The India-Burma-China line of British military effort is now demanding immediate attention on their part - and will absorb its proportion of our (U.S.)

munitions - in addition to the munitions which we are committed to furnish to China.[18]

The American Combined Chiefs also made an assessment of Churchill's observations. They:

- Supported the recommendations made by Churchill on 4 and 5 March.
- Noted a possible threat of a German thrust through the Caucasus to meet the Japanese in India. This made the Middle East a vital theater.
- Noted the British public was

  ◦ Disappointed that promised victories in Egypt were not realized
  ◦ Profoundly shocked by

    - The bold escape from Brest of the [German ships] Scharnhorst and Gneisenau through the English Channel, and
    - The fall of Singapore.
    - Observed that Churchill's government was being bitterly criticized. It was possible Churchill could be voted out of office.[19]

Calls for Churchill's resignation increased after the fall of Singapore. This must have received the attention of United States military leaders as they noted Churchill's political vulnerability. They realized that a new British government might negotiate a peace with Germany.

HEAVY BOMBER PERFORMANCE

Nearly every war plan generated by the United States, including those made before its entry into the war had a major plan component dependent upon heavy bomber operations. By early March, the United States had nearly three months of operational experience to test her theories, most of which took place in Southeast Asia. The reality was that the plans assumed operations from well constructed and supplied bases. This assumption was invalid for this period. The reality was that over 300 sorties[20] were attempted by 37 B-17s and 12 LB-30s. Of these, 40% failed to reach the target. Of those completing their missions, the bombers claimed the sinking of one destroyer, eight transports, and two ships of unknown identity. They also claimed that their gunners shot down 23 enemy fighters. All of this was accomplished at the cost of 38 bombers.[21] This was a rather inauspicious beginning to America's entry to the air war. The Japanese were losing 1 fighter plane for every 1.6 heavy bombers they shot down.

10TH AIR FORCE

Brereton assumed command of the 10th Air Force on March 5[22], which basically existed only on paper. He had eight heavy bombers; six of which he had brought with him from Java and two recent arrivals from the States. Three days later these bombers would begin operations; not as bombers, but as transports. By March 13, they had ferried 474 troops and 29 tons of supplies to Magwe, Burma. (see Figure 29) They evacuated 423 civilians on the return leg.[23]

HALPRO

As the new 10th Air Force commander, Brereton issued his first directive. Buried in that directive was the following statement regarding HALPRO:

> *The Halverson detachment, enroute from* the States for a special mission in China, equipped with 30 B-24s. Its original mission was to operate from heavy bomber bases located in Chekiang Province, eastern China. Its targets were to be Tokyo and other Japanese industrial centers.[24]

Brereton's use of the past tense is interesting in hindsight.

REINFORCEMENTS

Bill Mayhew and his fellow travelers continued their sojourn:

> *On 5 March 1942 at* 9:00 A.M. (0900) we dropped anchor in the harbor of Colombo [Ceylon]. We were not allowed on shore, due to an epidemic there at the time, but native traders were in small boats all around our ships continually, so the time passed quickly. These natives were selling everything from bananas to ebony elephants. Colombo has always been one of the stops on the round-the-world cruises, so most of the traders could speak some English. Little naked boys would paddle out to dive for pennies or anything else one would throw overboard to them. From what we could see of the town, it appeared rather clean, but naturally the waterfront area was rather dirty. There were no docks at Colombo, so we had to be refueled from clumsy barges while we were riding at anchor. About sundown (7:00 P.M--1900) of our second day in Colombo (7 March 1942) we put to sea again, headed for Bombay, India. However, an epidemic was raging there also, so we proceeded to Karachi, India.[25]

EAST INDIES

Allies were reported to have left Batavia.[26] (see Figure 28)

1942 MARCH 6

BURMA

General Alexander had just arrived in Rangoon on March 5 to assume command of the Burma Army. His first order the next day was to evacuate Rangoon.[27] (see Figure 29)

GLOBAL STRATEGY

Churchill's notes to Roosevelt and the apparent agreement by the CCS were having a ripple effect, in particular with Arnold and his staff. Fearing a collapse on the Russian front, they could envision a successful thrust by Germany towards the Middle Eastern oil fields. Some Allied action was needed to force Germany to pull troops from the Russian front. This "action" was what Stalin had been screaming for, i.e. the so-called Second Front. America's heavy bombers were the only "tool" Arnold had. He, therefore, argued for the concentration of United States planes, rather than the dispersal into smaller groups.

In particular, Arnold and his staff noted:

1. American plans centered on a main thrust to the continent from UK bases, to be delivered only

   - after a powerful force had been established there, and
   - after Germany weakened by an intensive bomber offensive.

2. Current plans to divert resources threaten that plan.[28]

Eisenhower authored a strategy paper supporting Arnold's position.[29]

BOMBING JAPAN

Arnold forwarded a memo to Roosevelt on potential Chinese bases and Japanese targets.[30]

1942 MARCH 7

BURMA

The evacuation of Rangoon was completed. The Burma Army would now have to be supplied by air.[31] (see Figure 29)

EAST INDIES

Allied communication with Bandoeng ceased, indicating that the Japanese had overrun the city. Japan now controlled Java.[32] (see Figure 28)

REINFORCEMENTS

Harry Holloway had joined the Army Air Corps on September 1, 1939, the day Germany invaded Poland. He attended radio school in 1940, and graduated from gunnery school on March 7. Then he joined John Lavin's crew.[33]

1942 MARCH 8

BURMA

Japanese troops entered Rangoon. Meanwhile, 10th Air Force began a five-day effort of flying troops and supplies from India to Magwe, Burma.[34] (see Figure 29)

NEW GUINEA

The Japanese landed at Lae and Salamaua. At Lae, the Japanese faced no opposition. At Salamaua, Allied forces destroyed some of the facilities and then withdrew into the surrounding hills.[35] (see Figure 19)

1942 MARCH 9

NEW GUINEA

Japanese success in New Guinea removed the last land barrier to the northwest coast of Australia. An invasion of the mini-continent now seemed a viable threat. The Joint Chiefs wrote that the Allies had three choices:

1. To send strong reinforcements to the Pacific at the cost of sacrificing the hope of an early and vigorous offensive against Germany;
2. To concentrate forces against Germany with acceptance of the possibility of losing all the Southwest Pacific;
3. To reinforce the Southwest Pacific and related areas to a point sufficient to maintain a defensive position while building up in the United Kingdom the forces required for assumption of the offensive at the earliest possible date.[36]

GLOBAL STRATEGY

To an impartial observer, it might appear that the Allies were reacting to events and not causing them. While the average American was not privy to the intricacies of the strategy debates, there was an awareness of the seemingly endless successes of the enemy and the continual withdraw of Allied forces. The populace was beginning to question if and/or when the Allies would strike. The Roosevelt administration's answer to that question was delivered in a *Time* article, *Let's Begin to Strike*. The last sentence of that article read, "'The time has now come,'" said General Marshall, "'when we must proceed with the business of carrying the war to the enemy.'"[37]

UNITED STATES CHIEF OF STAFFS

Roosevelt's administration announced a major shakeup in the United States military leadership. The Army's General Headquarters (GHQ) was replaced with three departments:

1. Army Ground Forces under Lt. Gen. Lesley J. McNair,
2. Army Air Forces under Lt. Gen. Henry H. Arnold, and
3. Services of Supply (later designated as Army Service Forces) under Maj. Gen. Brehon B. Somervell

These departments operated in the Zone of Interior (ZI) under Gen. George C. Marshall, the Army Chief of Staff.

Admiral Ernest J. King replaced Admiral Harold R. Stark as Chief Naval Officer (CNO).[38]

1942 MARCH 10

MIDWAY

A Japanese reconnaissance plane was shot down southwest of the island.[39] (see Figure 10)

MIDDLE EAST

Albert Wedemeyer, who was on the planning staff, was returning from England with Stimson. He noted that Stimson said in mid-March:

> *The Middle East is very* last priority of all that are facing us, we have foreseen for months that the British would be howling for help here, that we should not give in to them, and I think now is the time to stand pat.[40]

NEW GUINEA

The United States carriers *USS Lexington* and *USS Yorktown* were stationed in the Gulf of Papua off the southern coast of New Guinea. They launched aircraft to attack the Japanese positions at Lae and Salamaua. (see Figure 19)

Following this attack, eight B-17s from the 40th Squadron of the 19th Bombardment Group attacked the same targets. They were operating from Townsville, Australia.[41] Meanwhile, the Japanese captured Finschhafen, New Guinea.[42] (see Figure 30)

HALPRO

General M. F. Hammon issued a memo for Air War Plans Division in which he stated:

> *HALPRO called for the creation* of a squadron detachment, equipped with 24 B-24s under the command of Col. Harry A. Halverson. When mobilized and trained, it was to be ordered for special duty to the Far Eastern theater of operations, or on such other combat mission as might be determined by OPD.[43]

Now, for the first time, HALPRO's mission to fulfill Roosevelt's request to bomb Japan was seemingly in jeopardy. And the Doolittle mission was still training.

1942 MARCH 11

THE PHILIPPINES

MacArthur's entourage left Corregidor in four PT boats. His first stop was the island of Mindanao.[44]

1942 MARCH 12

REINFORCEMENTS

Bill Mayhew's ship arrived in Karachi. (see Figure 27)

> *We arrived there at 2:00* P.M. (1400) on 12 March 1942. We had been aboard a ship for 36 days on this voyage. In all, we had spent 68 days aboard ships, and traveled 18,045 miles from San Francisco (much of that in a zig-zag fashion). We left the ship for the last time here, and began to set up a permanent camp at the Karachi Airport (seven miles east of Karachi). The trip from Brisbane covered seven weeks and 7,764 miles. While we were setting up the tents for the camp, the troops were divided between New Malir (seven miles east of the airport) and the giant dirigible hangar at the airport. This hangar had been built in 1930 to house the British dirigible R.101. However, the R.101 crashed and the hangar was never used. Those troops that stayed in the dirigible hanger slept on cots under mosquito netting. Some troops put clothes over the netting to keep

pigeon droppings from coming through the netting. The 88th Reconnaissance Squadron personnel were located at New Malir in new barracks.

B-17Es (Flying Fortresses) began arriving from the States, so once again we took on the appearance of a fighting outfit. We were the beginning of the 10th Air Force. Some of these planes were in "Project Aquilla", which was the code name for a planned attack on Tokyo from China by B-17Es. Colonel Caleb Haynes was to lead the attack. He left the United States early to make arrangements for the mission. He was followed on 30 March 1942 by six B-17Es leaving Florida for India. These planes were flown by Lt. Col. Robert Tate, Lt. Col. Robert Scott, Lt. Col. Jerry Mason, Lt. Col. Torgas Wold, Major Max Fennell, and Major Donald Struther. It was determined after they reached India that B-17s did not have the range to do the job from the available bases in China. The other early B-17 arrivals (Projects X [10 planes] and 157 [7 planes]) were planes headed for Java that were stopped in India because the route was broken when Palembang, Sumatra fell to the Japanese. These planes were the latest model of B-17 (B-17E). They were equipped with tail guns and most had lower ball turrets (some were equipped with remote controlled lower turrets). All guns on board were .50 caliber machine guns except the .30 caliber guns in the bombardier's compartment. As we had never seen that type of B-17 before leaving the States (we had flown in B-17C and B-17D planes before going overseas), those of us that volunteered to go on a combat crew naturally had to receive instructions on the devices that had been added to the ship. For example, I read the technical orders to learn how to get into the ball turret. Once I was in, our tail gunner read the technical orders over the plane's intercom to tell me what to do to operate the turret. That is how I learned to be a ball turret gunner. It was all on the job training. A ball turret was about four feet in diameter, with a window between the feet that was 15 inches in diameter and two inches thick. A .50 caliber machine gun was mounted on each side of the turret about a foot from the gunner's head.[45]

1942 MARCH 13

GLOBAL STRATEGY

The prospect of a Japanese invasion of Australia caused Allied planners to do worst-case scenarios. In Washington, the CCS thought the Japanese might attempt to occupy Darwin, Wyndham, or Townsend. In Australia, their chiefs thought the Japanese might attempt a landing at Port Moresby (see Figure 25) by the end of the month, and then Darwin by early April.[46] (see Figure 20)

MIDDLE EAST

As if he needed reminding about the Middle East, Roosevelt received a communiqué from William Bullitt. In it, Bullitt pointed out that Turkey was in a key position to either thwart or enable German advance to the Middle East. If there was an Allied presence in the area, then Turkey might fight. But if there were none, then Turkey would most certainly cave to German pressure and allow German troops to march through their country.[47]

BOMBING JAPAN

Roosevelt had forwarded Arnold's March 6 memo to Currie two days earlier. Currie responded that potential Japanese targets had been identified for six months.[48]

1942 MARCH 14

THE PHILIPPINES

MacArthur and his party arrived at Mindanao.[49]

GLOBAL STRATEGY

As the possibility of a German-Japanese link-up somewhere around present day Pakistan became a planning nightmare for the Allies, it became impractical for planners to ignore one theater over the other. So the Joint Chiefs concluded that the Middle East and Far East theaters were interdependent. They stipulated that plans for their reinforcement should remain flexible in order that units might be shifted on short notice to whichever area appeared to have thfe greater need.[50]

This change to a more flexible strategy explained the previous comment made about HALPRO on March 10.

COMMITMENT TO CHINA

Despite the military realities, Hopkins listed for Roosevelt his "Matters of Immediate Military Concern." In it, he noted the following about China:

1. China – We must keep that line to China open and get it along. Believe Army needs to be jogged on this regularly
2. "The second phase of the Chinese business is to get a springboard from which to bomb Japan itself. For morale reasons, this is extremely important and the sooner it can be done the better.[51]

AUSTRALIA

Remnants of the 88th Reconnaissance Squadron of 7th Bomb Group, currently operating out of Townsville, (see Figure 30) were transferred to the 40th Squadron, 19th Bombardment Group. Three of these B-17s were sent to rescue MacArthur. One would ditch off the coast of Mindanao; the other two would arrive at Del Monte on March 16.[52]

1942 MARCH 16

GLOBAL STRATEGY

Two days after the JCCS admitted that the Middle East and India were intertwined and the key to preventing a Japanese-German linkup, the Joint Chiefs issued a memo restating the March 9 options:

1. To send strong reinforcements to the Pacific at the cost of sacrificing the hope of an early and vigorous offensive against Germany;
2. To concentrate forces against Germany with acceptance of the possibility of losing all the Southwest Pacific;
3. To reinforce the Southwest Pacific and related areas to a point sufficient to maintain a defensive position while building up in the United Kingdom the forces required for assumption of the offensive at the earliest possible date.[53]

Nowhere did they mention the Middle East or India.

PRESS

Time placed its finger on the key to both Allied and Axis strategies. In an article titled *SUPPLY: Oil Can Lose the War*, the author wrote:

> Allied experts who once boasted that oil would win the war began to realize last week that the oil may get into the wrong hands. It was a rude awakening.
> Before Pearl Harbor the U.S. and Britain's fleets drew on the vast oil fields of the Western Hemisphere. Soviet Russia and the Armies of the Middle East had Baku and Batum. Australia, Hong Kong and Singapore were next door to the Dutch East Indies. The anti-Axis powers of the world controlled 97.5% of world production. It was as simple as that.
> But the Japanese have closed the United Nations' filling station in the far Pacific. Adolf Hitler, if he drives into the Middle East, may capture Baku and Batum. Then the Axis would not only have oil enough for its war machine (after destroyed mills and refineries are repaired), but would force the United Nations into complete dependence on the Western Hemisphere.[54]

In the same issue, *Time* ran another article, *BATTLE OF THE PACIFIC: Fall of Java*. This article told the story that the Allied command probably wished had remained confidential:

> The Japanese conquered Java in eight days. They surged inward from the northern coasts, took the capital at Batavia, chopped up the long island into isolated bits, reduced the naval base at Surabaya and the Army's mountain stronghold at Bandung. On Java there was no Bataan. The Japanese victory was complete and swift.
>
> How Did It Happen? Recriminations by the bitter Dutch defenders told part of the story, but only part: promised Allied reinforcements did not arrive; an Allied High Command was imposed upon the Dutch, without the Allied troops, planes and ships which would have made the joint command . . .[55]

It seemed that the press had a pipeline into the JCCS inner sanctum.

1942 MARCH 17

THE PHILIPPINES

MacArthur flew to Darwin, Australia.[56] (see Figure 20) The news that he had left the Philippines was a cause for celebration. This news became the banner headline in *The New York Times*.[57]

GLOBAL STRATEGY

Arnold told Marshall, who in turn told Roosevelt, that:

- Portal made another plea for air reinforcements for Egypt,
- Portal thought that the British might furnish American aircraft types to Cairo from the American production allotted to them; the AAF would furnish personnel.[58]

HALPRO

Acting on his beliefs that a concentration of force was needed and that India and the Middle East shared a common strategic objective, Arnold (via Gen. Harmon) told Air War Plan Division:

- HALPRO mission was to send the planes into the CBI and bomb Japanese targets. There would be 37 planes in the project, and would have priority on B-24s coming off the line. It was to operate in the theater without replacement of aircraft. When it reached the point of diminishing raids, what was left of it,

will, as far as could now be anticipated, be passed to the control of the Commanding General of the AAF in India.
- Halverson was directed to report to Arnold without delay.[59]

1942 MARCH 18

GLOBAL STRATEGY

Roosevelt had asked Marshall to review Bullitt's assessment of the Turkish situation. On March 18, Marshall responded, saying that the urgency reflected in Bullitt's communiqué was

> *justified. From a military point* of view, the region invites attack, and its loss would permit junction by sea between the Japanese and the Germans with disastrous consequences for the United Nations implied by such an eventuality.

Marshall then went on to say that, despite the current view that the Middle East was a British theater, he had already told the British that the United States would supply air units to the theater. He also reminded Roosevelt of the need for planes in every theater.[60]

1942 MARCH 19

BURMA

The Japanese began their campaign against Toungoo, Burma.[61] (see Figure 29)

1942 MARCH 20

10TH AIR FORCE

Stilwell sent a status report to Marshall who, in turn, reported it to Roosevelt: "Stilwell says prompt action to build up (the) U.S. Air Force in India is essential. American Volunteer Group's equipment is exhausted."[62]

MIDDLE EAST

A memo from the Air War Planning Division read:

- Upon analysis, the table of production had shown that the request for reinforcements in the Cairo area could not be met without disruption of the United States schedule of commitments and detriment to the American training program.

- Under the circumstances, deliveries to the British represented the only source from which diversions to the Middle East could come.⁶³

What this memo meant was that sending reinforcements to the Middle East was becoming a military necessity. The Middle East was becoming a recognized theater of operations. However, the "Arsenal of Democracy" could not meet this new demand without diversion from some current task.

1942 MARCH 21

HALPRO

Harmon sent another memo regarding the purpose and logistics of HALPRO.

Subject: Halpro.

1. The organization so designated will, when mobilized and trained, be ordered to the Aquilla [India] theater of operations for special duty under the code designation of "Halpro" which has been assigned as a special mission designation of the Aquilla command or to such other combat mission as may at that time be determined by the War Plans Division, General Staff.
2. The airplanes for this project, B-24D's, will be assigned on first priority from production aircraft to the 3rd Air Force for use in the organizational training for this project. The total number or airplanes to be assigned is 37. Three are now available and assigned - the remaining 34 have been designated by number for such assignment.⁶⁴

More hints are evident that HALPRO's mission was in jeopardy. And its size had grown to 37, as stated in the March 17 memo.

Patterson Field issued a status report on the various projects underway. On page two of that report, it listed the Halverson Project. At this time, the work was focused on installing the extra set of fuel tanks. Three sets of tanks had been installed; seven sets were on site; seventy-three were on order from a local (Dayton, Ohio) company.⁶⁵

1942 MARCH 22

BURMA

The Japanese bombing of the Magwe airfield was so destructive that the AVG was forced to move to Loiwing and the RAF to Akyab.⁶⁶ (see Figure 29)

1942 MARCH 23

GLOBAL STRATEGY

The reality of too few assets chasing too many targets did not escape the press. *Time* published an editorial, *World Battlefronts, STRATEGY: Too Many Fronts?*

In it, the author wrote:

> Wrote Pundit Walter Lippmann last week: "We and the British are trying to make everything secure, which means that we are dispersing our forces and fighting on lines of communication which are so monstrously extended that our navies and our shipping cannot possibly be adequate. At the same time, for political reasons, we are not concentrating our force for effective action in the one theater of the war, namely western Europe, where our communications are the shortest, the strategic position the most favorable and the gains to be obtained by strong action in conjunction with Russia the largest and the most immediate."

He then proceeded to cite various facts and arguments for the current strategy. However, his summary made the critical point:

> All these were chill facts. Colder was the fact that the U.S. and its allies must concentrate somewhere. The U.S. lacks the forces to concentrate everywhere. A cold and offensive warrior, the U.S. Fleet's Commander in Chief Ernest Joseph King, said last week: "We've got to have more planes, more warships, more guns, more of everything. Meanwhile . . . we are going to do the best we can with what we've got."
>
> If the U.S. best is dispersed too widely and too far, Admiral King may never get his additional planes and warships and guns. Perhaps he and the U.S. will have to choose a place where they can win with what they have in 1942.[67]

INDIAN OCEAN

The British defenders surrendered their base on the Andaman Islands.[68] (see Figure 32)

1942 MARCH 24

HALPRO

Stilwell and Brereton discussed the War Department's project of heavy bomber operations against Japan. Stilwell asked:

What the hell are we doing trying to go to Japan to bomb when we have a war on our hands right here?

I [Brereton] agreed with him that support of operations in Burma should have first priority. General Stilwell asked when I would be ready for operations. My answer was that, depending on arrivals from the west, the heavy bombers would be ready about 1 May.[69]

It is unstated what Brereton thought would take a month. Plus, the theater commanders doubted the necessity of the mission to bomb Japan.

DOOLITTLE RAID

On March 24, 1942, a terse message announced that a flight of North American B-25 medium bombers, under the command of a lieutenant colonel named James Doolittle, would be arriving at McClellan AFB. His requests for service and supplies should be accorded the highest priority. Such orders seemed fairly routine at McClellan, formerly known as the Sacramento Air Depot. Situated seven miles northeast of Sacramento, it was the major west coast air depot for arming and preparing bombers and fighter planes for shipment overseas. McClellan's maintenance crews worked every day on hundreds of airplanes destined for bases throughout the western United States and the Pacific.[70]

1942 MARCH 25

DOOLITTLE RAID

Doolittle and his own B-25 arrived at McClellan Field. He immediately met with the field's engineering officer and two of his lieutenants to brief them on the work that needed to be done. "This project is a highly confidential one," he told them. "As a matter of fact I am having to notify General Arnold in code that I arrived."[71]

1942 MARCH 26

GLOBAL STRATEGY

Both MacArthur, in Australia, and Chiang, in the CBI, demanded more heavy bombers. Arnold asked Eisenhower for his opinion. Demands for heavy bombers exceeded production. This was Arnold's view of the problem:

- The present priority provided for building up MacArthur before India. There were 41 heavy bombers in Australia with two enroute. They were being sent out to Australia at the rate of two a day.

- In India, there were six heavy bombers on hand and four enroute with no more to be sent until the Australian heavy bomber force was built up to full strength.
- Should we alternate heavy bombers to India then Australia until both were built up to full strength?[72]

PLOESTI

The staff of the War Plans Division issued a study titled *The Danube as an Objective for Air Attack*. Data had been gathered on the oil industry of Rumania as well as on rail and water communications between the Balkans and Germany.[73] Interestingly, the United States was not at war with Rumania in the first months of 1942. Both the United States military attaché and the United States oil industry had representatives in Rumania, who were gathering information about the Ploesti oilfields.

1942 MARCH 27

BURMA

The Japanese bombing of the Akyab airfield forced the RAF to withdraw to India.[74] (see Figure 29)

10TH AIR FORCE

With continued success in Burma, it appeared that it was only a matter of time before Japan might focus on India. If true, this operation would be another fatal step in the highly feared link-up of German and Japanese forces. Brereton and the meager 10th Air Force was all that stood in its way. He wrote the following in his diary:

- Possibility of an invasion of India through the mouth of the Ganges River is indicated. Action is needed at once to delay the movement of supplies to lower Burma.
- RAF planes haven't the range to undertake Rangoon or the Andaman Islands as targets. The only force available, which could reach these two targets with an adequate bomb load, is the 9th Squadron, Tenth Air Force.
- My directive from the War Department had been modified to the extent that I was authorized to support General Wavell's action for the defense of India on either the Bay of Bengal or the Arabian Sea.[75]

(see Figure 32, showing the 1,000-mile range of a B-24 operating from Ceylon.)

CROSS CHANNEL INVASION

While Brereton was being ordered to thwart a possible Japanese invasion of India with essentially a non-existent air force, Washington continued to devise plans for a cross-channel invasion into Northern France. This new plan, entitled "Plan for Operations in Northwest Europe," was released on March 27. The invasion was to take place in the fall of 1942 or spring of 1943.[76]

1942 MARCH 28

DOOLITTLE RAID

The final planning ingredient for the Doolittle mission was the preparation of a Chinese airfield for the B-25s to land. However, the secrecy surrounding the project was so great that the Chinese were never fully informed about the details of the project. Nevertheless, Chiang gave a reluctant approval to the project on March 28.[77]

HALPRO

At Patterson Field, the final modification list for the HALPRO B-24Ds was issued. It contained 26 items.[78]

1942 MARCH 30

BURMA

The Allies withdrew from Toungoo.[79] (see Figure 29)

GLOBAL STRATEGY

Following the release three days earlier of "Plan for Operations in Northwest Europe" three days earlier, the Air planning staff released its supporting plan. Titled "Air Support of a Continental Invasion from the British Isles," it envisioned a four-phase process for that support:

1. Preparation
2. Strategic bombing
3. Close air support
4. Release from this support to resume strategic bombing of Germany.[80]

Since the United States did not have enough men and material to defend its present positions in Asia, the above plan raises the question about where the military planners assumed they would have sufficient resources to mount an air campaign in six months, let alone a year. After all, "by the end of March only nine of the bombers had reached Australia."[81]

10TH AIR FORCE

Rowan Thomas was now assigned to the 10th Air Force. On March 30, he, his crew, and five other planes took off for India.[82]

1942 MARCH 31

DOOLITTLE RAID

Doolittle's orders from Washington released his planes for takeoff on March 31 for the short hop to Alameda Naval Air Station, where the *USS Hornet* waited at a dock to hoist them aboard.[83]

HALPRO

The Operations Division of the Army Air Corps also received orders, this time concerning HALPRO. These orders read in part:

1. The organization so designated [HALPRO] will, when mobilized and trained, be ordered to the Aquilla theater of operations for special duty under the code designation of "Halpro" which has been assigned as a special mission designation of the Aquilla command or to such other combat mission as may at that time be determined by the War Plans Division, General Staff.
2. The airplanes for this project, B-24D's, will be assigned on first priority from production aircraft to the 3rd Air Force for use in the organizational training for this project. The total number or airplanes to be assigned is 37. Three are now available and assigned - "the remaining 34 have been designated by number for such assignment.[84]

Washington was now robbing its operational training unit to get enough planes. HALPRO was going to India, but not necessarily to bomb Japan.

FAR EAST

MacArthur sent a three-page communiqué to Marshall explaining what he was doing to reinforce Corregidor. He had only 12 B-17s, which were nearly exhausted. He reminded Marshall that when he [MacArthur] left for the Philippines, he was told that supplies would last until mid-May. Now he was told supplies would be consumed a month earlier. Marshall forwarded the message to Roosevelt on April 3.[85]

HEAVY BOMBER PRODUCTION AND DEPLOYMENT

During March, 156 heavy bombers were manufactured. Seventy were B-24s.[86] In the middle of March, Charles Hurd wrote an article entitled "Big Bombers Roll Into

Mass Output With 24-Hour Goal."[87] In it, he wrote that there were plans to shift to a 24-hour day production schedule.

By the end of March, only nine of the Project X B-17s had reached Australia. Recall that a week after Pearl Harbor, Stark had promised the delivery of 80 heavy bombers to the Philippines by mid February. While the AAF remained committed to the Southwest Pacific, only thirty bombers were scheduled to be deployed in April.[88]

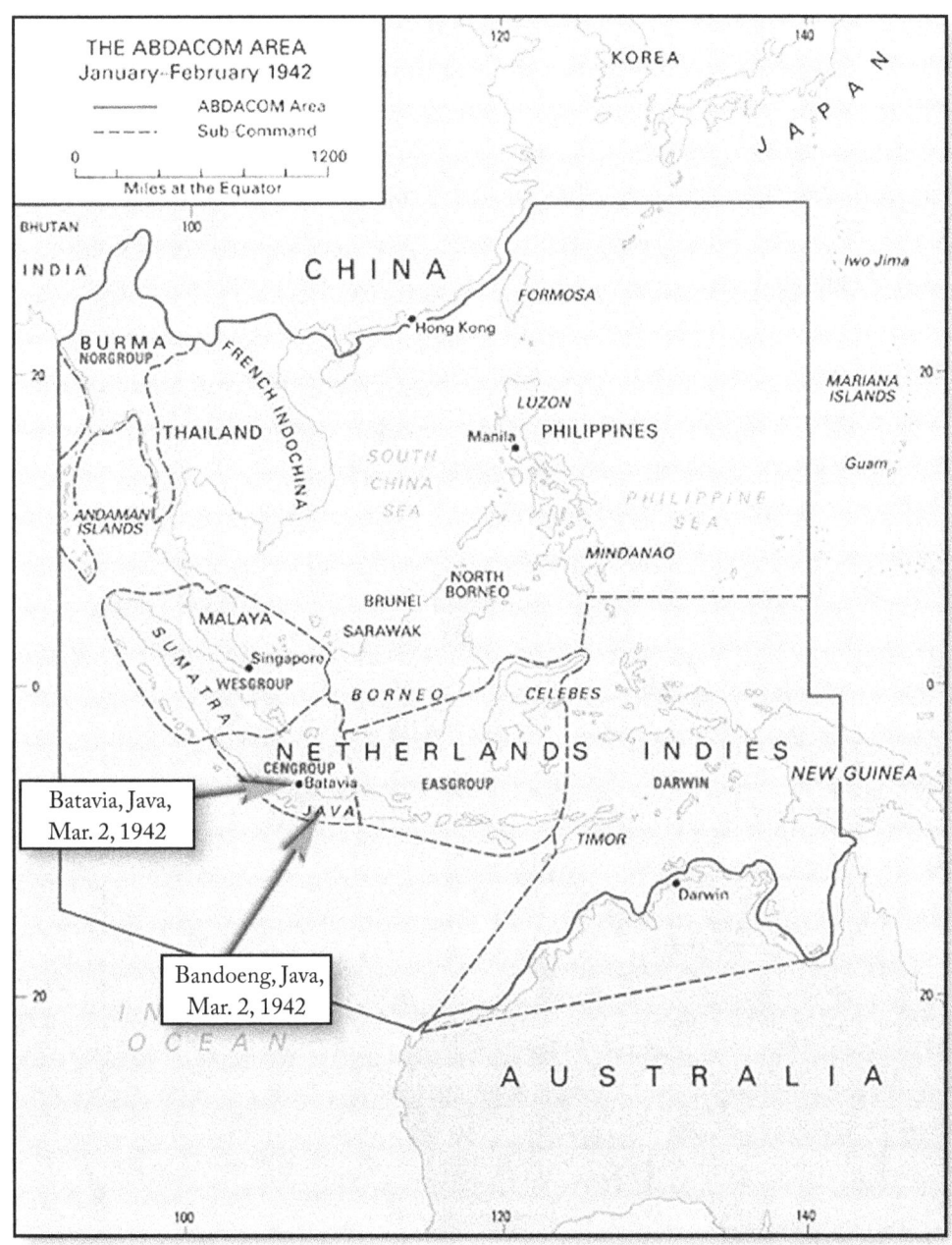

Figure 28 – ABDA – March 1942

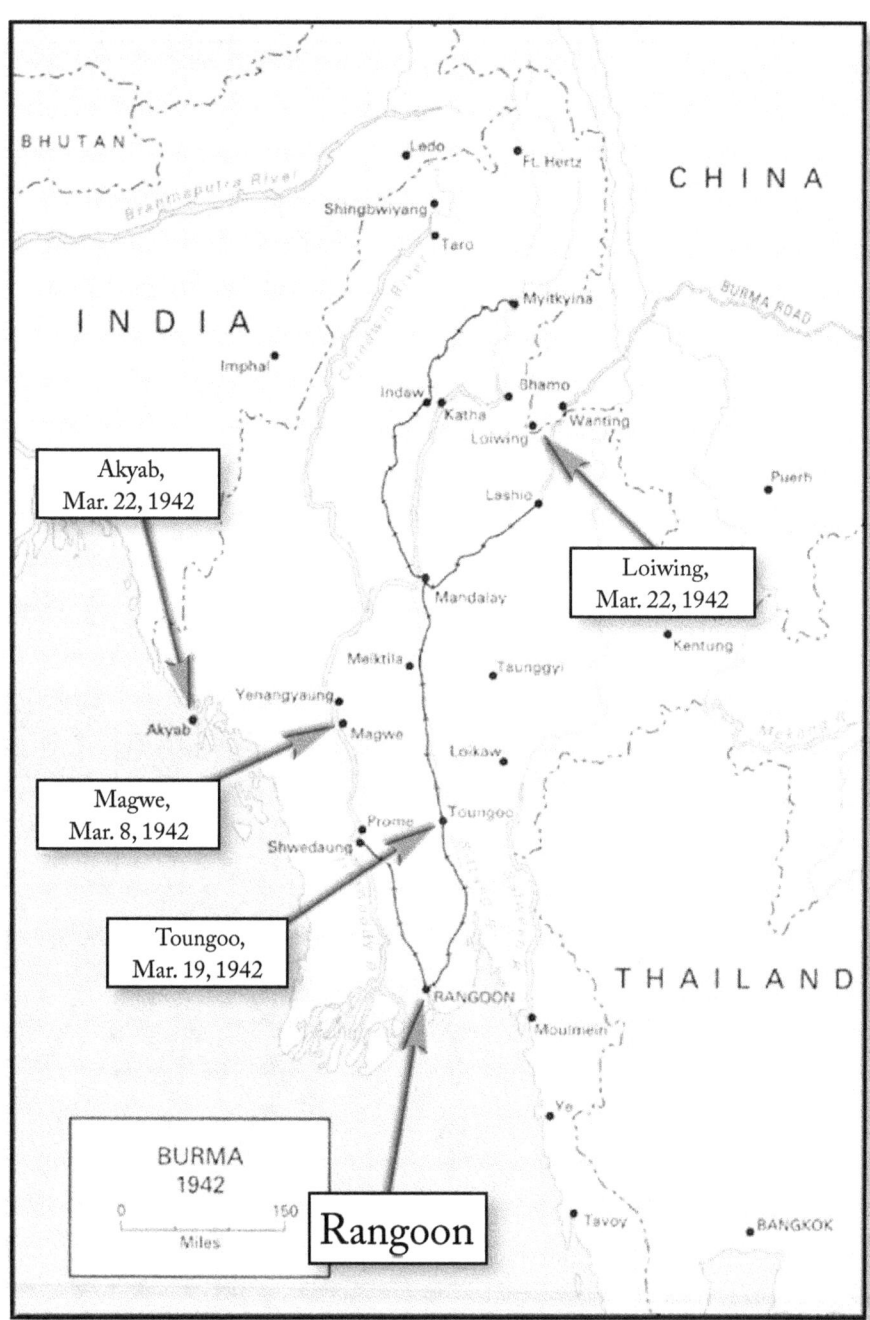

Figure 29 – Burma – March 1942

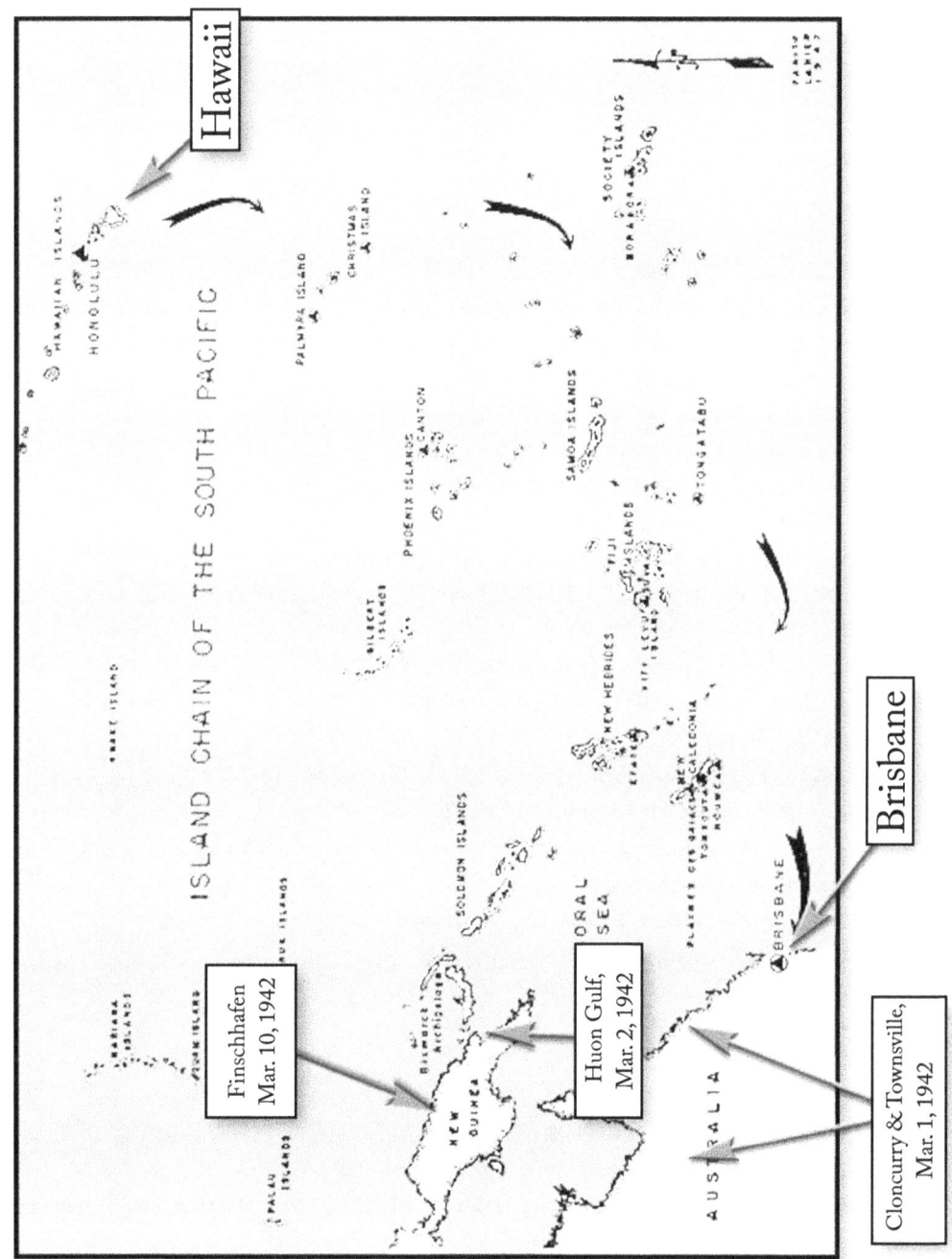

Figure 30 – Supply Chain – March 1942

ENDNOTES

1. Craven, *The Army Air Forces in World War II, Vol. 1*, pp. 398-399.
2. ----, "The "Kangaroo Squadron," http://www.ozatwar.com/usaaf/435th.htm , accessed August 11, 2015.
3. Williams, *Chronology 1941-1945 - 1942*, p. 26.
4. Walker, *The Liberandos*; Richardson, *The Forgotten Force*, p. 15.
5. Williams, *Chronology 1941-1945 - 1942*, p. 27.
6. Williams, *Chronology 1941-1945 - 1942*, p. 27.
7. Letter, King to Marshall, 2 March 1942. NRL, (footnote 15), ----, *US Marine Corps in World War II, Guadalcanal Campaign*, http://www.ibiblio.org/hyperwar/USMC/Guadalcanal/USMC-M-Guadalcanal-1.html , p. 4.
8. Assistant Chief of Air Staff, *The AAF in the Middle East*, p. 50.
9. Craven, *The Army Air Forces in World War II, Vol. 1*, p. 240.
10. Assistant Chief of Air Staff, *The AAF in the Middle East*, p. 49.
11. Brereton, *The Brereton Diaries*, p. 109.
12. http://www.historynet.com/countdown-to-the-doolittle-raid.htm .
13. ----, FDRL, PSF, Box 3, Marshall, George C., 1941 - 4/14/42 file, FDR to Marshall, March 4, 1942.
14. King, *Fleet Admiral King*, pp. 370-371.
15. Assistant Chief of Air Staff, *The AAF in the Middle East*, p. 57.
16. Williams, *Chronology 1941-1945 - 1942*, pp. 26 and 31.
17. Roskill, S.W., *War At Sea 1939-1945*, p. 5.
18. ----, FDRL, PSF, Box 3, King, Ernest J file, King to FDR, March 5, 1942.
19. Craven, *The Army Air Forces in World War II, Vol. 1*, p. 561.
20. A sortie is defined as one plane flying one mission. So, for example, if 5 planes were sent to attack a target, 5 attempted sorties were recorded.
21. Craven, *The Army Air Forces in World War II, Vol. 1*, p. 400.
22. Assistant Chief of Air Staff, *The Tenth Air Force*, p. 11.
23. Craven, *The Army Air Forces in World War II, Vol. 1*, p. 493.
24. Brereton, *The Brereton Diaries*, p. 109.
25. Mayhew, *transcript of interview*.
26. Williams, *Chronology 1941-1945 - 1942*, p. 27.
27. Williams, *Chronology 1941-1945 - 1942*, p. 27.
28. Craven, *The Army Air Forces in World War II, Vol. 1*, pp. 561-562.
29. ----, AFHRA, Hap Arnold file, Memo for HHA by D.D. Eisenhower, Asst Ch Staff, "Establishment of USAAF in the UK", March 6, 1942.
30. ----, FDRL, PSF, Box 98, Carter file, Arnold to FDR, March 6, 1942.
31. Williams, *Chronology 1941-1945 - 1942*, p. 27.
32. Williams, *Chronology 1941-1945 - 1942*, p. 27.
33. Holloway, letters to the author.

34. Williams, *Chronology 1941-1945 - 1942*, p. 28.
35. ----, *US Marine Corps in World War II, Guadalcanal Campaign*, http://www.ibiblio.org/hyperwar/USMC/Guadalcanal/USMC-M-Guadalcanal-1.html , p. 1.
36. Craven, *The Army Air Forces in World War II, Vol. 1*, pp. 409-410.
37. ----, *Time, Let's Begin to Strike*, March 9, 1942, http://www.time.com/time/magazine/article/0,9171,885884,00.html, accessed July 27, 2012.
38. Williams, *Chronology 1941-1945 - 1942*, p. 28.
39. Williams, *Chronology 1941-1945 - 1942*, p. 28.
40. Wedemeyer, *Wedemeyer Reports*.
41. ----, "The "Kangaroo Squadron," http://www.ozatwar.com/usaaf/435th.htm , accessed August 11, 2015.
42. ----, *US Marine Corps in World War II, Guadalcanal Campaign*, http://www.ibiblio.org/hyperwar/USMC/Guadalcanal/USMC-M-Guadalcanal-1.html , p. 1.
43. Assistant Chief of Air Staff, *The Ploesti Mission*, p. 14.
44. Williams, *Chronology 1941-1945 - 1942*, p. 28.
45. Mayhew, *transcript of interview*.
46. Craven, *The Army Air Forces in World War II, Vol. 1*, p. 408.

47. ----, FDRL, PSF, Box 6, West Africa file, Bullitt to FDR, March 13, 1942.
48. ----, FDRL, PSF, Box 98, Carter file, Currie to FDR, March 13, 1942.
49. Williams, *Chronology 1941-1945 - 1942*, p. 29.
50. Assistant Chief of Air Staff, *The AAF in the Middle East,* pp. 58, 66.
51. ----, AFHRA, Hap Arnold file, Hopkins to FDR, 14 March 1942.
52. ----, "The "Kangaroo Squadron," http://www.ozatwar.com/usaaf/435th.htm , accessed August 11, 2015.
53. Craven, *The Army Air Forces in World War II, Vol. 1*, pp. 409-410.
54. ----, *Time, SUPPLY: Oil Can Lose the War*, March 16, 1942, http://www.time.com/time/magazine/article/0,9171,801376,00.html , accessed July 27, 2012.
55. ----, *Time, BATTLE OF THE PACIFIC: Fall of Java*, March 16, 1942, http://www.time.com/time/magazine/article/0,9171,801368,00.html, accessed July 27, 2012.
56. Williams, *Chronology 1941-1945 - 1942*, p. 29.
57. ----, *New York Times, WASHINGTON HAILS 'BEST WAR NEWS'*, March 17, 1942, http://select.nytimes.com/gst/abstract.html?res=F70E10FB3F5D167B93CAA81788D85F468485F9, accessed February 18, 2014.
58. Craven, *The Army Air Forces in World War II, Vol. 2*, p. 8.
59. ----, AFHRA, Hap Arnold file, Maj. Gen. M.F. Harmon, C/AS to Col. Shumacker, 17 March 1942.
60. ----, FDRL, PSF, Box 6, West Africa file, Marshall to FDR, March 18, 1942.
61. Williams, *Chronology 1941-1945 - 1942*, p. 29.
62. ----, FDRL, PSF, Box 2, China file, Marshall to FDR, March 14, 1942.
63. Assistant Chief of Air Staff, *The AAF in the Middle East*, p. 54.
64. ----, AFHRA, Hap Arnold file, Harmon to Director of Military Requirements, March 21, 1942.

65. ----, Experimental Engineering Section, Memorandum Report, March 21, 1942.

66. Williams, *Chronology 1941-1945 - 1942*, p. 30.

67. ----, *Time, World Battlefronts, STRATEGY: Too Many Fronts?*, March 23, 1942, http://www.time.com/time/magazine/article/0.9171.802259.00.html, accessed November 16, 2009.

68. Williams, *Chronology 1941-1945 - 1942*, p. 30.

69. Brereton, *The Brereton Diaries*, p. 113.

70. http://www.historynet.com/countdown-to-the-doolittle-raid.htm .

71. http://www.historynet.com/countdown-to-the-doolittle-raid.htm .

72. ----, AFHRA, Hap Arnold file, Arnold to Eisenhower, March 26, 1942.

73. Assistant Chief of Air Staff, *The Ploesti Mission*, p. 13.

74. Williams, *Chronology 1941-1945 - 1942*, p. 30.

75. Brereton, *The Brereton Diaries*, p. 114.

76. Craven, *The Army Air Forces in World War II, Vol. 1*, p. 563.

77. Craven, *The Army Air Forces in World War II, Vol. 1*, p. 440.

78. ----, *Necessary Modifications to B-24D on HALPRO as of March 28, 1942*.

79. Williams, *Chronology 1941-1945 - 1942*, p. 31.

80. Craven, *The Army Air Forces in World War II, Vol. 1*, p. 563.

81. Air Corps plans had called for 2 B-17s to leave the U.S. each day until a total of 40 planes were in Australia. Craven, *The Army Air Forces in World War II, Vol. 1*, p. 413.

82. Thomas, *Born in Battle*, p. 47.

83. http://www.historynet.com/countdown-to-the-doolittle-raid.htm .

84. ----, AFHRA, Hap Arnold file, Harmon to Director of Military Requirements, March 31, 1942.

85. ----, FDRL, PSF, Box 3, Marshall, George C., 1941 - 4/14/42 file, MacArthur to Marshall, March 31, 1942; Marshall to FDR, April 3, 1942.

86. ----, Army Air Force Statistical Digest, http://www.ibiblio.org/hyperwar/AAF/StatDigest/aafsd-3.html , accessed January 22, 2014, p. 112; B-24 production from Author's personal papers.

87. Hurd, Charles, *New York Times, BIG BOMBERS ROLL INTO MASS OUTPUT WITH 24-HOUR GOAL*, March 16, 1942, http://select.nytimes.com/gst/abstract.html?res=F10F17F83F5D167B-93C4A81788D85F468485F9, accessed February 18, 2014.

88. Craven, *The Army Air Forces in World War II, Vol. 1*, p. 413..

APRIL 1942

On April 10, *The New York Times* announced the surrender of Bataan, one of the bleakest days in United States Army history.

A week later, on April 18, the United States celebrated its first victory. The Doolittle raiders had bombed Japan.

HALPRO's original reason for its existence was now gone. But the HALPRO men did not know this, and they continued their training. But for the United States military leaders, they now had a strike force, albeit small. How best to employ it was the question. Which reflected the overall strategic question – where do we place our scarce resources?

The bomber force that was originally sent to the Philippines to stop any Japanese attack was a mere shadow of itself. Three successive withdrawals now placed those forces in India, hoping to retard a Japanese advance into India. For all intents and purposes, they were the only Army Air Corps force facing the enemy.

1942 APRIL 1

DOOLITTLE RAID

At the Alameda Naval Station, 16 B-25 were loaded onto the deck of the *USS Hornet*. She would put to sea the next day.

HALPRO

Arnold's staff documented HALPRO's progress:

Subject: HALPRO

1. In connection with . . . above subjec . . . the following action has already been taken:

    a. Personnel movements . . . from Patterson Field, Ohio, to Fort Meyers, Florida, have been accomplished.
    b. Ammunition, Bombs, and Gasoline

- There was shipped from an Atlantic port the latter part of January per SS Dona Aniceta, 168,000 pounds of maintenance material for the B-24D airplanes, labeled for HALPRO.
- The Commanding General, Services of Supply, AQUILLA, has reported the receipt of this material and placed it in storage at Karachi Air Depot.
- There was also shipped, per SS Bering, approximately _____ rounds of .50 cal. Ammunition and 180 tons of appropriately selected bombs for HALPRO.[1]

The entire HALPRO contingent moved to Fort Myers, Florida. There, they would complete their final training prior to their deployment. The HALPRO personnel were temporarily assigned to the 98th Bombardment Group for administrative purposes. Halverson assumed command of the 98th.[2]

THE PHILIPPINES

Roosevelt awarded MacArthur the Medal of Honor. The citation read in part:

> ...*for conspicuous leadership in* preparing the Philippine Islands to resist conquest, for gallantry and intrepidity above and beyond the call of duty in action against invading Japanese forces, and for the heroic conduct of defensive and offensive operations on the Bataan Peninsula. General MacArthur mobilized, trained, and led an army which has received world acclaim for its gallant defense against a tremendous superiority of enemy forces in men and arms. His utter disregard of personal danger under heavy fire and aerial bombardment, his calm judgment in each crisis, inspired his troops, galvanized the spirit of resistance of the Filipino people, and confirmed the faith of the American people in their Armed Forces.[3]

This award appeared in sharp contrast to the label that the United States servicemen gave to MacArthur: "Dugout Doug." Several versions of a parody song (sung to the "Battle Hymn of the Republic") made the rounds among the GI's.

1942 APRIL 2

BURMA

The Allies withdrew from Prome.[4]

10TH AIR FORCE

The 10th Air Force flew its first offensive combat mission attacking the Andaman Islands.[5]

Chiang and Chennault had been reluctant to support the transfer of the AVG into the regular United States Army Air Corps. The men were contract fighters, i.e. mercenaries, who were under no obligation to enlist. The withdrawal from Prome resulted in the abandonment of much needed supplies. This action was the latest step in a series of withdrawals. The two leaders changed their minds.[6] (see Figure 32)

GLOBAL STRATEGY

The issue of priorities - Australia vs. India – was elevated to the JCS. After they made their decision, Arnold communicated it to General Millard F. Harmon, Chief of Air Staff.

> *The Australian heavy bomber force* was to be built up to 80 operational aircraft, with 40 in reserve. The India heavy bomber strength was to be built up to 50. Australia was to be given a heavier priority until its operational strength was reached.[7]

This was a continuation of the February-March decisions. As noted, only nine planes had reached Australia. The hidden hand of MacArthur was demanding reinforcements when there was none to be had.

SECOND FRONT

Following the release of the WPD's "Plan for Operations in Northwest Europe," Marshall sent a summary of that plan to Roosevelt under the title, "Plan for Invasion of Western Europe." He appeared to have down played the tiny fact that such a plan would preclude any major action in 1942.[8]

DOOLITTLE RAID

In Asia, Chiang asked that the Doolittle raid be delayed until late May as he could not provide supplies for the 25 B-25s he was told would be landing at his bases at Kweilin, Kian, Yushan, Chuchow, and Lishui. Also, he could not prevent Japanese troops from occupying the Chuchow area.[9] (see Figure 34)

1942 APRIL 3

HALPRO

Kenneth DeLong became the engineer on Therman Brown's crew. He related what happened at Fort Myers:

> *On 3 April, I was* ordered to report to Captain John Kane (98th squadron commander). I had no idea why. Captain Kane was very friendly and wanted to interview me about my personal goals in the Army Air Force. Originally, my goal had been to be a pilot. However, my right eye did not check out to 20/20. Being an aircrew member was my second choice. I told Captain Kane I would like to be an aerial engineer on a bomber crew. He asked me if I was interested in serving overseas in a combat zone, and my answer was yes.[10]

1942 APRIL 4

10TH AIR FORCE

Subject: 10th Air Force

That the following Secret Message in code be transmitted to Major General Lewis B Brereton, New Delhi, India, by most expeditious means.

Information as to assignment of HALPRO or any special project to your command will be furnished you at such time as it may be decided by War Department for you to assume such command.

In addition to the 10 heavy bombers now in India and en route, you will receive the balance of the 24 heavy bombers now being set up for HALPRO or which may be left at the time HALPRO is assigned to your command.[11]

Brereton was now given command of HALPRO when it reached India. The 37 planes had now been reduced to 24. Earlier, Halverson was given command of the Indian forces. This would have personal repercussions later.

MIDDLE EAST

Military Intelligence cabled its Cairo office that they needed detailed information concerning:

- The Balkans, with emphasis on the Rumanian on refineries,
- Rail communications between the oil fields and the Ukraine, and

- Vulnerable points in rail and water communications throughout the whole area.[12]

Ploesti as a target was appearing more and more evident.

DIEPPE

Following discussions within the British High Command, Lord Mountbatten ordered his staff at Combined Operations to draw up plans for an invasion of Dieppe. The British High Command had defined two different concepts for the invasion.[13]

1942 APRIL 5

INDIAN OCEAN

The Japanese carriers *Akagi, Soryu,* and *Hiryu* launched an air strike against Colombo, Ceylon, which damaged the port facilities. Later in the day, they attacked the British cruisers, *HMS Cornwall* and *HMS Dorsetshire,* which sank off the coast of Ceylon. Four days later, the carrier *HMS Hermes* was lost. The Japanese Navy had entered the Indian Ocean. (see Figure 32)

1942 APRIL 6

BURMA

The Japanese landed reinforcements at Rangoon.[14] (see Figure 31)

10TH AIR FORCE

Brereton received a communiqué from Stilwell via Marshall requesting a status on the 10th Air Force. Four days earlier, the 10th Air Force had conducted a raid on the Andaman Islands, located near in the center of the Bay of Bengal. (see Figure 32) Stilwell and Brereton had previously agreed that the 10th Air Force, which were not supposed to be ready until May 1, would be used to support Allied forces in Burma. Stilwell wanted the status report so "he might plan for its use in support of critical ground operations." Brereton believed heavy bomber operations were separate from air-ground support.[15]

This difference of opinion regarding the use of heavy bombers would perplex Allied leaders throughout the war.

CBI

Earlier, Roosevelt had sent Louis A. Johnson to India to evaluate the seriousness of the Japanese threat there. On April 6, Johnson sent his first report to Hull regarding his meeting with Wavell. Wavell told Johnson that he, Wavell, could hold India if the

United States could send 120 P-40s, 120 B-25s, 80 Hudsons, and 40 DC-3s by June 1942. Hull forwarded the request to Roosevelt on April 7, who in turn forwarded it to Arnold. Roosevelt was not sure if Arnold had seen it and was not sure if he, Roosevelt, should reply. He wanted Arnold's opinion.[16]

In his response two days later, Arnold told Roosevelt:

Subject: Request for Additional Airplanes in India.

- In connection with request made by General Wavell of Mr. Louis Johnson for additional airplanes in India, it is my opinion that we should handle this matter with care.
- India is in the war theatre for which the British have full responsibility. When this same matter was brought up before the Combined Chiefs of Staff, the British members stated that it was a British responsibility and that they would handle it. Accordingly the document concerning additional airplanes to India is now being considered by the British Joint Chiefs of Staff in London.
- In view of the above, I recommend that no further action be taken on this request.[17]

What was interesting about this development was that the Indians themselves did not consider the Japanese a threat – India was already "occupied." There were rumors that Gandhi was ready to negotiate a peace settlement with the Japanese. Churchill appeared to want United States intervention when it helped him, but was opposed to that help if it interfered with "internal issues" of the British Empire. It was this type of activity that always raised the perception among certain United States military planners that Churchill's overarching goal was preservation of the Empire, not the defeat of the Axis.

GLOBAL STRATEGY

While United States planners were contemplating action in India, Churchill was receiving a briefing from British intelligence on possible German plans for 1942. It was their opinion that if Germany failed to acquire oil, its reserves of oil would be depleted, causing severe military and economic damage.[18]

Churchill must have leaked his concerns to the press. *The New York Times* ran an article with a map depicting one version of that very strategy. (see Figure 35)

HALPRO

The HALPRO crews were training hard, but they were not told why, according to Ed Cave:

Ordered out of 98th Bomb Group to Ft. Myers Florida April 6, 1942 to what was later known as Page Field. We called it Palmetto field because it had no name. There were two runways - a parking ramp - enlisted personnel lived in tents and had tar paper and stand up screen mess structures with sand floors. Officers were permitted to live in town. Flew day and night putting in 97 hours flying time trying to learn combat tactics - had no idea what it was all about.[19]

1942 APRIL 8

HALPRO

Kenneth DeLong was interviewed by Feldmann and Richard Sanders:

Both interviewers had my complete records in front of them. They wanted to find out if I was satisfied with my job assignment (aerial engineer) and whether I had any reservations about serving overseas in a combat zone.[20]

1942 APRIL 8-10

GLOBAL STRATEGY

A series of major strategy meetings took place in London the second week of April. Marshall led the American delegation to secure British consent to Eisenhower's plan for a cross channel invasion in April 1943. The plan had three components –

1. The buildup of bases, supplies, and troops (Codename BOLERO),
2. An invasion in the Le Havre-Boulogne area (Codename SLEDGEHAM-MER), and
3. A breakout towards Germany (Codename ROUNDUP).

The United States promised 1,000,000 men and 2,550 aircraft by April 1, 1943.[21] At first, the British Chiefs of Staff were reluctant to accept the plan. Eventually, they came around to Marshall's arguments, stating that "in the broadest terms General Marshall's proposal was in line with our strategy."[22]

One historical analyst, Tuvia Ben-Moshe, wrote the following about this conference:

Here for the first time was a strategic plan to engage the German army directly at its present strength in order to defeat it in the field according to a clear timetable, a plan that did not depend on an extreme weakening or collapse of the German military machine. It was built on what the Allies might have been able to do and not on internal factors on the Continent over which they had little control. This was the fundamental difference between the American and British approaches.[23]

While a strategic plan may have been verbalized, it was never clear how the United States could train and supply the required men and planes within one year. Right now, they could not supply the air power necessary to stop any Axis advance on any front.

1942 APRIL 9

THE PHILIPPINES
General King surrendered the Bataan garrison.[24] (see Figure 6) The front page of *The New York Times* late April 10 edition announced:

JAPANESE CAPTURE BATAAN AND 36,000 TROOPS [25]

INDIAN OCEAN
Japanese aircraft struck Trincomalee, Ceylon.[26] (see Figure 32)

EASTERN FRONT
The Germans successfully thwarted Russian attacks in both the Crimea and near Moscow.[27]

1942 APRIL 10

GERMAN OIL
The March 26 report *The Danube as an Objective For Air Attack*, together with supporting materials, was submitted to the WPD, which pronounced it "sound tactically and strategically."
However, due to the scarcity of planes and commitments to other theaters, the plan was filed away for future use.[28]

1942 APRIL 11

DOOLITTLE RAID
Prior to his departure from the United States, Stilwell was informed of a projected bombing mission, known as the First Special Bombing Mission. Stilwell was probably not given a date for the mission. That changed on April 11, when Bissell sent a radiogram to Stilwell.[29]

1942 APRIL 12

SECOND FRONT

From London, Marshall cabled Roosevelt that Churchill had agreed to the cross-channel invasion plan.[30]

COMMITMENT TO CHINA

Madame Chiang Kai-Shek sent a message to Currie describing a desperate situation:

Situation in Burma "unspeakably dangerous." Possible complete disorganization of front and rear; panicky population; no fighting morale. British "hopeless and helpless"; Burmese "antagonistic and the countryside is honeycombed with fifth columnists."

CKS told Alexander that in China "heads would have been chopped off" to allow such conditions. Called for large scale air support. Otherwise "sacrifices useless . . . would be impossible to hold on."[31]

10TH AIR FORCE

Rowan Thomas and crew landed at Karachi. (see Figure 32) Brereton briefed the new arrivals:

General Brereton told us, quite frankly, that the crisis approaching India was of greater importance at that moment than the bombing of Japan. To all intents and purposes the Japs were not going to stop with the conquest of Burma, but were preparing to push on into India. India could keep the Japs going for years once they could take over her war potential and use her vast population for slave labor.

At that time General Brereton had only fourteen Fortresses under his command, and we could understand that our six brand new ships looked like manna from heaven to him.[32]

1942 APRIL 13

COMMITMENT TO CHINA

Currie forwarded Madame Chiang's message to Roosevelt:

Chinese army fighting gallantly against overwhelming obstacles but must have immediate large scale air support of hundreds of planes otherwise our sacrifices are useless and it will be impossible to hold on.

I should like, if you have no objection, to follow up this request for additional planes (from CKS) with Generals Arnold and Eisenhower (then ACS/G-3).[33]

DOOLITTLE RAID

General Joseph T. McNarney advised Roosevelt that Bissell, air advisor to Stilwell, was informed that the Doolittle mission was too far advanced for cancellation. Bissell was informed that the Second Mission [HALPRO], arriving at a later date, would be under Stilwell's command.[34]

It appeared that the Theater Commanders in the CBI would now be allowed to divert HALPRO from its original intended mission to bomb Japan. Halverson and his men were not informed of this change in mission.

PRESS

China had her supporters in the United States press. And her various representatives were not reluctant to use that support as leverage. On April 13, *Time* ran an article titled *Global Yang*:

> China needed U.S. arms. China was imperiled by India's peril. Yet one of China's foremost strategists last week raised his sights beyond the target of his own army's terrible tasks and achieved a global look at World War II. His conclusion: the U.S. and Britain can make their best contribution to the war in 1942 by opening a front in northern Europe. Said bullnecked, moon-faced General Yang Chieh, lecturing to the Chinese War College in Chungking: "In northern Europe . . . it would be easiest for Britain, America and Russia to cooperate."
>
> General Yang still wanted the U.S. to send all the material aid it could to China. Like other Chinese, he felt that some U.S. officers were still not sufficiently energetic in dispatching aid to China. But, said global General Yang, who was Chinese Ambassador to Moscow in 1938-39: if Hitler is crushed, "Japan's fate is sealed."[35]

1942 APRIL 14

BURMA

The Allies ordered the destruction of the oil fields at Yenangyaung. This was accomplished over the next two days.[36] (see Figure 31)

SECOND FRONT

The British and American conferees formally accepted Marshall's plan for the cross-channel invasion.[37]

However, Churchill added one caveat:

One broad reservation must however be made - it was essential to carry on the defence of India and the Middle East.[38]

Whether they wanted to admit it or not, the United States was now defending the British Empire.

GLOBAL STRATEGY

Churchill's addition of India to the Allied effort was disclosed to Hull. He received communication from London that Churchill was about to inform Roosevelt about the situation in the Indian Ocean:

The enemy have apparently moved very powerful forces in that direction and the implications of this will readily occur to you.[39]

In response to both Britain's and China's pleas for airplanes, Arnold told Roosevelt:

1. There were limiting factors to sending planes to the CBI theater

   - There was a limited number of landing fields within striking distance of the Japanese Air Force.
   - There were not more than 25 transports available at that time,
   - The AAF aircraft were already headed for the CBI Theater.

2. More airplanes would be sent as the facilities became available.

Following the meeting, Arnold sent a memo to Kuter, assistant chief of Air Staff. Kuter was to get in touch with all concerned and

tell them all to step on the gas consistent with safety of crews and to ensure that the planes will arrive. In other words, take up all slack possible in connection with this movement.[40]

10TH AIR FORCE

Rowan Thomas and crew took off for Rangoon.[41] (see Figure 31)

1942 APRIL 15

10TH AIR FORCE

Earlier in the month, the 10th Air Force had conducted its first mission against the Andaman Islands. Stilwell was a bit peeved and implied, via Marshall, to Brereton that he, Stilwell, was in charge of the 10th Air Force. Meanwhile, the British had lost two cruisers and a carrier near Ceylon. On April 15, Stilwell was informed that the 10th would be used to help with the British plans for the defense of the waters around Ceylon.[42] (see Figure 32)

HALPRO

Following four months of training, Al Story found himself in the 44th Bomb Group as a trained gunner. He had just completed gunnery school in Las Vegas, Nevada.

On April 15, 1942, our group was shipped by train back to the 44th, which had moved to Fort Myers, Florida. By then they had some airplanes, B-24s! It was shortly after this that I heard of HALPRO.

One day about twenty or thirty of us were told to go to a certain tent for a meeting. We had no idea what it was about. We all congregated in front of the tent, which, judging by a sign in front appeared to be Group Headquarters. I, by happenstance, was standing closest to the door and was asked to come in. Inside was a Colonel that I assumed to be Colonel Halverson. I later learned that it was Colonel McGuire. He said that we had been chosen for a very important mission, that he could not say what it was but it was important and they only wanted volunteers for this mission. Of course, I volunteered. He thanked me and I went outside where all the others were anxiously waiting to find out what was going on. It developed that every one of the guys volunteered. We were now HALPRO!

After that, things really began to move. I was assigned to crew #23. I was the tail gunner. We were all promoted one grade, so I was now a Corporal. Our pilot, Lt. Cave, took a picture of the crew as we stood before our plane. He obviously was not in the picture. It was shortly after this photo was taken that we became known as the Silver crew. Since we had no tow targets for gunnery practice, we had to fly out over the Gulf and throw out bags of aluminum powder that burst and spread out over the water. While we were shooting at these targets, someone accidentally stepped on one of the bags and the powder coated every one of us so that we were completely aluminized, face, hands, arms, and legs. A few days later the same thing happened to another crew so we sort lived down our reputation. After much practice and refitting our planes were finally ready to head overseas.[43] (see Figure 36)

1942 APRIL 16

DOOLITTLE RAID

Washington received two important messages regarding the Doolittle mission. The first came from Arnold, who said that the mission would not be executed for three to four days. The second was that Chiang had reluctantly agreed that Doolittle could use the airfields at Yushan, Lishui, Kian, Keweilin, and Henyang but not Chuchow.[44]

BOMBING JAPAN

A fifth set of Japanese targets was identified.[45]

1942 APRIL 17

GLOBAL STRATEGY

Churchill informed Roosevelt of his approval of the April 1943 cross channel invasion; his sole proviso was that the Japanese and Germans must be prevented from joining forces in the Middle East.

If that could be forestalled, he advocated a crescendo of activities against the continent, "starting with an ever increasing air offensive both by night and day."[46]

Churchill agreed, but he had conditions.

FAR EAST

Stilwell was not in agreement with the orders that the 10th Air Force would be supporting the British defense of Ceylon. (see Figure 32) He sent two messages, one on April 17 and the second, the next day. Both stated that the Chinese were suspicious of British intents. Plus, he and the Chinese were upset that a decision had been made without consultation with the theater commander. He refused to send the appropriate orders on to Brereton.[47]

While Stilwell was arguing with Washington, Arnold sent a telegram to the Air Corps Commander in London that B-17s, LB-30s, and B-25s were being diverted to the 10th Air Force for reconnaissance in the Bay of Bengal. And Brereton did not know when he would be getting these planes.[48]

PLOESTI

United States Intelligence sent its first report to Washington regarding the Balkans. One of the reported items was:

There are at least 100 AA Batteries in Constanza, showing great importance attached possibly from view point relating to the refining of present and anticipated captured oil supplies.

There is no undue cause for alarm but above movements indicate that Axis is preparing bases in Greek Islands for jumping off points for attack on South Eastern Mediterranean, should main campaign warrant it, or for diversion.[49]

Constanza, on the Black Sea, is the closest port to the Ploesti refineries.

HALPRO

The first set of travel orders was issued to Halverson. And, he was officially told that he might be held in Egypt for emergency operations.[50]

REINFORCEMENTS

Harry Holloway and crew began their overseas odyssey.

Approximately 17 April 42, John Neal Lavin's (B-17) crew left the States on a "secret" mission.

We left Florida to Borinquen Trinidad, to Belem Brazil, where we had an accident – damaging our aircraft and killing about 4 civilians on the ground. The airplane was repaired from parts of other crashes and flown back to the states, by the crew that brought us a new plane.

This cost us a week or so, but we then took off for Natal Brazil, then from Natal to Roberts Field, Accra on the Gold Coast, then to Kano. From Kano to Khartoum, Khartoum to Aden, then Aden to Karachi (Pakistan today). Were quarantined then flew from Karachi to Allahabad, India. Because of our delay in Belem, our secret project was scratched. We flew 2 bombing missions from Allahabad against the Japanese, bombing Myitkyina Airfield in Central Myaumar (Burma).[51]

1942 APRIL 18

DOOLITTLE RAID

Early on the morning of April 18, the *USS Hornet* and her escorts encountered Japanese boats about 650 miles east of the Japanese Islands. The original plan was for the *Hornet* to steam within 400 miles of Japan. To avoid risking the carriers, the B-25s were launched. All pressure on HALPRO to attack Japan was about to be removed.

DIEPPE

Lord Mountbatten's staff reported back. They opted for a frontal assault, gave the operation the code name Operation RUTTER, and selected the early part of June for the execution.[52]

1942 APRIL 20

COMMITMENT TO CHINA

Currie cabled United States representatives in China:

- Am assured by Arnold that all planes sent are for GMO [Chiang] and Stilwell.
- Bombers set up for May delivery to China.[53]

1942 APRIL 21

HALPRO

HALPRO's twenty-four B-24Ds, which had been delivered to the Mobile Air Depot for modifications, began to arrive at Fort Myers, Florida.[54] Meanwhile, Brereton learned that HALPRO would operate in China independent of his command.[55]

PUBLIC OPINION

While Roosevelt and his team were planning the war against the Axis, he was also fighting a domestic problem – a considerable portion of the populace was against the war. The Federal Bureau of Investigation (FBI) was doing surveillance against perceived opponents of his policies. There was such resentment about America's participation in World War I that he went so far as to propose that the current conflict be called "War of Survival" at a press conference on April 3. Much to his surprise, his domestic opponents pointed out that the new name reflected a defeatist attitude. Since the news was filled with Axis victories, a negotiated peace with Germany appeared to be not all that unreasonable. Roosevelt dropped the idea.[56]

1942 APRIL 22

PLOESTI

Fellers sent a cable to Arnold, reporting that Tedder was anxious for the assistance of an American air force, especially heavy bombers. They were to be used against Rumanian oil, and he [Tedder] was willing to supply the necessary gasoline, oil, and bombs.[57]

REINFORCEMENTS

Bill Mayhew and the 10th Air Force were also busy in April.

From about the first of April 1942 we flew just over that section of India to grow accustomed to the planes. The same Japanese fleet that had attacked Pearl Harbor attacked Colombo on 5 April 1942, causing extensive damage. Therefore, we had four crews and two planes remain at Karachi to fly patrols over the Arabian Sea to look for Japanese ships and try to keep the sea lanes open for our supply ships. By the first part of April 1942 I was able to get assigned as an armorer gunner on one of our few Flying Fortresses. I was one of those that remained at Karachi. I flew as a tail gunner at that time. The rest of the outfit moved inland to Allahabad, India to be closer to the targets in Burma. Those of us that remained behind felt rather slighted, because one rarely gets much excitement on patrol flights. However, all of us got as much excitement as we wanted before we returned to the States. On 22 April 1942 the 88th Reconnaissance Squadron became the 436th Bombardment Squadron in the 7th Bombardment Group.

Many 7th Bomb Group personnel were involved in rescuing soldiers and refugees from Burma when that country fell to the Japanese in early 1942. At least 8,000 refugees were removed by air from Myitkyina (pronounced Mitch'enaw), Burma alone. One plane carried 74 people that was designed for 21. An English girl had a baby at the Myitkyina airfield while waiting for evacuation. Lt. Col. Jerry Mason had a woman give birth in his plane while he was evacuating refugees. Colonels Caleb Haynes and Robert Scott, both members of "Project Aquilla", were responsible for evacuating 4,500 soldiers and refugees, and over two million pounds of freight. Major Paul J. Long flew a B-17E loaded with Irish soldiers to Magwe, Burma. On the return trip he evacuated 48 British civilians. Colonel Haynes brought out 30 members of General Joseph Stilwell's staff when they were cut off in northern Burma.

No names were painted on the B-17s while we were in India. In fact, the serial numbers of these planes had been painted over. We had so few B-17s in India (20 at one time, usually more than out of commission) that the military authorities were afraid these numbers on the planes would allow the Japanese to determine how few bombers we had.[58]

SOUTH PACIFIC

The 40th Squadron was re-designated to the 435th Squadron.[59]

1942 APRIL 23

COMMITMENT TO CHINA

Apparently, the contest over where the planes were going and who was in charge was percolating up to Roosevelt. On April 22, Soong had pressured Roosevelt for answers. Roosevelt passed the question on to his military chiefs. On April 23, Arnold sent an answer to Captain John McCrea, Naval Aide to Roosevelt. In it, Arnold wrote:

> *...it was necessary, due to* conditions in the Indian Ocean and the Bay of Bengal, that General Brereton's mission be changed with a view of his operating against Japanese ships, which might be headed towards either Ceylon or the Indian East Coast ports. I believe it would be most unwise for the President to commit himself in any way that would tie down the operations of either the 10th Air Force under Brereton, or the AVG's directly under Stilwell, when the situation in that war theater is as fluid as it is now. As a matter of fact, conditions may change so that it will be necessary to send all of the Air Forces in that area against targets in East China. Accordingly, we should avoid stating positively and definitely where any Air Force in that theater should or should not be employed.[60]

1942 APRIL 24

COMMITMENT TO CHINA

Roosevelt replied to Soong:

> *We have commenced the building* up of a small but strong air force in India for ultimate employment in the China theater. Because of shipping and communication difficulties still existing and the loss of Rangoon, this force has been concentrated initially in India. Moreover, because of India's strategic importance as the gateway to China, an interim mission has been assigned to this force to operate against enemy naval units threatening India.[61]

PRESS

Fellers reported that the British had already worked out plans for bombing Ploesti.[62] This was reported in *The New York Times* on January 20, 1941, over a year earlier. The title of the article read:

BRITISH ARE READY TO BOMB RUMANIA.[63]

They were ready, but they didn't.

HALPRO

By April 24, Brereton had been informed that he would be receiving planes, other than the HALPRO planes, but he was not told how or why. He sent two cables to Washington, one on April 24 and the second on April 25, requesting clarification.[64]

1942 APRIL 25

COMMITMENT TO CHINA

The controversy over the use of the 10th Air Force caused Arnold to send a letter to the British contingent of the Combined Joint Chiefs. In it, he says:

> *The War Department fully appreciates* the need for and importance of air support in the campaign now being waged in India. In view of the present critical situation there, General Stilwell has been informed that, until further notice, Brereton's air force will, in conformity with British direction, conduct operations in the Bay of Bengal region in the general area from Ceylon to the northward.
>
> Regarding the 10th Air Force, action has been taken initiated to build the pursuit element to one group immediately and to strengthen the bombardment element by addition of eleven heavy and twenty-six medium bombers.[65]

It was now clear that Chiang was NOT the Supreme commander in the CBI. Stilwell was NOT the theater commander. And the just approved plan for the 1943 cross channel invasion was NOT the primary Allied objective.

MIDDLE EAST

United States Naval Intelligence received another communiqué from Cairo:

> *The continuing preparations Crete and* Dodecanese Islands present a growing potential threat at Suez and Eastern Mediterranean. Recommend U.S. augment immediately Allied Air Force this area to forestall Axis attack by destroying Axis preparations and to enable carry out offense against Ploesti and Bari refinery Italy, which is the only refinery capable of refining high sulphuric Albanian oil.
>
> Ploesti considered Achilles heel. Axis prosecution war there foremost viral target.
>
> RAF anxious for assistance of American Air Force especially heavy bombers.[66]

1942 APRIL 27

BURMA

General Alexander decided to focus on the defense of India, not Burma. Essentially, Burma was assumed to be lost.[67]

HALPRO

Stimson noted the following in his diary:

> *Later Arnold brought in Colonel* Harry A. Halverson who is leading a new special mission in the Far East. The query is whether we can put it through in the present situation. After a conference we started him off, for, even if events prove obstructive to his main mission he can be made useful on the way.[68]

From his travel orders, Halverson probably knew he might get temporarily diverted. He may have guessed that HALPRO would be "made useful on the way" to China. As to the remainder of his command, post war interviews indicate that they never suspected a diversion. They thought they were going to "China."

1942 APRIL 28

PLOESTI

In a note to Marshall, Lt. Colonel Chester Hammond, a military aide to Roosevelt, wrote: "FDR taking a personal interest in carrying out a raid over Ploesti."[69]

HALPRO

> *On April 28, 1942 we* [Ed Cave] flew to Mobile AL and picked up 23 new B-24Ds flown in from Long Beach CA.[70]

1942 APRIL 29

BURMA

Japanese captured Lashio, located at one end of the Burma Road.[71] The Burma Road was the major supply line into China. Thus, China could no longer be supplied by land and was essentially blockaded. (see Figure 31)

Marshall received a telegram from Stilwell, which he forwarded to Roosevelt on April 29. Stilwell informed Marshall that Chiang was not receiving any of the supplies currently sitting on Indian docks. If the supplies could not get to the men, then Chiang would bring the men to the supplies. Then they would march to Burma. And any offer of British help would be totally unappreciated. Obviously sensitive from the 10th Air Force affair, Stilwell said he was not about to embark on such a project without Washington's OK.[72]

Wesley F. Craven and James L. Cate, the two historians of the Army Air Corps, made the following observation:

> *If there were elements of* the comic opera in the continuing exchange of messages which preceded a general understanding that the Tenth -- whose largest mission to date comprised a total of six bombers would remain under theater control but with instructions for the time being to cooperate with the RAF, -- the exchange also revealed some of the extraordinary perplexities of the CBI that would add to the story of air operations in that theater an unusually complex chapter on administration. There was no escape from the necessity to base in India any effort for the support of air operations in China, and thus no possibility of indifference to the security of India itself. Yet, the forces available were wholly inadequate for either mission, and they necessarily served chiefly as a token of intent. Changes thus tended to acquire a significance altogether out of proportion to the forces involved.[73]

NEW GUINEA - SOLOMON ISLANDS

Admiral Chester Nimitz ordered Admiral Frank J. Fletcher to take the carriers *USS Lexington* and *USS Yorktown* to the Solomon Islands. Once there, he was to refuel and then begin to "operate in the Coral Sea commencing 1 May."[74] (see Figure 33)

PUBLIC OPINION

On the domestic front, having a Germany First strategy was not the majority opinion of Americans when they were asked

> *Granting that it is important* for us to fight the Axis every place we can, which do you think is more important for the United States to do right now: put most of our effort into fighting Japan or put most of our effort into fighting Germany?

Sixty-two percent responded Japan, and 21 percent Germany.[75]

HEAVY BOMBER PRODUCTION

During April, 171 heavy bombers were manufactured. Eighty were B-24s.[76]

Figure 31 – Burma – April 1942

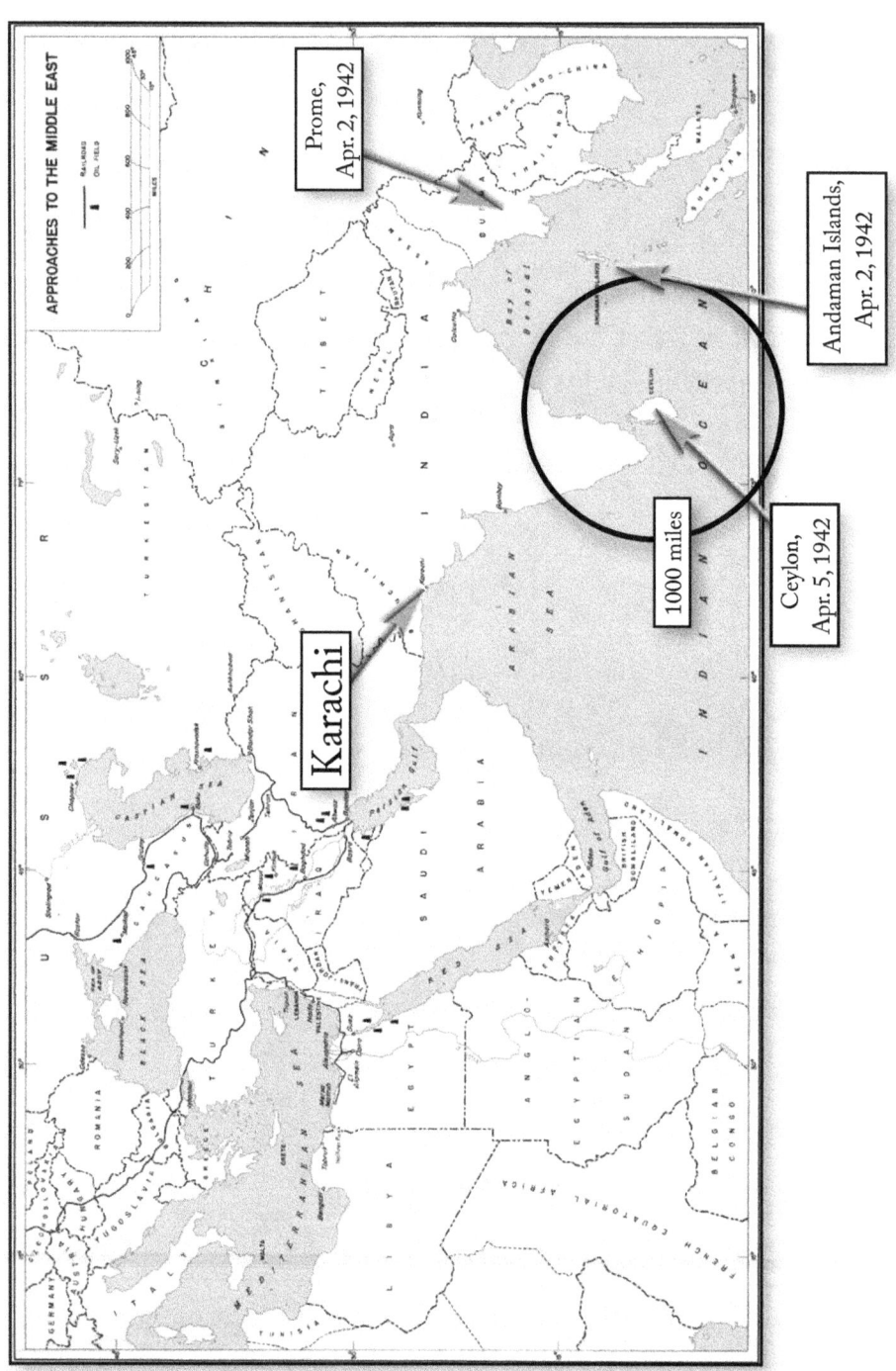

Figure 32 – India – April 1942

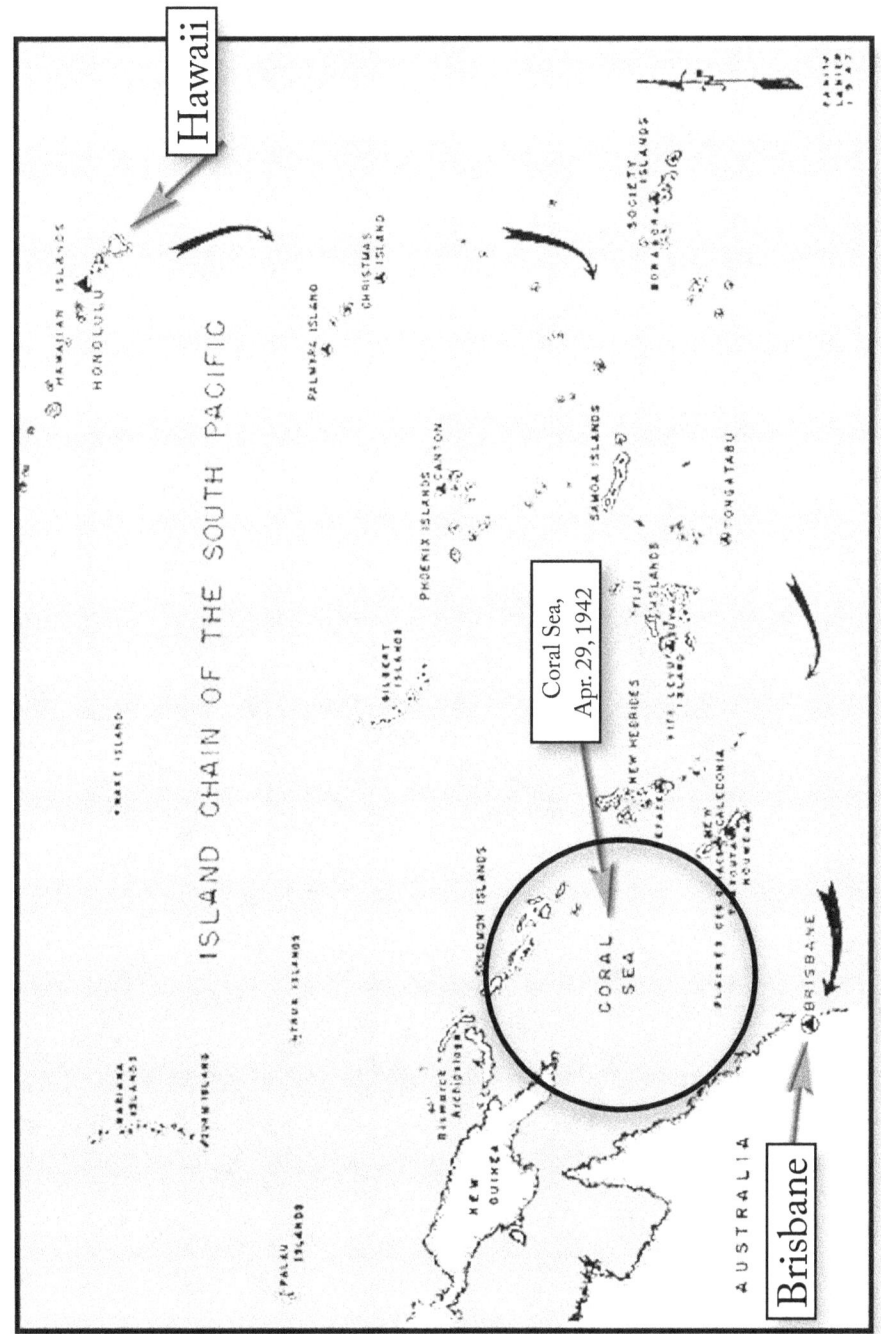

Figure 33 – Supple Chain – April 1942

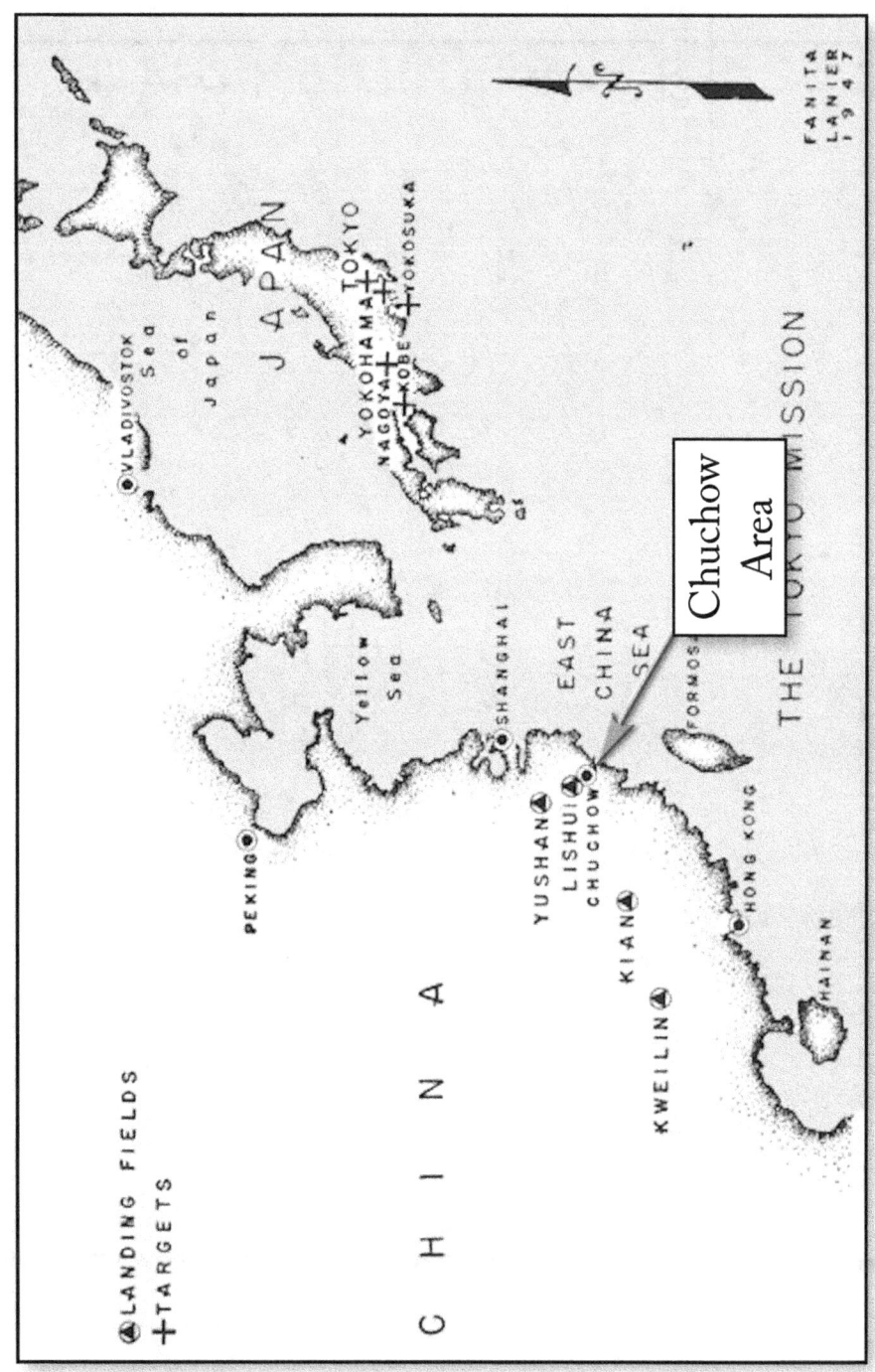

Figure 34 – Doolittle Landing sites – April 1942

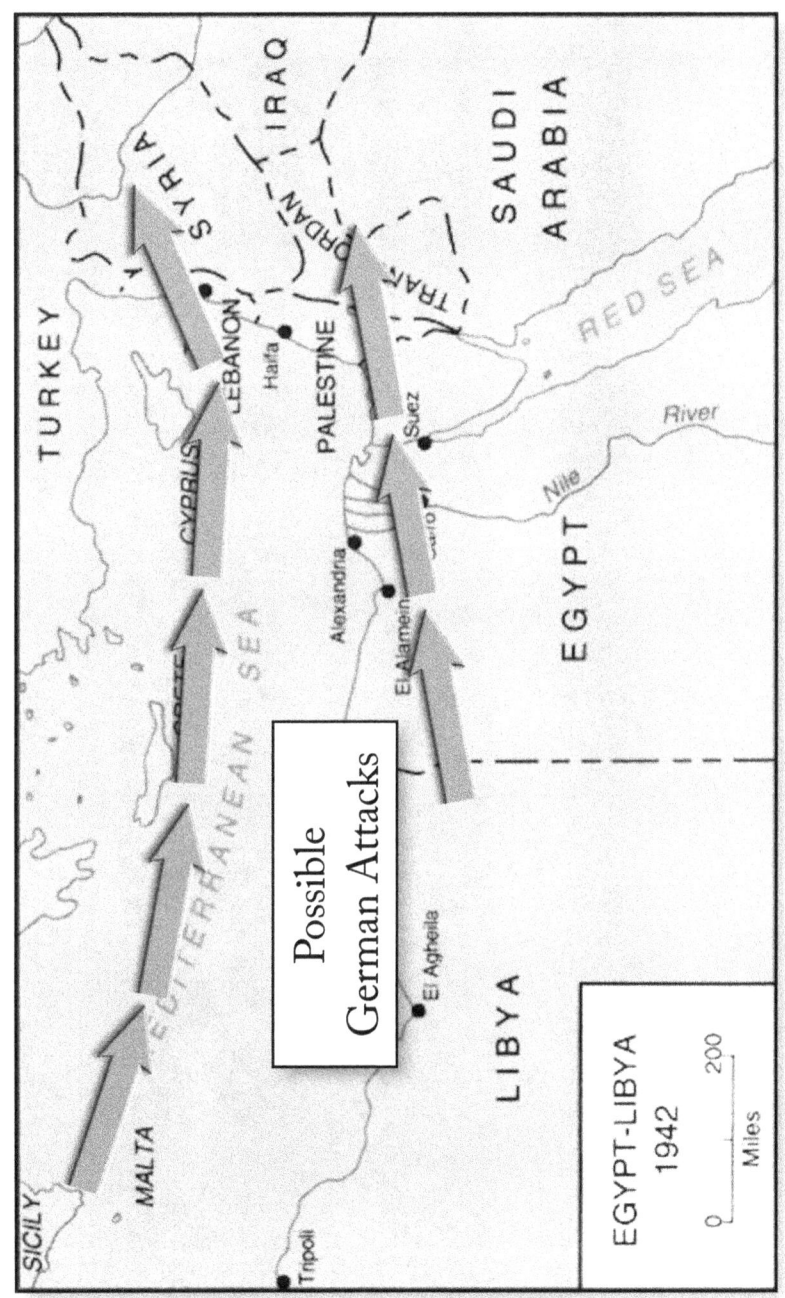

Figure 35 – Where the Powers Prepare for Spring – April 1942

Figure 36 – *Hellzapoppin* #23 and crew

ENDNOTES

1. ----, AFHRA, Hap Arnold file, Twining to Craig, April 1, 1942
2. Walker, *The Liberandos*; Richardson, *The Forgotten Force*, p. 16.
3. ----, Military Times Hall of Valor, http://valor.militarytimes.com/recipient.php?recipientid=676 , accessed August 11, 2015.
4. Williams, *Chronology 1941-1945 - 1942*, p. 31.
5. Craven, *The Army Air Forces in World War II, Vol. 1*, p. 503.
6. Xu, *The Issue of US Air Support for China*, p. 461.
7. ----, AFHRA, Hap Arnold file, Arnold to April 2, 1942.
8. Lowenthal, Mark M., *Roosevelt and the Coming War: The Search for United States Policy 1937-42*, p. 431.
9. Craven, *The Army Air Forces in World War II, Vol. 1*, p. 440.
10. Walker, *The Liberandos*.
11. ----, AFHRA, Hap Arnold file, Arnold to Brereton, April 4, 1942.
12. Assistant Chief of Air Staff, *The Ploesti Mission*, p. 13.
13. ----, *Operation 'Rutter'*
14. Williams, *Chronology 1941-1945 - 1942*, p. 32.
15. Craven, *The Army Air Forces in World War II, Vol. 1*, p. 440.
16. ----, FDRL, PSF, Box 3, India file, Johnson to Hull, April 7, 1942.
17. ----, FDRL, PSF, Box 3, India file, Arnold to FDR, April 9, 1942.
18. Ben-Moshe, *Winston Churchill and the "Second Front*, p. 512.
19. Cave, *transcript of interview*.
20. Walker, *The Liberandos*.
21. Cline, *United States Army in World War II, The War Department*, pp. 155-157.
22. Ben-Moshe, *Winston Churchill and the "Second Front*, p. 511.
23. Ben-Moshe, *Winston Churchill and the "Second Front*, p. 510.
24. Williams, *Chronology 1941-1945 - 1942*, p. 32.
25. ----, *New York Times, JAPANESE CAPTURE BATAAN*, April 10, 1942, http://www.freerepublic.com/focus/chat/2870163/posts, accessed February 18, 2014.
26. World War II Database, http://ww2db.com/ship_spec.php?ship_id=10
27. Williams, *Chronology 1941-1945 - 1942*, p. 32.
28. Assistant Chief of Air Staff, *The Ploesti Mission*, p. 13.
29. ----, FDRL, PSF, Box 2, China file, Acting Chief to FDR, April 16, 1942.
30. ----, FDRL, PSF, Box 83, Marshall file, McNarney to FDR, April 12, 1942.
31. ----, FDRL, PSF, Box 2, China file, Madame CKS to Laughlin Currie, April 12, 1942.
32. Thomas, *Born in Battle*, pp. 56, 64.
33. ----, FDRL, PSF, Box 2, China file, Currie to FDR, April 13, 1942.
34. ----, AFHRA, Hap Arnold file, McNarney to FDR, April 13, 1942.

35. ----, *Time, General Yang*, April 13, 1942, http://www.time.com/time/magazine/article/0,9171,766466,00.html , accessed July 29, 2012.
36. Williams, *Chronology 1941-1945 - 1942*, p. 33.
37. Cline, *United States Army in World War II, The War Department*, p. 159.
38. Ben-Moshe, *Winston Churchill and the "Second Front*, p. 511.
39. ----, FDRL, PSF, Box 3, Harry Hopkins file, Mathews to Hull, April 14, 1942.
40. ----, AFHRA, Hap Arnold file, Arnold to FDR, Arnold to Kuter, April 14, 1942.
41. Thomas, *Born in Battle*, p. 67.
42. Craven, *The Army Air Forces in World War II, Vol. 1*, p. 503.
43. Story, *Private Story*.
44. ----, FDRL, PSF, Box 2, China file, Acting Chief to FDR, April 16, 1942.
45. ----, FDRL, PSF, Box 98, Carter file, Carter to file, April 16, 1942.
46. Craven, *The Army Air Forces in World War II, Vol. 1*, p. 564.
47. Craven, *The Army Air Forces in World War II, Vol. 1*, p. 440.
48. ----, FDRL, Map Room (MR), Arnold to USFOR, April 17, 1942.
49. ----, FDRL, MR, Box 39, ALUSNA to OPNAV, April 17, 1942.
50. Assistant Chief of Air Staff, *The AAF in the Middle East*, p. 70; Assistant Chief of Air Staff, *The Ploesti Mission*, p. 14.
51. Holloway, letters to the author.
52. ----, *Operation 'Rutter'*.
53. ----, FDRL, PSF, Box 28, Currie In file, Currie to Segac, April, 20, 1942.
54. Walker, *The Liberandos*.
55. Craven, *The Army Air Forces in World War II, Vol. 1*, p. 504.
56. Steele, *American Popular Opinion*, p. 720.
57. Assistant Chief of Air Staff, *The Ploesti Mission*, p. 14.
58. Mayhew, *transcript of interview*.
59. ----, "The "Kangaroo Squadron," http://www.ozatwar.com/usaaf/435th.htm , accessed August 11, 2015.
60. ----, FDRL, MR, Box 168, McCrea to Arnold, April 22, 1942, Arnold to McCrea, April 23, 1942.
61. ----, FDRL, PSF, Box 2, Chiang file, FDR to Soong, April 24, 1942.
62. ----, AFHRA, Bonner Fellers file, Fellers' cable to G2, April 24, 1942.
63. ----, *New York Times, BRITISH ARE READY TO BOMB RUMANIA*, January 20, 1941, http://select.nytimes.com/gst/abstract.html?res=FB0D12F9345D1B7B93C2AB178AD85F458485F9, accessed July 29, 2012.
64. Craven, *The Army Air Forces in World War II, Vol. 1*, p. 503.
65. ----, AFHRA, Hap Arnold file, Arnold to Air Marshall DCS Evill.
66. ----, FDRL, MR, Box 39, ALUSNA to OPNAV, April 25, 1942.
67. Williams, *Chronology 1941-1945 - 1942*, p. 35.
68. Stimson, *Stimson Diaries*, entry for April 27, 1942.
69. ----, AFHRA, Bonner Fellers file, Hammond to C of S, April 28, 1942.

70. Cave, *transcript of interview*.
71. Williams, *Chronology 1941-1945 - 1942*, p. 35.
72. ----, FDRL, PSF, Box 2, China Folder, Marshall to FDR, April 29, 1942.
73. Craven, *The Army Air Forces in World War II, Vol. 1*, pp. 503-504.
74. Morton, *Strategy And Command: The First Two Years*, p. 276.
75. Steele, *American Popular Opinion*, p. 706.
76. ----, Army Air Force Statistical Digest, http://www.ibiblio.org/hyperwar/AAF/StatDigest/aafsd-3.html , accessed January 22, 2014, p. 112; B-24 production from Author's personal papers.

MAY 1942

The month of May would set the stage for the major events of the rest of 1942.

HALPRO would leave Florida for bases in China, which is where the crews anticipated they were going. Despite the "moral victory" of the Doolittle raid, Roosevelt continued to press for more bombing raids on Japan.

The Germans began their summer offensive. The Caucasus oilfields appeared to be the objective.

Admiral Nimitz combined two of his carriers into a carrier task force. They would engage the Japanese at the battle of the Coral Sea. Meanwhile, United States intelligence reported that a significant amount of Japanese message traffic was being read.

Rommel continued having success in North Africa. The British withdrew to Gazala, (see Figure 5) on the Libyan coastline, roughly 40 miles from Tobruk. The border with Egypt was less than 100 miles away.

Realizing that the Burma Road was the only means of supplying bulk quantities of materiel to China, the Japanese had been aggressively trying to sever the road. In late April, they captured Lashio, a major terminus on the Burma Road, forcing Stilwell to begin his infamous march out of Burma. But more importantly, there was now no way to supply Chinese forces over land. (see Figure 31)

1942 MAY 1

BURMA

The Japanese captured Monywa.[1] (see Figure 37)

SOUTH PACIFIC

The Japanese began their assault of Tulagi.[2] (see Figure 39)

PLOESTI

Fellers continued his message barrage on Washington, arguing that the United States must furnish the planes to bomb Ploesti. He suggested that the bombers use an RAF base at Khartoum, and stage at forward RAF bases for any mission. The Khartoum base would also allow for the bombers to defend eastern approaches to the Suez as well as an

Axis forays into northwest Africa. Fellers also suggested that if Ploesti was the target, the bombers could use a Russian base at at Taganrog.[3]

HALPRO

The size of the HALPRO complement was stabilized at eighty-two officers, three warrant officers, and one hundred twenty five enlisted men.[4]

Despite the whipsawing by upper management regarding their final destination and objective, the HALPRO crews still thought they were on a mission to bomb Japan. Hawk Cave wrote:

> *Went back to Ft. Myers* on the first day of May 1942 and flew to Wright Patterson Field to get overseas equipment. Took out the ball turrets in the belly since they didn't work properly and put a piece of metal skin under the bottom and filled the hole with gear. The B-24 was made for 48,000 lbs and we carried 67,000 lbs. when we left the States. They put in two 500 gal. rubber gas tanks and one in the right and one in the left side of the front bomb bays. I knew then we were in for some long flights to the unknown. Each of the 23 B-24s had two extra ground crew members to help take care of the aircraft and bombsights etc. We learned after leaving Florida that we were heading for China to make three raids on Tokyo and come home.[5]

1942 MAY 2

NEW GUINEA - SOLOMON ISLANDS

A Japanese invasion force was discovered approaching Tulagi Island. Allied defenders destroyed the installation and retreated to New Hebrides.[6] (see Figure 39)

MIDWAY

Admiral Nimitz visited the island to inspect defenses.[7] (see Figure 10)

1942 MAY 3

THE PHILIPPINES

Marshall was on an inspection trip to the southern United States when his staff informed him of Roosevelt's decision to send another 25,000 troops to MacArthur. Although he was respectful, Marshall resented that he was not consulted prior to the decision. Marshall rattled off a series of reasons why the decision should be reviewed:

1. Marshall was embarrassed during his early April trip to England when he could only promise two and one half divisions for Sledgehammer by September 15.
2. He opposed sending troops to Australia as it would commit vitally needed transports to a long sea voyage.
3. He preferred sending planes to Alaska or Hawaii rather than Australia.[8]

NEW GUINEA - SOLOMON ISLANDS

The Japanese invasion force seized Tulagi Island and made it into a seaplane base. Admiral Fletcher had taken the United States carriers *USS Lexington* and *USS Yorktown* on patrol near the Coral Sea. They were accompanied by two United States oil tankers, *USS Neosho* and *USS Tippecanoe* and six Australian cruisers, with two more Australian cruisers were enroute. Hearing the news of the assault on Tulagi, he ordered the Task Force to steam there.[9] (see Figure 39)

FAR EAST

The situation in India presented an interesting political and military dilemma for the United States. Gandhi and his movement viewed the British as occupiers. If the Japanese forced the British out of India, that would be fine with them. So, the more the United States appeared to be supporting the British, the more Tokyo worked that feeing into its propaganda. (See *The New York Times* article *Tokyo Radio Tries To Win India's Aid*,[10] published May 3, 1942)

1942 MAY 4

BURMA

The Allies evacuated Akyab. (see Figure 29) The AVG withdrew from Loiwing for Kunming, China.[11] (Kunming is not on Figure 29. It is to the east of Ft. Hertz, well into China.)

NEW GUINEA - SOLOMON ISLANDS

Aircraft from the *USS Yorktown* attacked the Japanese forces at Tulagi. After the attack, the *USS Yorktown* sailed away to rejoin the *USS Lexington*.[12] (see Figure 39)

HALPRO

Eighteen planes were already on hand. Five HALPRO B-24Ds arrived from the Mobile Modification Depot.[13]

COMMITMENT TO CHINA

In some respects, Stilwell's message to Marshall had some effect, as Roosevelt must have asked Arnold about ferrying supplies into China. Arnold responded on May 4. In essence, there were two routes. The existing route was about to be lost to the Japanese. The second had just been surveyed, but involved flying over mountain ranges, some up to 21,000 feet. During the rainy season from May 15 thru October 15, passage was hazardous at best.[14] "Flying the Hump" was about to begin.

1942 MAY 5

BURMA

Following the capture of Lashio by the Japanese, Stilwell began his march out of Burma. (see Figure 31) He originally headed to Myitkyina. Enroute, he learned that the Japanese were in Bhamo. He changed his destination to India.[15] (see Figure 37)

COMMITMENT TO CHINA

Apparently, Roosevelt did not like the answer he got from Arnold. In a rather terse message to Arnold, he stated:

> *I gather that the air* ferry to China is seriously endangered. The only way we can get certain supplies into China is by air.
>
> I wish you and Mr. Lovett would confer immediately with Dr. Soong and General Shen on alternative air routes.
>
> I want you to explore every possibility, both as to airplanes and routes. It is essential that our route be kept open, no matter how difficult.[16]

Arnold could read between the lines. He knew he was being asked to solve the problem. On the same day, he then told two of his staff members, Cols. Charles P. Cabell and Lauris Norstad, to "Let your imagination run wild and give me an answer – by the following Thursday."

Arnold suggested other ways to get aid to China by air,

- Through Nome, Alaska and eastern Siberia, or
- By way of Eastern India or
- Around the High Tibetan plateau, or
- Possibly through Basra in the Persian Gulf.[17]

May 5 was a Tuesday. They had two days to figure something out.

HALPRO
Another HALPRO B-24D arrived from the Mobile Modification Depot.[18]

1942 MAY 6

THE PHILIPPINES
General Wainwright surrendered the Corregidor garrison.[19] The front page of the late edition of the May 6 edition of *The New York Times* screamed:

> CORREGIDOR SURRENDERS UNDER LAND ATTACK
> AFTER WITHSTANDING 300 RAIDS FROM THE AIR [20]

GLOBAL STRATEGY
Roosevelt took umbrage with Marshall's May 3 chastisement of his "decision" about sending men and material to Australia. He responded on May 6, stating

> *I did not issue any* directive on May first regarding the increase of combat planes ... and ground troops. I did ask if it could be done. I understand now that this is inadvisable...
> I do not want "Bolero" slowed down.[21]

On Christmas Eve last, Roosevelt had made commitments to Churchill. He was then forced to retract them the next day when his advisors challenged him. He had a pattern and Marshall called him on it.

OPERATION TORCH
Apparently, Roosevelt must have decided that there was something to Marshall's basic complaint of a decision being made without an overarching plan. So, he issued his first policy statement on the conduct of the war. He circulated it to Secretary of War Stimson, Chief of Staff Marshall, General Arnold, Secretary of the Navy Knox, Admiral King and Harry Hopkins. The recipients were under strict orders to not discuss the memo with anyone other than the addressees.

In the first two pages, he outlined strategies for all of the major theaters. He then took to task those advisors, both American and British, who opposed actions before 1943:

> *I regard it as essential* that active operations be conducted in 1942.
> If we decide that the only large scale offensive operation is to be in the European area, the element of speed becomes the first essential.[22]

Roosevelt wanted the Joint Chiefs to pick an operation in any theater, but it had to be in 1942. And if it was in Europe, it had to be immediate. Roosevelt had just taken the first step towards Operation TORCH.

SOUTH PACIFIC

The Japanese continued into the Coral Sea. (see Figure 39) Both the Japanese and American fleets continued westward, both apparently unaware of the presence of each other.

In support of the American fleet, B-17s of the 435th Squadron flew reconnaissance missions. They spotted the Japanese fleet and made bombing runs on a carrier and a heavy cruiser.[23]

1942 MAY 7

COMMITMENT TO CHINA

Arnold responded to Roosevelt's order to "keep the supply lines to China open".

1. There were three possible routes for getting supplies to China.

   - The first was the existing route, but was very close to Japanese held territory. The other two made use of Russian air space. One of the Russian routes lacked sufficient facilities, and was not recommended. Both alternatives would require Russian permission.

2. Fifty B-24s could be used as transports. It would take ten days to modify these bombers to cargo planes, which could carry about 1200 tons per month.[24]

SOUTH PACIFIC

Both the American and Japanese Coral Sea fleet commanders suspected the other fleet was nearby. (see Figure 39) They sent out scout planes. About 8 a.m., the Japanese spotted the *USS Neosho* and her escort, the destroyer *USS Sims*. The Japanese launched two separate attacks; the second one sank the *Sims* and left the *Neosho* a drifting wreckage.

Meanwhile, United States scout planes found the Japanese carrier *Shoho* and four cruisers. The resulting strike force sank the *Shoho*.

As a result, the Japanese ordered the landing forces to withdraw, pending further developments.

PLOESTI

United States intelligence officers in Cairo continued sending their weekly reports to Washington. The May 7 report estimated that Germany had 15,000 tons of oil reserves. There was a pipeline from Ploesti to the loading docks at Constanza, a port on the Black Sea.[25]

1942 MAY 8

BURMA

The Japanese seized Myitkyina.[26] (see Figure 37)

NEW GUINEA – SOLOMON ISLANDS

The Battle of the Coral Sea (see Figure 39) occurred. Both the Japanese and American fleets launched scout planes before dawn. Both scouts found the enemy, but the Japanese were the first to get their aircraft airborne. By 9 a.m., the American aircraft had been launched. Both sets of aircraft attacked the other fleet by 11 a.m.. The Japanese carrier *Shokaku* was severely damaged and had to return to Japan for repairs. The carriers *USS Lexington* and *USS Yorktown* were also both severely damaged. The *Lexington* would eventually be scuttled; the *Yorktown* limped back to Pearl Harbor.

This was the first battle in naval history where the two opposing forces never saw one another. The main battle took place between aircraft launched from the opposing carriers.

While the battle was a tactical victory for the Japanese, it was a strategic success for the Allies. First, it stopped the Japanese advance in the southern Pacific at the Coral Sea. Second, two Japanese carriers, one sunk and one damaged, would not be operational for the Battle of Midway. Instead of having six Japanese carriers, the Battle of Midway would pit four Japanese carriers against three American carriers, since the *USS Yorktown* would be repaired.

EASTERN FRONT

The German Army Group South began a drive to reclaim Kerch.[27] (see Figure 38)

HALPRO

One of the HALPRO B-24 crashed during testing.[28] They now had 23 planes.

1942 MAY 9

COMMITMENT TO CHINA

Arnold was not particularly satisfied with the answers he had received from Cabell and Norstad. They had recommended continued use of the original air route, which Arnold had passed on to Roosevelt. Now, the Japanese were close enough to the route to seriously jeopardize those ferry flights. He needed another answer.

So, Arnold asked a similar question of Howard Craig, the head of the Air War Planning Division. Arnold knew that the only viable alternatives involved the use of Russian bases, a condition which Russia would object in order to avoid being dragged into the Pacific War. Even though he knew he would likely receive a rejection, Craig still made a visit to Russian General Alexander I. Belyaev, head of the Soviet Supply Mission to the United States. Anticipating Belyaev's rejection of his request, Arnold told Craig, "You will have to find ways and means of combating that line of argument."[29]

Arnold also made a request of another member of his staff, Col. Hoyt S. Vandenburg. He wanted Vandenburg to address the issue of the relative value of 50 B-24 bombers and to clarify if they were going to be used as cargo planes or bombers. Interestingly, Arnold suggested the answer he was looking for should agree with the desire of the president for the bombers to be used as cargo planes. So this should be the conclusion. But he did want Vandenburg to say that half of the 50 bombers should come from the American allotment and the other half should come from the British allotment.[30] In other words, he wanted the British to share the burden.

PLOESTI

India was not the only theater where differences of opinion existed. Ploesti was another. Roosevelt had already expressed an interest in Ploesti, so the various planning departments were asked to evaluate a mission to Ploesti. Air WPD (AWPD) was in favor of assigning planes for the bombing of Ploesti. The Strategy Section in OPD had objected to it, "due to other commitments."[31]

DIEPPE

The main components of Operation RUTTER were now finalized and approved by the British High Command. The operation was scheduled to take place between July 4 and 8.[32]

1942 MAY 10

HALPRO

The last B-24 from the Mobile Depot was delivered on May 10. That delivery was not without its own excitement. Scotty Royce and Major Ed Gavin had decided to fly it to Fort Myers. This was Royce's recollection of the flight:

Finally, in May, the last B-24 was finished and Ed Gavin and I flew it to Fort

Myers. I'll always remember that flight because we flew straight across the Gulf of Mexico, and right in the middle of our journey I was in the co-pilot's seat flying the airplane. Ed Gavin had gone back into the bomb bay to familiarize himself with the new fuel transfer system. He somehow had routed fuel from the manifold to all of the engines. He then connected the manifold to the bomb bay tank, which was empty. Thereupon, all four engines ran out of gas simultaneously, and there was an ominous silence.

I put the ship in a glide toward the water about two-thousand feet below, and Ed came charging up to the flight deck wanting to know what I had done. I told him I hadn't done anything, and directed the same question to him with increasing anxiety. He disappeared back into the bomb bay and started throwing valves and switches, and soon the engines roared back to life.[33]

1942 MAY 11

PLOESTI

Ray Smith, chairman of the Inter-Divisional Oil Commission of the Bureau of Economic Warfare,

declared that Ploesti was deserving of the highest target priority among all United Nations objectives. Since Germany's production of substitute and synthetic fuels was steadily on the increase, the sooner the attack was made the more damaging it would be to Axis oil economy. He recommended that the most urgent consideration be given to the immediate carrying out against Ploesti of the plan of attack, which has already been elaborated by our military authorities.[34]

COMMITMENT TO CHINA

Brereton noted the status of the Chinese airfields. His comments were not very positive from an Allied perspective:

Reconnaissance reports that Myitkyina airfield is in enemy hands. Our morning ferry to Kunming, three transports were turned back by Chennault because Kunming was heavily bombed this morning. Yunanyi cannot be used, and Laoshan is believed to be in Jap hands. If Chennault withdraws to Chungking for the protection of the capital, our ferry line over "the Hump" is out. Orders were issued for the 9th Heavy Bomb Squadron to hit Myitkyina with everything they have to prevent its use and increase the security of Dinjan.[35]

GLOBAL STRATEGY

The military necessity of keeping Russia in the war was obvious, and the potential reality that Hitler would strike again on the Eastern Front was equally obvious. The writers in *Time* were not blind to what might happen. In an article titled:

FOREIGN RELATIONS: TOUGH BABY FROM MOSCOW

the author wrote:

> The nation's greatest concern was far across the seas. Russia must not be allowed to fall if U.S. aid can prevent it. Russia's fall would turn loose on Asia and Africa a terrific Nazi army, an army of millions of men, thousands of planes and thousands of tanks, an army big enough to fight on a 2,000-mile front-as it is now doing.
>
> That was why last week U.S. eyes & ears were fixed on that great stretch of unknown land thousands of miles away; a land few Americans had ever seen, and whose place names few Americans could pronounce. On the reaches of Russia, from the Barents Sea to the Black Sea, World War II would be very nearly decided. Ahead lay six months of good military weather. In those six months probably lay the great decisions of the 20th Century.
>
> Throughout the United Nations, suspense mounted. All winter long, retreating here & there, Hitler and his troops had endured, had waited, had piled up strength for the decisive battle. Now spring spread northward in ever-widening circles. The zero hour was hard by. Hitler had feinted with one of the "peace offensives" which are always his last step before war of nerves turns to war of gunpowder.
>
> Only days, hours, minutes of this ticking silence - and the blow would fall in the greatest cataclysm of blood and steel the world had ever borne.
>
> Hitler has to destroy the Russian Army in 1942 or lose the war. And the U.S. has to keep Russia fighting or face a war that will be immeasurably longer and tougher.[36]

1942 MAY 12

EASTERN FRONT

With the Germans focused on Kerch, the Russians open up a campaign against Kharkov.[37] (see Figure 38)

AAF TASK FORCE BR

The first B-17s arrived at their base in England. They were part of the 97th Bomb Group. Much to the dismay of Arnold, it would take three more months for it to fly its first mission.

1942 MAY 13

HALPRO

Testing of HALPRO planes was completed. They were readied for departure.[38] All 23 HALPRO planes were parked on the ramp at Fort Meyers.[39]

1942 MAY 14

COMMITMENT TO CHINA

Eisenhower inherited the task given to Vandenburg, who was asked to evaluate the proposal to convert 50 B-24s into cargo variants. In his answer, Eisenhower made the observation that there were not enough planes to adequately supply the Chinese. Thus, this proposal was more for moral support. Since the 50 B-24s would have to come from the buildup of Allied forces in England, he asked: whether these B-24s would be more useful bombing Germany or delivering a morale boost to the Chinese.

Eisenhower also proposed delaying the departure of HALPRO until they found out from Stilwell and Brereton the possibilities of using this kind of plane in India during the next summer. He argued that the B-24s should be assigned to whatever project might appear best:

- For the defense of India or
- For carrying out the original HALPRO, or
- For the buildup in England.

If HALPRO was to be sent to India, Eisenhower wanted the British to be informed that this allotment, without replacement, "represents the entire bomber production that can be lent to India and China."

The British were informed that, henceforth, other than maintaining the strength of the 10th Air Force, the buildup of the CBI would be entirely a British responsibility.[40]

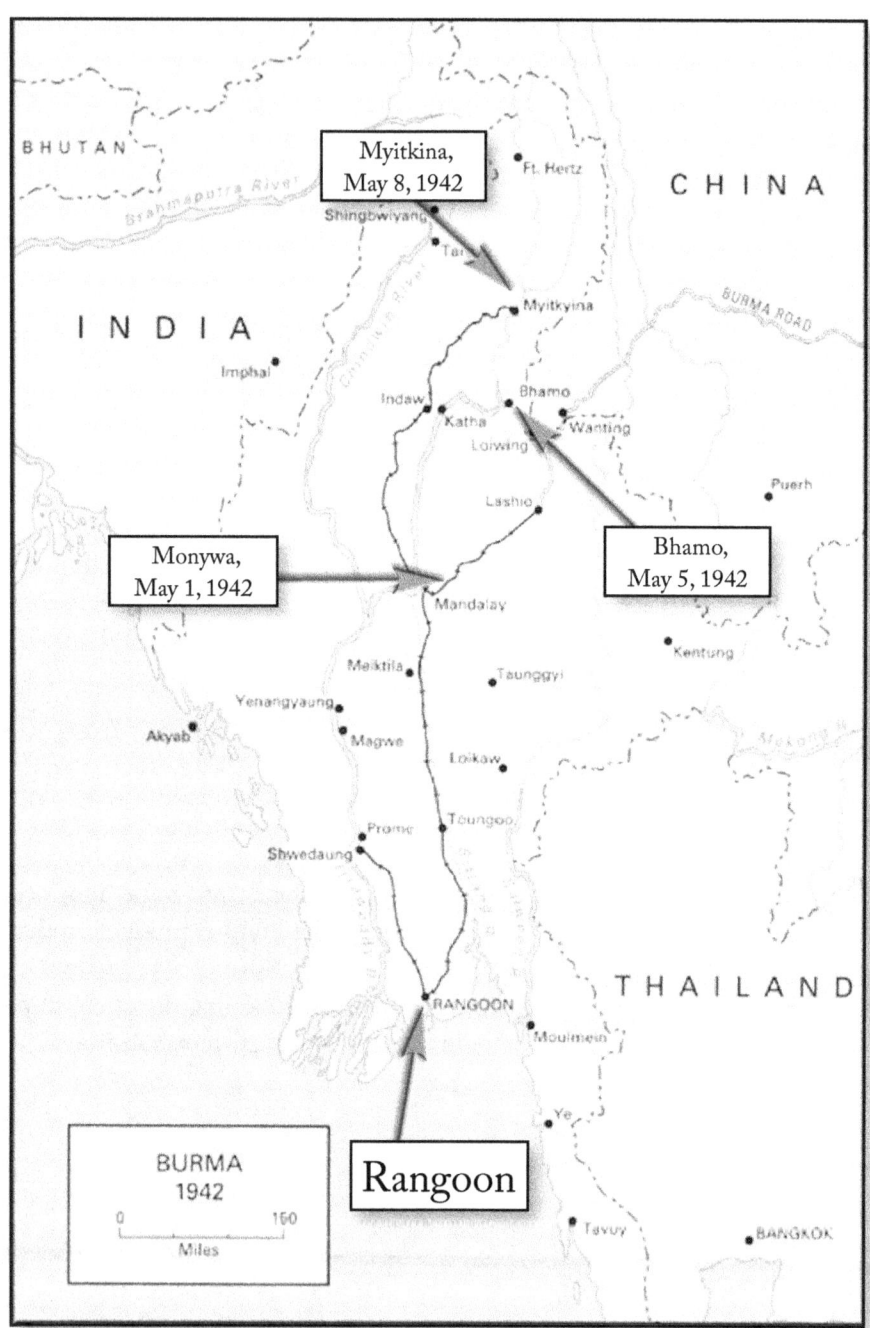

Figure 37 – Burma – May 1942

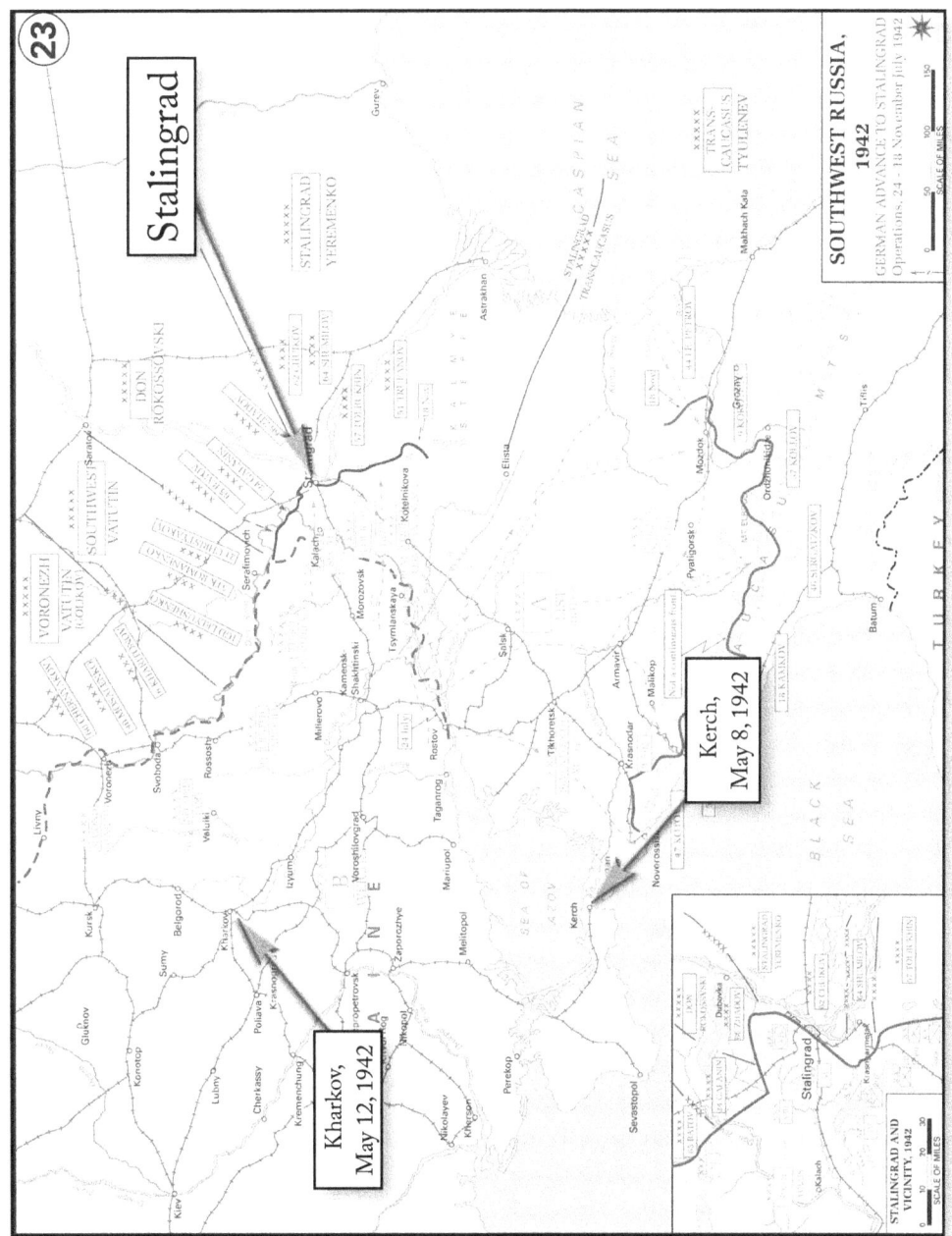

Figure 38 – Stalingrad – May 1942

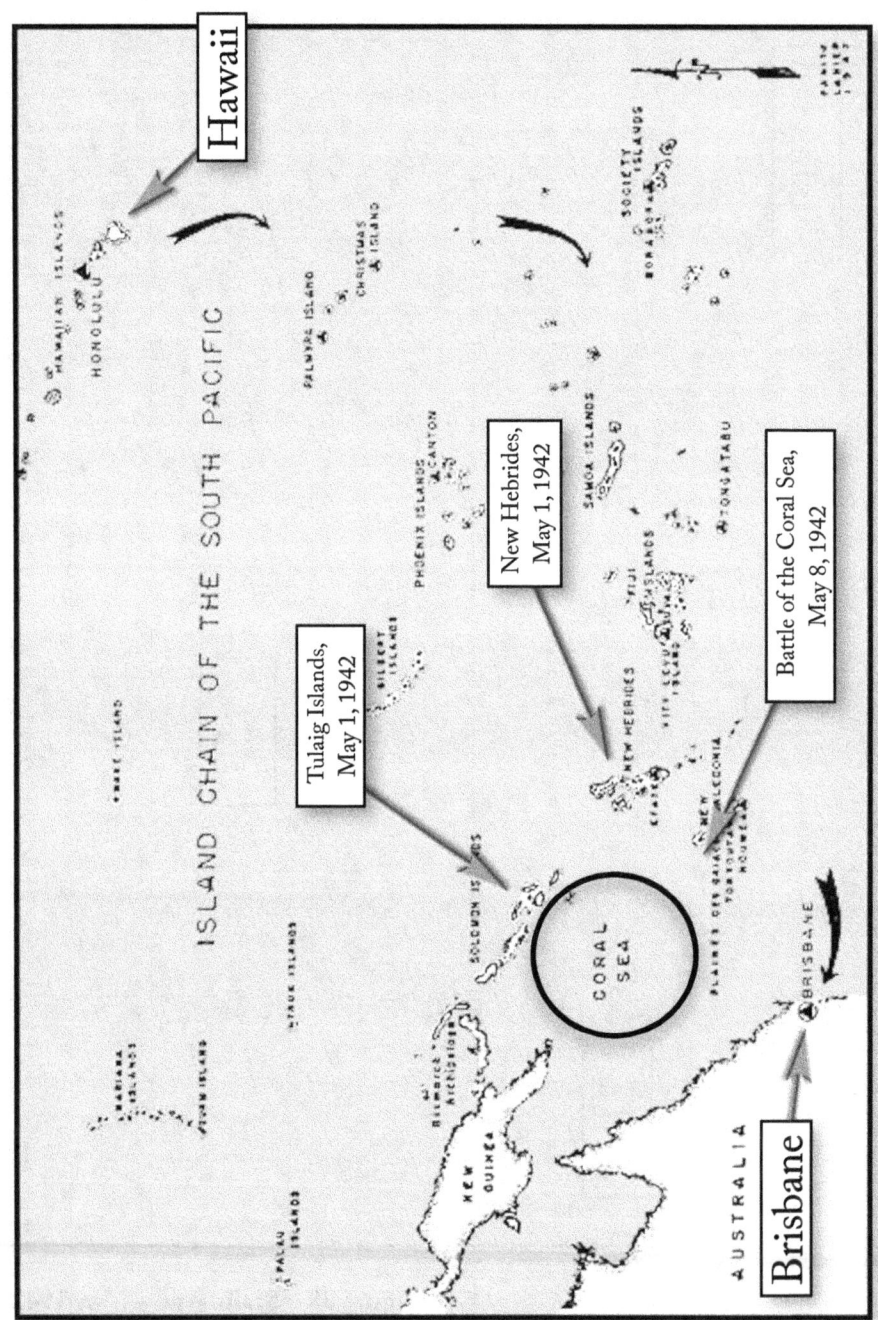

Figure 39 – Supply Chain – May 1942

ENDNOTES

1. Williams, *Chronology 1941-1945 - 1942*, p. 35.
2. ----, *US Marine Corps in World War II, Guadalcanal Campaign*, http://www.ibiblio.org/hyperwar/USMC/Guadalcanal/USMC-M-Guadalcanal-1.html , p. 2.
3. ----, FDRL, MR, Box 39, Fellers to G- 2, May 42, 1942.
4. Richardson, *Forgotten Force*, p. 17.
5. Cave, *transcript of interview*.
6. Williams, *Chronology 1941-1945 - 1942*, p. 36.
7. Williams, *Chronology 1941-1945 - 1942*, p. 36.
8. ----, FDRL, PSF, Box 4, Marshall, George C., 4/14/42 - 1944 file, Marshall to FDR, May 1, 1942.
9. Williams, *Chronology 1941-1945 - 1942*, p. 36.
10. Kluckhorn, Frank, *New York Times, Tokyo Radio Tries to Win India's Aid*, May 3, 1942, http://select.nytimes.com/gst/abstract.html?res=F50F10FB3A5E167B93C6A9178ED85F468485F9, accessed August 1, 2012.
11. Williams, *Chronology 1941-1945 - 1942*, p. 36.
12. Williams, *Chronology 1941-1945 - 1942*, p. 36.
13. ----, Daily Status Report, May 4, 1942.
14. ----, FDRL, PSF, Box 2, China Folder, Arnold to FDR, May 4, 1942.
15. Williams, *Chronology 1941-1945 - 1942*, p. 36.
16. ----, FDRL, PSF, Box 2, China Folder, FDR to Arnold, May 5, 1942.
17. ----, AFHRA, Hap Arnold file, Arnold to Cabell and Norstad, May 5, 1942.
18. ----, Daily Status Report, May 5, 1942.
19. Williams, *Chronology 1941-1945 - 1942*, p. 37.
20. ----, *New York Times, CORREGIDOR SURRENDERS UNDER LAND ATTACK* , May 6, 1942, http://query.nytimes.com/gst/abstract.html?res=9801E4DB153CE33BBC4F53DFB3668389659E-DE, accessed February 19, 2014. .
21. ----, FDRL, PSF, Box 4, Marshall, George C., 4/14/42 - 1944 file, FDR to Marshall, May 6, 1942.
22. ----, FDRL, PSF, Box 83, Marshall file, FDR to JCS, May 6, 1942.
23. ----, "The "Kangaroo Squadron," http://www.ozatwar.com/usaaf/435th.htm , accessed August 11, 2015.
24. ----, FDRL, PSF, Box 2, China Folder, Arnold to FDR, 7 May 1942.
25. ----, FDRL, MR, Box 39, ALUSNA to OPNAV, May 7, 1942.
26. Williams, *Chronology 1941-1945 - 1942*, p. 37.
27. Williams, *Chronology 1941-1945 - 1942*, p. 37.
28. ----, Daily Status Report, May 8, 1942.
29. ----, AFHRA, Hap Arnold file, Arnold to Craig, May 9, 1942.
30. ----, AFHRA, Hap Arnold file, Arnold to Vandenburg, May 9, 1942.
31. ----, AFHRA, Hap Arnold file, Nevins for Chief, S&P Croup, 9 May 42; OPD 381 Africa.

32. ----, *Operation 'Rutter'*.
33. Walker, *The Liberandos*.
34. Assistant Chief of Air Staff, *The Ploesti Mission*, pp. 14-15.
35. Brereton, *The Brereton Diaries*, p. 123.
36. ----, *Time, FOREIGN RELATIONS: Tough Baby from Moscow*, May 11, 1942, http://www.time.com/time/magazine/article/0,9171,790399.00.html, accessed November 29, 2009.
37. Williams, *Chronology 1941-1945 - 1942*, p. 37.
38. ----, Daily Status Report, May 12, 1942.
39. ----, Daily Status Report, May 13, 1942.
40. ----, AFHRA, Hap Arnold file, Eisenhower to McNarney, May 14, 1942.

REALITY OVERCOMES THE BEST LAID PLANS

MAY 1942

1942 MAY 15

BURMA

General Alexander moved his headquarters to Imphal, India. Meanwhile, Stilwell completed his exodus from Burma and arrived in India.[1] (see Figure 40)

NORTH AFRICA

The assessment of the Arnold's Assistant Chief of Air Staff was:
Since Rommel's divisions had succeeded in breaking through their defenses in Cyrenaica, it was obvious that the main British forces would be obliged to withdraw to the Egyptian frontier in order to avoid encirclement. Heavy bombers were therefore badly needed to slow down the German pursuit and to stave off the collapse of the Middle East.[2] (see Figure 41)

EASTERN FRONT

The Germans captured Kerch, and the Russians continued against Kharkov (see Figure 38).[3]

PLOESTI

Based upon Fellers' April 22 memo, Arnold asked the WPD to study the impact of a raid on Ploesti. They released their study with the title "Strategic Targets within Range of Middle East Bases." Their major findings were:

- Rumanian oil produced in or near Ploesti amounted to six million tons a year of refined products, which was 90% of all the oil the Germans could get out of the entire area. The output was going mostly to Germany and to the Ukrainian front.
- An interruption of just 100,000 tons per month would provide maximum assistance to Russia.
- If HALPRO could not get to its present destination (China) then it should be diverted to perform a bombing mission against Ploesti.[4]

HALPRO

Three memos were then released which effectively changed the HALPRO mission. Since Chinese bases were no longer an option, the WPD began looking for alternative uses for the HALPRO bombers. Ploesti and the Middle East offered the best alternative.

- The loss of Burma to the Japanese made it extremely doubtful that HALPRO could accomplish its mission.
- Therefore, Ploesti should be substituted as HALPRO's target, a move that would assist in the defense of Russia.
- British authorities were asked to provide facilities for advance airdromes and to cooperate in the work of preliminary planning for the Ploesti mission.
- HALPRO was ordered to proceed as far as Khartoum, and to wait there for further instructions.[5]

Orders were issued for HALPRO to begin its overseas deployment on May 16. The first step was to fly from Fort Meyers to Morrison Field, West Palm Beach.

Special Orders Number 45

Pursuant to authority contained in Secret Letter, the following named officers and enlisted men of the Halverson Detachment will proceed on or about May 16, 1942, via military aircraft, commercial aircraft (WD Cir. 12,1942) and by rail where necessary, from Fort Meyers, Florida to Morrison Field, West Palm Beach, Florida, thence via the ***********, reporting upon arrival at ********* to the ********* for further instructions.[6]

DOOLITTLE RAID

While the Doolittle raid boosted morale in the United States, tactically, it had three negative effects,

- The Allies lost the future use of the sixteen badly needed bombers
- The Allies soon lost their eastern airfields to the Japanese, who advanced upon Chuchow from the Hangchow area on 15 May
- A short time later, the fields at Chuchow, Yushan, and Lishui fell into enemy hands.[7]

There were now no forward airfields from which HALPRO could operate.

MIDWAY

Naval Intelligence learned from their interception of Japanese messages of the planned Japanese attack on Midway Island. (see Figure 10) They informed Nimitz of their analysis on May 15. While they did not know the exact date of the attack, they thought the Japanese fleet would leave their bases from Japan and Saipan by May 29. Army Air Corps began sending B-17s to Midway to perform reconnaissance missions.[8]

If this speculation was true, then the Japanese naval forces would be withdrawn from the Indian Ocean. Such a withdrawal eliminated a possible link-up between German and Japanese forces in the Middle East.

10TH AIR FORCE

By May 15, Rowan Thomas had returned to Karachi. (see Figure 32) He was on a reconnaissance mission over the Indian Ocean looking for an American convoy. As the sun began to set, they found and then "buzzed" the ships.[9]

1942 MAY 16

NORTH AFRICA

The British ground commander, Maj. Gen. Neil Ritchie, issued orders for a defensive stand at Gazala.[10] (see Figure 5)

PLOESTI

The United States asked the Russians about the use of Soviet airbases for a Ploesti mission.[11]

1942 MAY 17

COMMITMENT TO CHINA

Stilwell notified Washington that, subject to further instructions from Washington, the 10th Air Force would be in charge of all preparations for reception of HALPRO.[12] So despite a decision two days earlier that it was useless to send HALPRO to India, Stilwell had yet to be informed.

GLOBAL STRATEGY

At a White House meeting on May 17, Arnold recorded the decisions Roosevelt made:

1. Combat planes (i.e. B-24s) would not be used to ferry cargo into China.
2. DC-3 planes would be used as long as possible to ferry cargo into China.

3. Efforts would be made to base US operating units in Russia for operations against German units from Caucasian bases.
4. Endeavors would be made to make possible additional raids on Tokyo or Japan Proper.
5. In general, nothing would be done to weaken the BOLERO effort.[13]

1942 MAY 18

MIDWAY

The 7th Air Force was alerted to a possible Japanese invasion of Midway or Hawaii.[14] (see Figure 10)

PLOESTI

In the May 17 memo where Arnold recorded Roosevelt's decisions, Arnold asked Harmon to explore the Ploesti mission in greater detail. His inquiry was based upon the three key assumptions made in the recently released study about Ploesti:

1. Stalin had to approve giving logistical support, which, thus far, had been "denied all foreign nations"
2. The Russians would stabilize their front with the Germans
3. Any United States logistical buildup would have to come out of the buildup for BOLERO.[15]

BOMBING JAPAN

Arnold also asked Cabell and Norstad to explore the ways and means of bombing Japan, noting:
- Distances for the moment had us licked unless we could go into Siberia.
- You fellows use your imagination and see what ideas on the subject you can present me. I realize that the distance from the Aleutian Islands probably makes that prohibitive, but there might be some other way to get at this proposition.[16]

The HALPRO mission had already been changed, but Roosevelt and Arnold were still looking for ways to bomb Japan.

1942 MAY 19

BOMBING JAPAN

Following a conversation with Soong, Roosevelt asked Arnold to consider the following:

1. A very large amount of Japanese equipment is being made in Shanghai.
2. This equipment is dependent on power from the Shanghai power plant, which is the only source of power.
3. That if this power plant were destroyed it would slow up Japanese production for six months.

He suggests that we bomb the Shanghai power plant. His thought is that our planes should take off from India, thus constituting putting the power plant out of business, to return to the Chengtu field near Chungking.[17]

OPERATION VELVET

An aerial attack on Japan was not the only mission of concern to Roosevelt. Earlier in May, he had expressed an interest in Air Corps operations from the Caucasus to help defend the Soviet oil fields. In anticipation of a White House meeting the next day, Arnold asked his staff to investigate the feasibility of such a project.[18]

1942 MAY 20

CAUCASUS

Allied speculation about German intent always contained "ifs". One of the "ifs" happened. A German drive to the Caucasus oilfields was now obvious. If the Germans were successful, not only would the Germans secure a badly needed supply of oil, but also they would threaten the Allied supply line into Russia. In an article entitled "Kerch Mopped Up," the author hypothesized that the quickest route to the oil fields was via Crimea. (see Figure 43)

BURMA

For all intents and purposes, the Japanese had captured Burma.[19]

MIDWAY

United States code breakers, reading Japanese messages, now knew of the upcoming Japanese invasion of Midway and the Aleutians. Supplies and reinforcements were rushed to both.[20] (see Figure 10)

COMMITMENT TO CHINA

Arnold addressed the bombing of Shanghai:
- In connection with the Dr. Soong proposal for bombing the Shanghai power plant, a study is being made in my office.
- Preliminary study indicates that it may be possible to bomb this power plant

from bases in India, stopping long enough, either enroute or returning, at some airport in China for refueling.[21]

While Roosevelt and Arnold were dealing with Soong's Shanghai idea, a Chinese government spokesman in Chungking was telling reporters that the Japanese appear to be mounting an offensive. "The 'situation was fraught with serious possibilities.' The spokesman requests 'all practicable increased American air support.'"[22]

In addition, another chapter in China's continuous public relations plea for help appeared in the press. Japan had been waging an offensive to remove potential Chinese air bases that were within striking range of its homeland. Combined with their success in severing the Burma Road, China was becoming more and more isolated. They made their case in an article of the first page of the *The New York Times*:

CHINA CALLS ALLIES TO AVERT KNOCKOUT.[23]

OPERATION VELVET

At a White House meeting on May 20, Roosevelt reaffirmed his interest in operations from the "Caucasus regions."[24]

HALPRO

Early in the morning, "A" flight lifted off the Fort Myers runway bound for Morrison Field. The flight was short, and all seven aircraft had landed by 9:45. The stay at Morrison had been projected not to exceed twenty-four hours, during which additional supplies would be loaded and some last minute work performed on the B-24s.[25] HALPRO began its overseas deployment despite uncertainty over its objective.

1942 MAY 21

BOMBING JAPAN

John Carter, who was responsible for collecting information about potential Japanese targets, had lunch with Harvey Davis, the Director of the Stevens Institute. Davis suggested that the Air Corps drop a few bombs on Japanese volcanoes. Carter forwarded the idea to Roosevelt.[26]

Davis must have been serious about his idea. Even more amazing, Carter forwarded it to Roosevelt. It seemed that someone was looking for any reason, regardless of how far-fetched, to get HALPRO to China. Roosevelt must have considered it worthy of study as he forwarded it to Arnold the next day.[27]

Cabell and Norstad respond to Arnold's May 18 request for means to bomb Japan with three ways and means of bombing Japan:

1. A modified HALPRO under the 10th Air Force direction and using aircraft of that organization.
2. Operations from Russian bases should be considered so that we could take advantage of a favorable change in the political situation.
3. A repetition of the Jimmy Doolittle project. While this plan may have lost some element of surprise, it was still sound and practical if the equipment were available.

They also recommended the development of a glider that would be capable of carring 1,000 gallons of fuel. Heavy bomber operations could then be conducted from Alaska with landings in Russia and China.[28] Arnold sent these brainstorm ideas of Cabell and Norstad to Craig the next day and asked him to evaluate.[29]

COMMITMENT TO CHINA

Hull responded to the May 20 message from China that she needed bombers and pursuits, first and last, within the next three months. In his message to Roosevelt, Hull wrote: "We should give the Chinese proof at an early date that China is looked upon as a vital theater of war by our Government."[30]

Now, we were getting to the crux of the issue: if China folded, Japan would have a free hand. HALPRO represented a commitment to China.

CONVOY P.Q. 16

Convoy P.Q. 16, a British supply convoy to Russia, left from Hvalfjord, Iceland for the ports of Archangel and Murmansk.

1942 MAY 22

HALPRO

"A" Flight left Morrison Field to begin their departure from the continental United States. They planned to arrive at Borinquen Field, Puerto Rico. "B" Flight of HALPRO left Fort Meyers for Morrison Field.[31]

1942 MAY 23

EASTERN FRONT

Following a German counter attack near Kharkov (see Figure 38), the Germans had surrounded the Russian forces in the Izyum area west of the Donets.[32] (see Figure 18)

1942 MAY 24

COMMITMENT TO CHINA

United States intelligence speculated that the Japanese had turned their focus towards Midway. (see Figure 10) One of the supporting pieces of evidence were indications that the Japanese fleet, currently in the Indian Ocean, was withdrawing to Singapore. Accordingly, on May 24, the order committing the 10th Air Force to support RAF operations was rescinded by the War Department. Stilwell, who had just concluded his march from Burma, promptly replied that the 10th Air Force would be recommitted to a mission primarily in support of China.[33] (see Figure 32)

HALPRO

James Sibert was the pilot of the HALPRO plane *Queen B*. As the lead plane of "B" flight, he began their departure from the continental United States.

> *Up at 04:30. Ate a* good breakfast and met the crew at the ship. I crawled in the Pilot seat and was the first ship off since we were considered the lead ship of "B" Flight. We waved farewell to our ground crews and taxied to the end of the runway and gave the old girl the needles, circled and headed out across the blue water of the Caribbean enroute to Puerto Rico. After 6 hours of water flying we set down at Borinquen I made the landing and drug in over some land on the runway, which shut off the reconstruction area. (One runway 8000 ft). Spent the night - was briefed and each ship bought a case of cigarettes at 55 cents a carton.[34]

"C" Flight left Fort Meyers for Morrison Field. Shortly after takeoff, Martin Walsh's B-24, nicknamed *Wash's Tub*, experienced an equipment mal-function and he had to return to Morrison Field. Upon landing, the nose wheel collapsed. This was a problem that would plague the early models of the B-24.[35]

Al Story, aboard *Hellzapoppin*, thought HALPRO was going to the Far East.

> *We all left Ft. Myers* on the same day, May 23, 1942. [Author's note – he is off by one day] Our first stop was Miami Air Force Base. There I first suspected we

were headed for Japan when we were told to exchange our tracer ammunition for incendiary. The only use for incendiary bullets is for starting fires and we all had images of Japanese houses made of paper.[36]

1942 MAY 25

COMMITMENT TO CHINA

Chiang warned the United States that unless China saw visible evidence of United States help, "Chinese confidence in their Allies will be completely shaken."

During a meeting with Brereton and Wavell, Stilwell referred to a message he received from Marshall. Quoting Chiang, Marshall said that:

1. the British failure in Burma,
2. the departure of high-ranking American officers from Chungking, and
3. the lack of air support and other aid from America "had reduced the Chinese morale to a desperate state."

In view of Japanese advances in the Chuchow area and their success in Burma, the Generalissimo stated that immediate air support was necessary. If it was not forthcoming, Chinese resistance would be in danger of collapsing.[37]

10TH AIR FORCE

Brereton presented his plan for the 10th Air Force :

The heavy bombers would continue to operate from bases in India until such time as fuel and ammunition could be made available in China.

As fuel accumulated in China, the heavies could be moved to Chinese bases. The fighters and the mediums, the latter based at Kunming, could begin their operations immediately, provided fuel and ammunition could be put in place.

It was hoped this plan would increase American operations considerably and give Chinese morale an immediate boost.[38]

Even Brereton thought he was getting the HALPRO bombers.

PRESS

A *Time* article announced the beginning of the German summer offensive for 1942:

BATTLE OF RUSSIA: PUSH WITH A DIFFERENCE

... The spot the Nazi chose was the Crimea, where his troops had held on through the winter with Russians in front of him and behind him (at Sevastopol). (see Figure 8)

In the Crimea, then. Field Marshal Dieter Wilhelm von Mannstein, Junker-born apostle of the swift and crushing thrust, slogged east toward Kerch. Before his power drive the Russians fell back toward the end of the Crimea, and at week's end were fighting in Kerch. With water at their backs they were in a tough spot, but so since last October had been the great naval base at Sevastopol. And Sevastopol was still in Russian hands.

What the German was after in the Crimea seemed reasonably plain. If he cleared out Kerch, he would find it easier to have another try at Sevastopol, chief base of the Russian Black Sea fleet. And east of Kerch, across only four miles of water, lay the Caucasus and its oil. If the German could get across, he would have something more than the fuel and lubricating oil he bitterly needs. He would also be in a position to lance down into Persia and cut the roads over which U.S. and British supplies are flowing into Russia.[39]

1942 MAY 26

NORTH AFRICA

Rommel began his attack on Gazala.[40] (see Figure 5)

MIDWAY

USS Kittyhawk delivered additional supplies.[41] (see Figure 10)

10TH AIR FORCE

Brereton was not the only one who thought HALPRO would join the 10th Air Force. Currie sent the following message to the American team: "10th Air Force is being strongly reinforced by planes now enroute."[42]

HALPRO

James Sibert and *Queen B* were completing the last leg of the Southern American portion of the flight.

> *We flew all night and* weather was bad. On instruments for about 3 hours. Over the Atlantic the whole time except when we flew over Vichy, France territory, which we were supposed to avoid. Queen B humming along and the enlisted crew getting plenty of sleep. Ebert, my Navigator quiet but sure of himself I'm satisfied. Flew along coast all morning. Landscape of S.A. was beautiful. Much like North Texas. Landed at Natal Brazil at 15:00 and found another ship from A flight with leaking tanks. Had a nice meal with the strongest coffee ever en-

countered. Also introduced to the "Giant sized Bottled Beer". (Cervesa) To bed happy.[43]

Behind him, six of the eight "C" flight aircraft departed for Puerto Rico following in the wake of A and B flights.[44] The last of the HALPRO planes were about to begin their trans-Atlantic journey. Cave's plane was in "C" flight.

On May 26, 1942 we ["Hawk" Cave] took off for Porto Rico, Trinidad to Natal, South America - Accra Africa across the ocean - to Khartoum Egypt.[45]

Before reaching Belem, the stop before Natal, Al Story took notes on their trans-Caribbean flight.

Next morning we were off again for Georgetown, British Guyana. The main thing I remember about this part of the flight was that we flew mostly through very rough weather, thunder and lightning, with up and down drafts and the rough weather was at night. I was required to sit in the tail turret through all this because our route took us near the Island of Martinique. This was a French possession and we had to guard against attacks by French fighter planes from there. By this time the French had sided with the Germans so this was a concern. Other than the rough weather, everything was routine. We landed at the airfield near Georgetown early in the morning.[46]

The attitude of the Vichy French was unknown and it was unclear if they would support the Allies. Therefore, HALPRO crews needed to be wary of possible aggressive action by the French.

SOUTH PACIFIC
An internal Allied report indicated that the Allies were intercepting 60 percent of the Japanese Imperial Navy's message traffic. And of these, 40 percent were being read. Two days later, the Japanese changed their communications.[47]

1942 MAY 27

HALPRO
After a brief stay in Natal, the *Queen B* was readied for her trans-oceanic flight.

I made the take off loaded full for a 16-hour trans-ocean flight to another continent. This long heard of "Dark Africa". Took off at 21:00.[48]

With "B" Flight heading over the Atlantic, "C" Flight departed for Waller Field, Trinidad. Upon landing, the various crews noted that one plane from "A" flight, *Ole Faithful*, was parked on the tarmac. Alfred F. 'Kal' Kalberer, the original pilot of *Ole Faithful*, had apparently performed

> *a near miracle. He had* taken off at night, flown a short while when the oil pressure of engine fell and he turned for base with the huge overload. He found the field with the flare path extinguished for some cause and had come on in with three engines.[49]

Since Halverson was a passenger on *Ole Faithful*, he elected to take one of the "B" Flight planes, *Arkansas Traveler*, and proceeded on. The Homer Adams' crew, who had been flying *Arkansas Traveler*, was left standing on the tarmac watching the engine on *Ole Faithful* being replaced. They were somewhat dismayed that Halverson had not simply transferred his luggage to *Arkansas Traveler* and flown on with them.

"C" Flight lost a second plane when *Malicious* developed gas tank leaks. Richard Sanders, the pilot, elected to return to Patterson Field for the needed repairs.

While the "C" Flight crews rested, they heard another story about "A" Flight. Colonel George McGuire was Halverson's deputy commander, and was onboard the plane *Town Hall*. Piloting another "A" Flight bird, *Ole Rock*, was George Uhrich. For reasons unknown, Uhrich always managed to reach the first two destinations ahead of McGuire and his plane, *Town Hall*. Uhrich thus landed before his leader, McGuire. For this achievement, Uhrich earned the nickname "Speedball."

1942 MAY 28

MIDWAY
Admiral Raymond Spruance's Task Force departed Pearl Harbor for its position off Midway.[50] (see Figure 10)

NEW GUINEA – SOLOMON ISLANDS
A United States force arrived at Espiritu Santo, New Hebrides to build a runway for heavy bombers. These bombers would support the planned invasion of the Solomon Islands.[51] (see Figure 39)

EASTERN FRONT
The battle of Kharkov ended as the Germans eliminated the Russian forces west of the Donets.[52] (see Figure 38)

HALPRO

Queen B flew on into the night.

Didn't have to get up for I flew all night: Perfect weather and the usual head winds at 12,000 ft. Kept careful check on my power settings and fuel consumption. Eb. [Ebert] used celestial all the way until 100 miles from the coast of Africa where we hit a front and varied our course to the left and let down to 2,000 ft. Intercepted the coast at Tackeradi and on up to Accra and Pan Am. Approached field and set up glide at 105 and slowed it up to 90 when she set in (light) [sic]. Runway 4,000 ft but bad room to spare. Treated swell by Pan American.[53]

At midnight, "C" Flight left for their last Western Hemisphere destination, Natal, South America. They arrived that afternoon. "C" Flight temporarily lost another plane. *Jap Trap* had to divert to Recife, Brazil to refuel. The crew would quickly catch up.

For a young kid from Georgia, Al Story must have felt out of place.

We refueled there [Belem] and headed on out for Natal, a small city on the easternmost tip the country. This leg of the flight was particularly interesting because we flew low over the jungle and looking down, we could see, among other things, Indian villages and huts in clearings carved out of the forest. We also crossed the Amazon here and it was impressive. It seemed to be at least twenty miles wide where we crossed it. We landed at Natal in late afternoon and spent the night there.[54]

The "C" Flight crews asked about the McGuire-Uhrich rivalry; they wanted to know who landed first at Natal. The answer: Uhrich.[55]

Meanwhile, Hawk Cave, in *Hellzapoppin*, was about to cross the ocean.

I remember very vividly when we left Natal South America to fly across the ocean we took off at night in a hell of a thunderstorm. Got up to about 10,000' and it cleared off like the top of a table. I don't know what time it was but some time during the early morning hours Wicklund called me on the intercom and said "I've got some bad news and some good news" I said well let me have the bad news first". He said, "We've reached the point of no return". I said, "What's the good news?" He said, "I think I know where we are".[56]

Al Story, also on board *Hellzapoppin*, recorded his impressions.

Next afternoon we headed out over the Atlantic on our way to Accra in what was then called The Gold Coast. The plane droned on all night and sometime after the sun had risen we could see land. It was the Gold Coast and there was a city. It was Accra! Our navigator hit it right on the nose. Once again, as we approached the African Coast, we had to be on guard against the possibility of attack by French fighter planes stationed at an air base at Dakar in Senegal. Nothing happened though, and we landed without incident.[57]

Apparently, the navigator knew where they were. And the HALPRO crews remained concerned about the French.

1942 MAY 29

BOMBING JAPAN

Arnold had passed the suggestion to attack Shanghai to Col. Robert L. Walsh. Walsh responded on May 29, saying that Shanghai was an important logistical center and had a textile industry. If the Air Corps managed to destroy the electric power plant, the plant would be out of commission for six months, with most of the damage to the public morale.[58]

HALPRO

"C" Flight departed Natal at 10 o'clock in the evening on their trans-Atlantic leg – destination Accra, Africa. They arrived about noon on May 30. Upon landing, "C" Flight learned that they had lost another plane. Francis Nestor, piloting *Mona The Lame Duck*, could not get his landing gear to lock in its stowed position, and had returned to Natal. The plane had also developed gas tank leaks. If the tanks could not be repaired in Natal, he would have to return to Patterson Field.[59]

Following another day of rest, James Sibert and *Queen B* were ready for their final leg to Khartoum.

> *Briefed for trip across the* wilds of Africa to Khartoum and got our sandwiches and I made the takeoff and as we were loaded to capacity I had to pull it off when I had used all of the 4000 ft of runway. Headed out to the coast (Gold Coast). Followed coastline to Lagos and then turned on a course for Khartoum.
>
> After two hours out I ran into a line of thunderstorms approximately 100 miles long, tried to go around, over, but finally settled for 14,000 ft and headed into it on instruments. Not too rough but lightning was flashing all around us and rain was so hard it took paint off of the props and bare leading edge of the wing. Then instruments for 3 1/2 hours. After getting out of rain, never saw stars until early morning.[60]

REALITY OVERCOMES THE BEST LAID PLANS, MAY 1942 | 245

COMMITMENT TO CHINA

Most retrospective reviews about HALPRO often speculate on when the crews knew about the decision to hold HALPRO in the Middle East; and not proceed on to China. Richard Miller made the following entry in his diary:

> *Have talked with many in* the Ferry Command here [Accura] and was set back to learn that the Japs have some really hot pilots and very good equipment. There is a great deal of speculation as to how we will come out - most agree that half of us will not return after the first mission - i.e., that our outfit can't last for more than 3 trips - who knows?
> But by God, it will be a swell scrap while our props continue to turn.[61]

In answer to the above speculation, at least he did not know on May 29 that they would stay in the Middle East.

1942 MAY 30

MIDWAY

Admiral Fletcher's Task Force departed Pearl Harbor for its position off Midway Island.[62] (see Figure 10)

MIDDLE EAST

Arnold and Rear Admiral John Towers traveled to London on May 26 to confer with RAF and Royal Navy (RN) officers regarding allocation of aircraft. The discussions made progress until the subject of the Middle East arose. The United States was faced with two equally unattractive options. Either acquiesce by allowing the RAF to have large quantities of aircraft and stores to maintain the RAF or send its own combat units, replacing altogether an equivalent RAF strength and utilizing aircraft previously allotted to the British.

Since American production was beginning to gain pace and American training units were increasing its output, Arnold and Towers chose the second option. On May 30, they committed to getting nine groups to the theater:

- One heavy group completed by 1 October 1942;
- Two medium groups completed by 1 March 1943;
- Six pursuit groups, two available in the theater by September 1942, two by December 1942, and two by April 1943.[63]

HALPRO was the only heavy bombardment group that would be in position to meet this schedule.

HALPRO

Kalberer had no better luck with his confiscated plane, *Arkansas Traveler*. "C" Flight crews found him in one of the hangers working on one of the landing gears. He had pulled off another flying miracle – the gear had dropped from its stowed position and Kalberer had flown the plane with one gear down part way across the Atlantic. He had landed at Accra with only 30 minutes of fuel remaining.

And as to the McGuire-Uhrich rivalry, who landed first? Uhrich. And apparently, McGuire was not taking Uhrich's disobedience very well.

Following a meal and rest, the crews boarded their planes for the next leg – Khartoum.[64]

Queen B continued on into the night.

> *As dawn approached I saw* the most beautiful sunrise of the trips. The Col. [Feldmann] relieved me and took over. I sprawled out on the flight deck and went to sleep. Woke up about 3 hours later over desert sands and the Col. was feathering No's 1 and 4 engines ~ out of gas. Plenty in Nos 2 & 3 so we transferred fuel and went on. Was sweating out our fuel supply when we spotted the Blue Nile on the left and ahead Khartoum at the intersection of the Blue and White Niles. Good navigating for Eb. The Col came in hot and set her down with a jolt but everyone was so tired we didn't care as long as we were on the ground at our destination. As Khartoum was our rendezvous point for all three flights "A" flight ships were there and Col. McGuire met us. S/Sgt Milliren opened a bottle of champagne that sent the cork "flying high". Everyone laughed and we celebrated the good luck on the trip. Reported in and was taken to barracks when I saw all the fellows Nesbitt, Smith etc.[65]

Therman Brown in *Draggin Lady* was one of the last HALPRO planes to arrive at Khartoum.

> *On May 30, 1942, our* plane landed at Khartoum. We rejoined all twenty-three of the Halverson Detachment (HALPRO's) B-24s. All had flown from Ft. Myers, Florida, down to Natal, Brazil, non-stop across the South Atlantic to Accura, Ghana, and non-stop across Africa to Khartoum. NO planes were lost. This was quite a feat for the time.
>
> We did not know it yet, but there was a change of plans. Rommel's Afrika Korps had pushed the British Eighth Army back to El Alamein – only 90 miles

from Alexandria. Our mission to China was permanently delayed while we were thrown into the breech.

We stayed at Khartoum for about ten days. The Chinese general [Ping Han Whang] who had been traveling with us since we left Ft. Myers, Florida didn't look happy. Then he never did. He left us not long after and went on to China. He was to have been the liaison officer for the Chinese when we got there.[66]

According to HALPRO veterans who were later interviewed, when the crews left Florida, they still thought their mission was to bomb Japan. The Chinese General already knew otherwise.

PLOESTI
Roosevelt asked Vyacheslav Molotov what the Soviet attitude would be towards the HALPRO forces landing on Russian bases after bombing the Rumanian oil fields. Molotov was enthusiastic about the Ploesti bombing raid. It was likely his idea to attack Ploesti via shuttle bombing operations. The Russians gave permission to the United States to use two bases - at Armavir and Krasnodar.[67] (see Figure 42)

CONVOY P.Q. 16
The bulk of Convoy P.Q. 16 arrived in Murmansk; eight ships made it to Archangel the next day. Of the 35 ships that began the voyage, only eight were lost.

1942 MAY 31

NORTH AFRICA
Axis forces began their attack on Sidi Muftah.[68] Sidi Muftah is southwest of Tobruk. (see Figure 41)

MIDDLE EAST
Arnold and Portal had met the previous day to discuss the commitment of American units to the Middle East. At this meeting Arnold realized that there were still a few unresolved issues:

1. the British desire for reinforcement of the Middle East;
2. the RAF Coastal Command's request for long-range bombers for the Battle of the Atlantic;
3. and the demands for light bombers, which were needed for any air-ground operations.

Portal noted on May 31, "what a very real effort you have made to meet our point of view."⁶⁹

Portal's comment was an example of the British arrogance that some United States officials found offensive.

COMMITMENT TO CHINA

Madame Chiang sent a rather interesting note to Currie, who passed it to Roosevelt. It did not deal with China, but rather India. In essence, she observed that Mahatma Gandhi was using the war to further his Independence Party objective of overthrowing British rule. The British were threatening to quell the civil disobedience actions, thus furthering tensions. She thought that the only way to avoid total disruption of the war effort was for the United States to recognize Indian independence.⁷⁰

HALPRO

The next morning, on the way to Khartoum, "C" Flight ran into a rainstorm. Five hours later, they were still in it. After a fourteen and a half hour flight, *Black Mariah II* was the first "C" flight plane to land, and the fourteenth HALPRO plane to arrive. Kalberer arrived, dangling the one gear.⁷¹ But something was amiss. Instead of continuing their flight to China, all the planes remained in Khartoum for further orders.

Al Story, on *Hellzapoppin*, watched the first planes of 'C' flight takeoff. Then it was their turn.

> *We stayed there [Accra] for* two days and a night, waiting for our stragglers to catch up. On the evening of our second day there, we were all lined up beside the runway, ready for our flight across Africa. Our next destination was to be Khartoum in The Sudan. Khartoum is located in the forks of the White and Blue Nile Rivers. We were lined up according to our plane numbers. Being No. 23 we were, as usual, next to last in line for take-off. While those ahead of us taxied out for take off I was sitting atop our plane refueling. I noticed that as each plane lifted off the ground some of our crew were laughing. Then I realized what the laughter was about. At the end of the runway there was a stand of elephant grass about fifteen to eighteen feet high. As each plane lifted off the ground the wheels and landing gear dragged through this grass!
>
> By the time our turn came, darkness had fallen so we taxied down and lumbered along the runway in the dark. We made it into the air and headed eastward. Old Hellzapoppin purred along all night long. Finally as it began to get lighter next morning I could see that something was trailing out from our left wing. I reported this to the pilot who asked if we could tell what it was. None of us could. It was a mystery. Anyway, the plane seemed to by flying well and as we had no other

options, we kept on. As we approached Sudan a large weather system had sprung up in our path so we tried to fly south around it. As it turned out, the system was so large that by the time we got around it we were running low on fuel. Our navigator located a small auxiliary British airfield somewhere in southwestern Sudan and we landed. It was at this field that our pilot managed to contact the rest of our group who had made it to Khartoum before the storm. That is when we were informed that the line we were dragging from our wing was the landing lights for the field back at Accra.

This emergency field was surrounded be a fence of acacia thorns and was about fifteen feet high. On the outside of this fence we could see hundreds of natives walking by. They were all, the men at least, carrying long spears.

The natives, all carrying their long spears, continued to walk past and look at the huge plane until darkness fell. Our pilot was busy making arrangements for fuel and arranging for food and a place to sleep for the night. Conrad Pierce and I were assigned to guard the plane while the rest went off to eat. After they returned from eating, Conrad and I were taken to the mess hall such as it was. About the only thing I remember about going to eat were the numerous kangaroo rats in the headlight beam of the vehicle. After our return, the pilot said that two of us were needed to guard the plane during the night. He did not want to leave it unguarded. I was nominated. Our radio operator, Durfree, also volunteered . So while the others went off to the nearby barracks, we were left on guard.

There was no moon that night. It was pitch dark as we sat under the wing of the plane and talked. All the natives had gone and the only sound we heard was the incessant sound of drums coming from the nearby village, and barking dogs. That sound went on most of the night. In the blackness we did make out a small animal, perhaps a jackal or a big-eared fox trotting by. No other animals were seen. Finally we decided to get into the plane and into our sleeping bags for the rest of the night. Next morning as we were preparing to take off I was talking to one of the British soldiers about the numerous natives walking past. I asked him why they all carried spears. "Its for protection against the lions." He said.[72]

It was interesting that he was focused on lions. Concerns about natives, spears and lions showed there were other interesting dimensions to their lives.

HEAVY BOMBER PRODUCTION
During May, 178 heavy bombers were manufactured. Ninety were B-24s.[73]

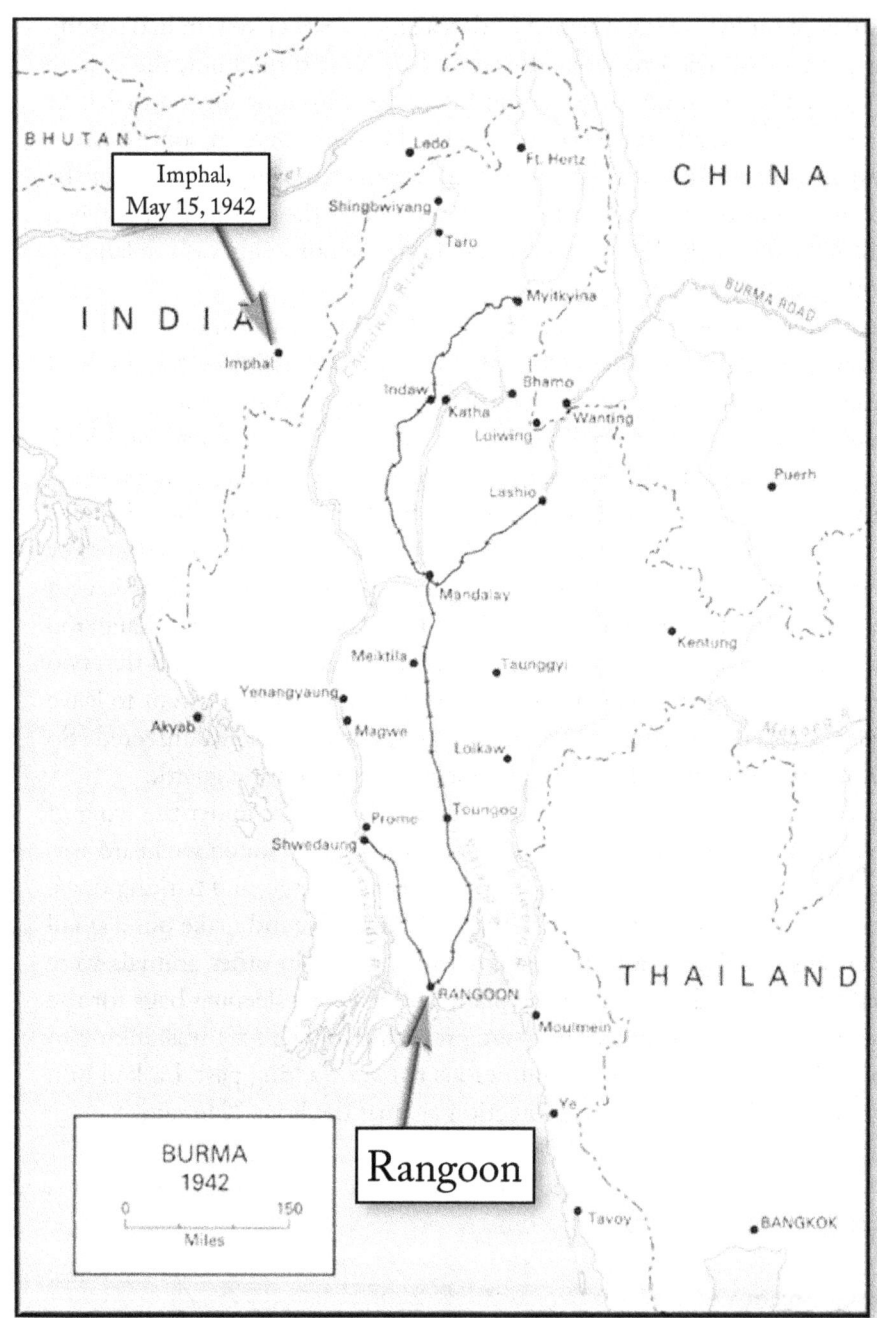

Figure 40 – Burma – May 1942

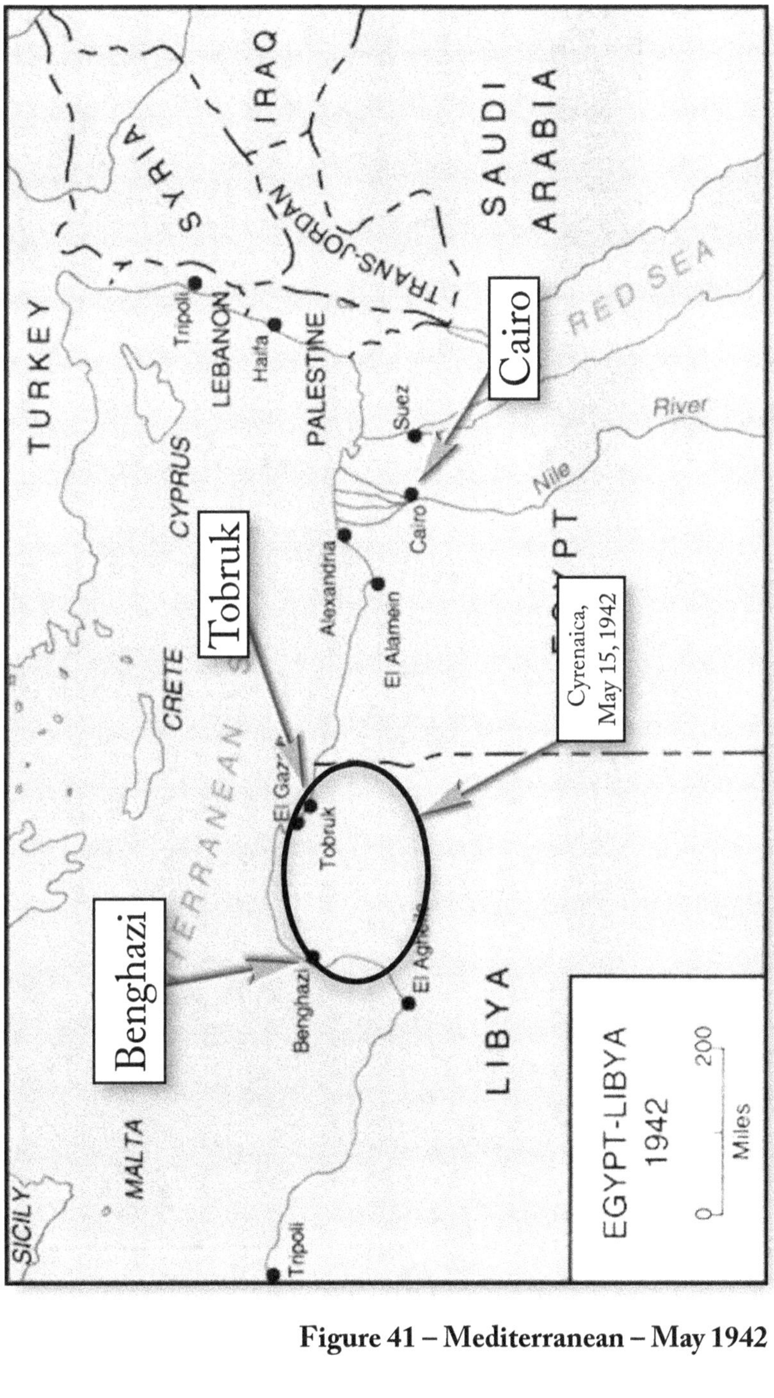

REALITY OVERCOMES THE BEST LAID PLANS, MAY 1942 | 251

Figure 41 – Mediterranean – May 1942

252 | THE OTHER DOOLITTLE RAID

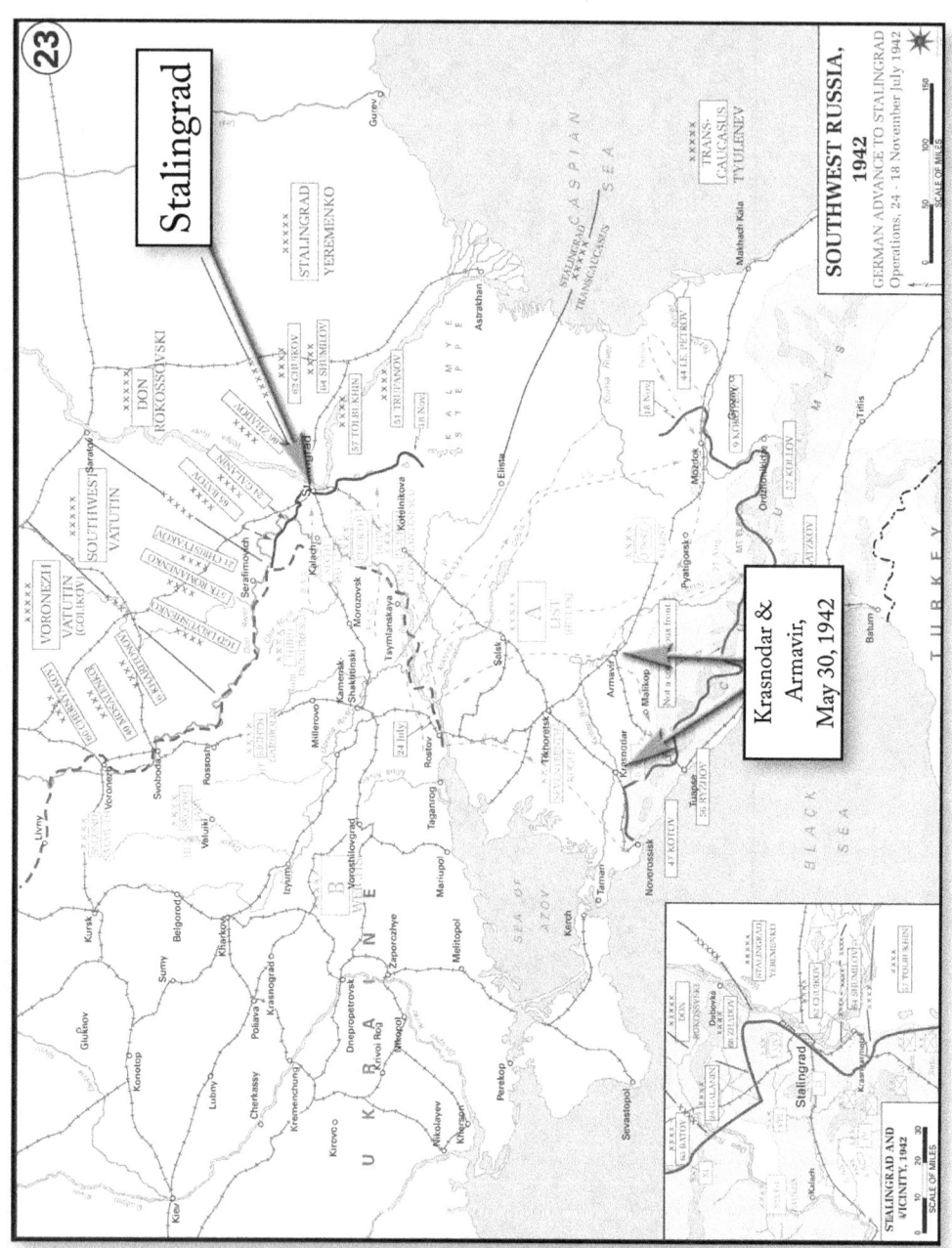

Figure 42– Stalingrad – May 1942

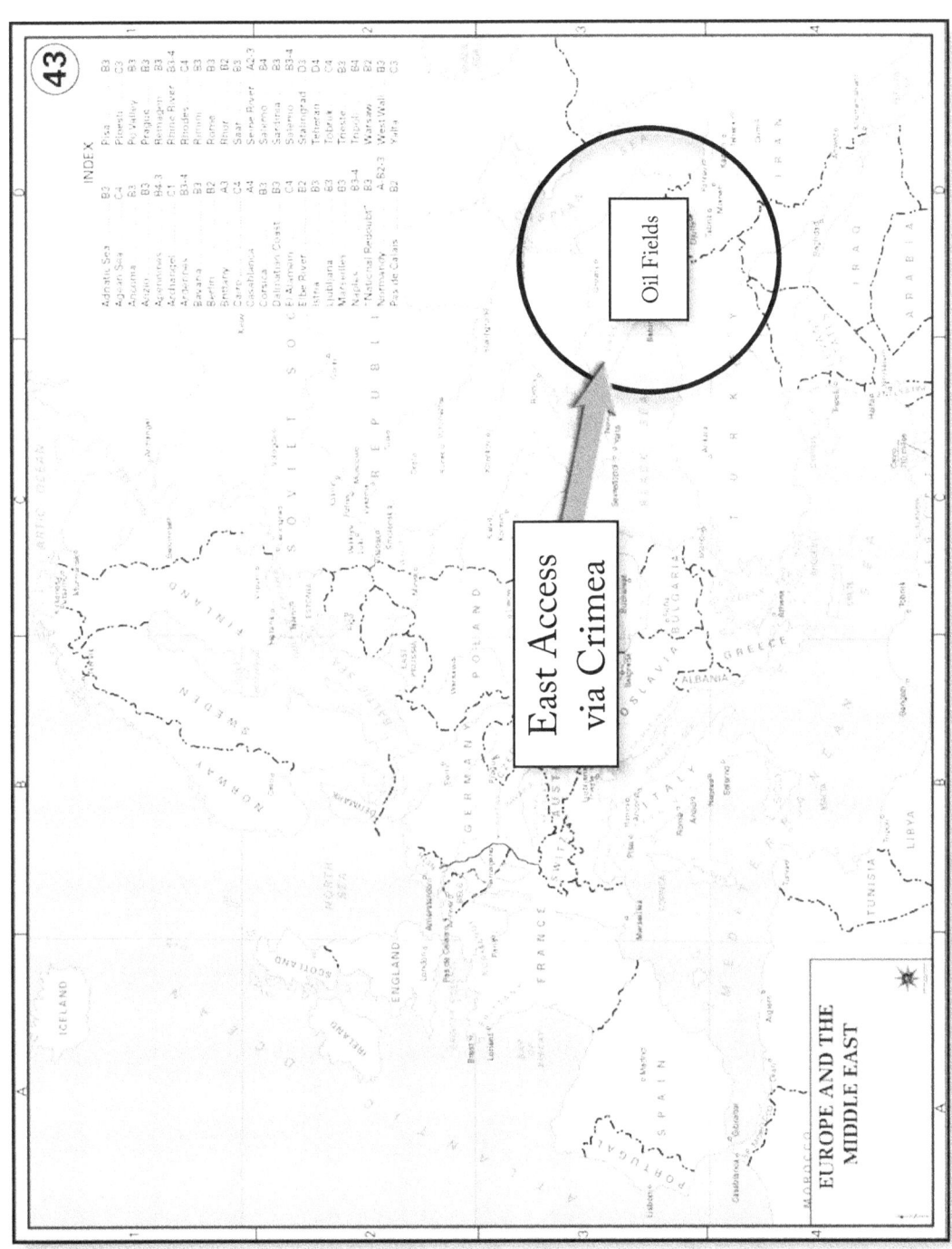

Figure 43 – Possible Strategy – Eastern Front – May 1942

ENDNOTES

1. Williams, *Chronology 1941-1945 - 1942*, p. 38.
2. Assistant Chief of Air Staff, *The AAF in the Middle East*, p. 70-71.
3. Williams, *Chronology 1941-1945 - 1942*, p. 38.
4. Assistant Chief of Air Staff, *The Ploesti Mission*, pp. 14-15; ----, AFHRA, Hap Arnold file, Craig to Arnold, May 15, 1942.
5. Assistant Chief of Air Staff, *The Ploesti Mission*, p. 15.
6. Walker, *The Liberandos*; Richardson, *The Forgotten Force*, p. 22.
7. Craven, *The Army Air Forces in World War II, Vol. 1*, p. 444.
8. Craven, *The Army Air Forces in World War II, Vol. 1*, p. 452-457.
9. Thomas, *Born in Battle*, pp. 98-99.
10. Jackson, *The Battle For North Africa*, p. 202.
11. Craven, *The Army Air Forces in World War II, Vol. II*, p. 10.
12. Craven, *The Army Air Forces in World War II, Vol. I*, p. 504.
13. ----, AFHRA, Hap Arnold file, Arnold to Harmon, May 18, 1942.
14. Williams, *Chronology 1941-1945 - 1942*, p. 38.
15. ----, AFHRA, Hap Arnold file, Arnold to Harmon, May 18, 1942.
16. ----, AFHRA, Hap Arnold file, Arnold to Cabell and Norstad, May 18, 1942.
17. ----, FDRL, PSF, Box 3, Japan Folder, FDR to Arnold, May 19, 1942.
18. Lukas, *The Velvet Project*, p. 146.
19. Williams, *Chronology 1941-1945 - 1942*, p. 38.
20. Williams, *Chronology 1941-1945 - 1942*, p. 38.
21. ----, FDRL, PSF, Box 3, Japan Folder, Arnold to FDR, May 20, 1942.
22. ----, FDRL, PSF, Box 2, China Folder, American Ambassador to FDR, May 20, 1942.
23. Forman, Harrison, *New York Times, CHINA CALLS ALLIES TO AVERT KNOCKOUT*, May 19, 1942, http://select.nytimes.com/gst/abstract.html?res=F70611FC3858167B93C2AB178ED-85F468485F9, accessed February 19, 2014.
24. Lukas, *The Velvet Project*, p. 146.
25. Walker, *The Liberandos*; Richardson, *The Forgotten Force*, p. 28.
26. ----, FDRL, PSF, Box 98, Carter Folder, Carter memo, May 21, 1942.
27. ----, FDRL, PSF, Box 82, Arnold Folder, FDR to Arnold, May 22, 1942.
28. ----, AFHRA, Hap Arnold file, Cabell and Norstad to Arnold, May 21, 1942.
29. ----, AFHRA, Hap Arnold file, Arnold to Craig May 22, 1942.
30. ----, FDRL, PSF, Box 2, China Folder, Hull to FDR, 21 May 1942.
31. Richardson, *The Forgotten Force*, p. 32, 34.
32. Williams, *Chronology 1941-1945 - 1942*, p. 38.
33. Craven, *The Army Air Forces in World War II, Vol. I*, p. 504.

34. Sibert, *personal diary*.
35. Richardson, *The Forgotten Force*, p. 37-8.
36. Story, *Private Story*.
37. ----, AFHRA, Hap Arnold file, Chiang to FDR, May 25, 1942.
38. Brereton, *The Brereton Diaries,* pp. 125-126.
39. ----, *Time, BATTLE OF RUSSIA: Push With a Difference*, May 25, 1942, http://www.time.com/time/magazine/article/0,9171,766559,00.html, accessed July 30, 2012.
40. Mitcham, *Rommel's Desert War*, p.46.
41. Williams, *Chronology 1941-1945 - 1942*, p. 38.
42. ----, FDRL, PSF, Box 28, Currie In Folder, Currie to Chunking, May 26, 1942.
43. Sibert, *personal diary*.
44. Walker, The Liberandos.
45. Cave, *transcript of interview*.
46. Story, *Private Story*.
47. Frank, *Guadalcanal*, pp 38-41.
48. Sibert, *personal diary*.
49. Richardson, *The Forgotten Force*, p. 43-4.
50. Williams, *Chronology 1941-1945 - 1942*, p. 39.
51. Williams, *Chronology 1941-1945 - 1942*, p. 39.
52. Williams, *Chronology 1941-1945 - 1942*, p. 39.
53. Sibert, *personal diary*.
54. Story, *Private Story*.
55. Richardson, *The Forgotten Force*, p. 43-5.
56. Cave, *transcript of interview*.
57. Story, *Private Story*.
58. ----, AFHRA, Hap Arnold file, Walsh to Arnold, May 29, 1942.
59. Richardson, *The Forgotten Force*, p. 47-9.
60. Sibert, *personal diary*.
61. Miller, unpublished diary, p. 7.
62. Williams, *Chronology 1941-1945 - 1942*, p. 39.
63. Craven, *The Army Air Forces in World War II, Vol. II*, p. 8.
64. Richardson, *The Forgotten Force*, p. 49-50.
65. Sibert, *personal diary*.
66. Brown, *War Stories*. Actually, 2 B-24s were still in Florida.
67. ----, AFHRA, Hap Arnold file.
68. Williams, *Chronology 1941-1945 - 1942*, p. 39.
69. Craven, *The Army Air Forces in World War II, Vol. I*, p. 567.
70. ----, FDRL, PSF, Box 3, India Folder, Madam CKS to Currie.
71. Richardson, *The Forgotten Force*, p. 50-2.

72. Story, *Private Story*.
73. ----, Army Air Force Statistical Digest, http://www.ibiblio.org/hyperwar/AAF/StatDigest/aafsd-3.html , accessed January 22, 2014, p. 112; B-24 production from Author's personal papers.

JUNE 1942

June began with a monumental victory for the Allies at the Battle of Midway and ended with major Allied defeat with the surrender of Tobruk. These two events would seal the fate of HALPRO.

Midway caused the Japanese to cease her major drive into India. And with the collapse of the Burma Road, there was now no way for HALPRO to operate from Chinese bases. The fall of Tobruk would force the United States to buttress the British defense of Egypt. The United States was now committed to the Mediterranean Theater of Operations (MTO).

With this new emphasis on the MTO, HALPRO began air operations against Axis targets. Realizing that one air group was not enough, Arnold committed a second air group, the 98th Heavy Bombardment Group, to the fray. In addition, the surviving bombers of MacArthur's air force (the 9th Squadron) were also committed to the Middle East. India and China were now left on their own.

Initially operating from a base in Egypt, the apparently unstoppable Afrika Corps forced HALPRO to withdraw to Palestine.

Politically, the fall of Tobruk presented another major dilemma for the Allies – the British press began questioning Churchill's leadership. With the coming mid-term elections, Roosevelt realized his position was also precarious. While he could not be removed from office, he could find himself with severe Congressional opposition.

And if the all of this was not enough, the Germans began Case Blue, the attack on the Caucasus oil fields.

REINFORCEMENTS

After flying reconnaissance missions over the Arabian Sea, Bill Mayhew joined a crew:

> *In early June, I was* assigned as the permanent ball turret gunner on Major Max R. Fennell's crew. Our crew consisted of Major Fennell, pilot; Captain William (Jumbo) Stewart, co-pilot; Captain Lewellyn Daigle, navigator; Master Sergeant Joseph (Joe) Taulbee, bombardier; Master Sergeant Joseph (Joe) Rose, flight engineer and upper turret gunner; Sergeant Keith (Mac) McJunkins, radio operator

and waist gunner; Sergeant Augustus (Pat) Patrick, tail gunner; and Sergeant Wilbur (Bill) Mayhew, ball turret gunner. Captain Stewart had trained with Jimmy Doolittle for the Tokyo raid from the aircraft carrier "Hornet" as a spare pilot, but he did not go on that mission. Instead he came to India and became Major Fennell's co-pilot. We flew shy one gunner because our pilot was somewhat superstitious. He had to kick one gunner off our crew earlier and he was afraid to possibly add another "weak link" to the crew. Our B-17E was eventually named "Fennell vs. Rommel" (Serial number 41-9029).[1]

1942 JUNE 1

HALPRO

Marshall sent the following to the Cairo mission:

- The War and State departments solicited the cooperation of the Soviet government in permitting the use of landing fields in the Caucasus-Ukraine area following the attack on Ploesti.
- The possibility of a return trip raid from Soviet bases was also suggested.[2]

1942 JUNE 2

MIDWAY

Admiral Fletcher's Task Force joined Admiral Spruance's Task Force. Fletcher ordered the combined fleet to a station 200 miles north of Midway. (see Figure 10)

Meanwhile, reconnaissance PBYs, flying from the Aleutians, found the diversionary Japanese invasion force.[3]

EASTERN FRONT

Germany renewed her attack against Sevastopol with a five-day artillery barrage.[4] (see Figure 8)

COMMITMENT TO CHINA

Soong reminded Arnold that two weeks had elapsed since their meeting on air transport. Arnold had agreed to review the situation and had asked for two weeks to study the problem. Soong had two questions:

1. If further DC planes were being sent in accordance with previous commitments?
2. If the B-24 transport type (C-87) would be sent and if so, in what numbers?

Soong also sent a copy of his letter to Harmon saying, "I shall look forward with interest to the substantive reply you hope to give us within the next few days."[5]

1942 JUNE 3

MIDWAY

By June 3, 17 B-17s and four B-26s were on Midway Island to assist the 30 Navy PBYs in reconnaissance missions. At 0904, one of the reconnaissance planes found two Japanese ships 470 miles west of Midway. An attack was mounted by nine B-17s from the 431st Bomb Squadron on a transport force 570 miles west of the Island.[6] (see Figure 10)

1942 JUNE 4

MIDWAY

The Midland Island PBYs and B-17s resumed the search for the Japanese fleet. At 0545, a PBY crew reported "many planes heading Midway" 150 miles north west of the Island. By the end of the day, planes from the three United States carriers attacked and sank two of the four Japanese carriers. The remaining two Japanese carriers would be sunk the following morning. (see Figure 10)

HALPRO

Halverson radioed Marshall that "it would be of great assistance to use a field in Krasnodar-Rostov area."[7]

Marshall, in turn, told the Soviet ambassador:

> *I understand from Pres. Roosevelt* that his Excellency, Commissar Molotov, had stated that there would be no difficulty in arranging for the reception at an airport in southern Russia of the United States Army planes shortly to take off from the Middle East with a mission of bombing the refinery and other oil installations at Ploesti and Rumania.

Marshall then asked:

- Is this arrangement possible?
- Can the planes be refueled and re-armed so they could attack on the return trip to the Middle East?

He stated that 24 B-24 planes would be ready to take off from Aleppo (Syria) within ten days.⁸ Marshall was contemplating a flight path for the HALPRO planes that would take them west from Aleppo along the southern border of neutral Turkey, then north up the western edge of neutral Turkey to Ploesti, then head straight east to Krasnodar. (see Figure 46)

While Marshall was busy trying to negotiate the use of the Russian base, Arnold had fired off a series of questions to the Cairo mission. First, he wanted to know why the British had not attacked Ploesti in October of 1941. The answer was that the British lacked the planes to execute the mission. Second, Marshall wanted a status report. During the winter of 41-42, Germany had built up her stock of oil supplies. Presently, all Rumanian output was being sent to the Eastern front. There was also a comment that while the Rumanian government favored the Axis, it was not certain that the general populace did.⁹

1942 JUNE 5

GLOBAL STRATEGY

The United States of America declared war on Bulgaria, Hungary and Rumania. The United States had to declare war first, prior to violating their respective air space and/or dropping bombs. Otherwise, she would be guilty of repeating the same "offense" that the Japanese had committed.¹⁰

HALPRO

Halverson called a staff meeting and reported that he had been summoned to Fayid, Egypt, the next stop on HALPRO's original itinerary. Halverson's party was immediately summoned to a conference. At this meeting, the HALPRO attendees were briefed on the critical state of the desert war in North Africa, and Halverson was informed that the Japanese had overrun the advanced refueling bases in China. Major General Russell Maxwell, Commanding General, United States Forces in Middle East, produced a wire from Arnold directing HALPRO to hold indefinitely at Fayid and participate in a bombing mission against the oil refineries at Ploesti, Rumania.¹¹

A letter was sent requesting Russian approval for the use of one or more airdromes in the Stalingrad-Astrakahn area to serve as refueling bases for a B-24 squadron. It was anticipated that about 20 B-24s would participate in the mission.

Belyaev was requested to query the Soviet government about granting approval of a repetition of the mission from Russian bases and that arrangements for the mission would be made between Halverson and the local Russian commander.¹²

1942 JUNE 6

NORTH AFRICA

Auchinleck assured Churchill that Tobruk was strong and was prepared for an 80-day siege.[13]

Axis forces concentrated their attack on the Knightsbridge area and Bir Hakeim, southwast of Tobruk.[14] (see Figure 47) It appeared Rommel was trying to encircle Tobruk.

HALPRO

Arnold sent a message to Halverson, noting that while discussions were occurring, approval had not yet been received. If approval was received, the objective would be to hit Ploesti twice, once going in and another coming back.[15]

Meanwhile, the British had additional ideas on how to use the HALPRO bombers. Malta is an island in the Mediterranean and a British outpost, situated squarely in the middle of the German supply lines to Rommel's forces. As such, it was a constant threat to the flow of men and material.

All British efforts to supply Malta were harassed by German forces. The British decided to split the harassing German forces by sending in two supply convoys, one from the east; the other from the west. The convoy from the west faced greater prospect of German interference due to the British withdrawal in the face of Rommel's progress in Libya. When Tedder was informed that HALPRO would be in the Middle East, he had requested their assistance in fighting the Italian fleet if the fleet should depart to intercept the convoy. After some hesitation, the War Department approved the request on 10 June.[16]

1942 JUNE 7

MIDWAY

The front page of the late edition of the June 7 issue of *The New York Times* headlined:

2, PERHAPS 3, JAPANESE CARRIERS SUNK WITH ALL THEIR PLANES, NIMITZ REPORTS [17]

EASTERN FRONT

After a five-day artillery barrage, German forces marched on Sevastopol.[18] (see Figure 8)

1942 JUNE 8

SOUTH PACIFIC

Following the success at Midway Islands, MacArthur saw an opportunity to strike Japanese forces. He proposed that if he were given a division of landing trained assault troops, he could re-take Rabaul and drive the Japanese back 700 miles. He also requested two carriers to provide air cover for the landings. Since this plan required the use of Navy assets, the Navy needed to agree to the plan. Needless to say, the Navy was not about to release any of its assets to the Army. To justify its opposition to the plan, the Navy stated they were reluctant to operate its carriers within an area surrounded by Japanese air bases. They counter with a proposal to take the outlying Japanese bases prior to any assault on Rabaul.[19] (see Figure 19)

While the Army and Navy were debating what the strategic plans should entail, the first Japanese troops landed on Guadalcanal.[20] (see Figure 45)

MIDDLE EAST

The WPD considered the United States commitment to the Middle East so important that it proposed the creation of a United States Middle Eastern Theater Commander equal in rank to the British counterpart.[21]

HALPRO

HALPRO received orders that they were to proceed to the air base at Fayid, Egypt.[22] James Sibert and *Queen B* were the sixth crew to leave Khartoum.

Up at 04:30 and a "Haboob" Sand storm was in progress. Ate breakfast and went out to the ship to take off. Was ship #6 to take off and after 15 min. delay due to lowering of visibility I finally got off. I made an instrument takeoff as I could only see half way down the runway, I watched my gyro compass and airspeed, when I had enough A.S. and saw the end of the runway directly ahead I pulled it off and stayed on instruments until I had climbed to 7000 ft before seeing the outline of the sun. At 7000 we were on top of the dust level and set our course for Fayid. 50 miles out we were out of the dust area and flew across the Sahara until we hit the valley of the Nile. Flew up the "Green Valley" until we were almost to Cairo then went on to our base. Landed. Hauled our baggage into barracks. Had 2 cans of cold American beer. Ate supper and went to bed.[23]

Richard Miller, *Queen B* copilot, recorded the flight also.

Dust and More Dust!! Left today for Fayid - an R.A.F. Station 60 miles east of Cairo.

We are to be right on Bitter Lakes which is part of the Suez Canal _ 20 miles from the main part of Suez canal. Took off on instruments because the dust was so heavy - in fact, 9 planes had to stay in Khartoum to await clearer weather. Flew across the Libyan Desert. Gad - that would be a rugged place to have to set down. Followed the general course of the Nile for a few hundred miles. Saw the Pyramids - fascinating piece of engineering. Cairo looked mighty good and green on the way up, we got into the British Defense corridor - a very good system. For example, we flew around one town at a certain distance from it and at a designated altitude. On entering the vital zones we had to fire Varey signals. Likewise on the approach to the field, we fired the Signal. Looks like we are going to really get into the war for keeps.[24]

1942 JUNE 9

HALPRO

We were not unhappy when word finally came that we were to move on. We were flying to a British base, Fayid, on the Suez Canal just north of Great Bitter Lake. On June 9, 1942, we flew down the Nile River to Cairo and east to Fayid, our new home for twenty days.[25]

Winds reached gale strength during the takeoff process, reducing visibility to below minimums. Those who had not left remained at Khartoum.[26]
Miller was informed for the first time about HALPRO's first mission:

Tonight we had a meeting and found that we haven't stripped our ship for nothing. We are to raid a spot that the British have wanted to hit for two years - must be in Italy - who knows. It is exciting as hell - this preparing for a raid - wish to hell we could go out every day!![27]

1942 JUNE 10

NORTH AFRICA

British commander Ritchie ordered the withdrawal from Bir Hakeim.[28] (see Figure 47)

COMMITMENT TO CHINA

Soong had asked Arnold when China could expect delivery of the promised planes. Arnold's reply pointed out the obvious:

> *It is to me a* source of regret and great concern that I cannot give you a more favorable picture of the transportation of supplies.
>
> It is fruitless to throw into this activity more transport planes than can be profitably operated when they are so vitally needed in other theatres.

Arnold promised to take measures to build the forces up to a total of 75, but

> [I]t is impossible for me in the face of existing combat demands for heavy bombardment airplanes to divert B-24s to this project. I feel that you will agree with me that the cause of China will be better advanced by their direct application in combat against our common enemy.[29]

Finally, the United States was admitting that the defense of China was secondary.

HALPRO

Eighteen HALPRO B-24s were on the ramp at Fayid. Fourteen were considered combat ready; the remaining four required minor maintenance after the long flights from the USA. At noon, a meeting was called and HALPRO's operations officer George McGuire announced the fourteen crews for the Ploesti mission.

Three meetings were held regarding the impending mission. Bernie Rang assembled the other navigators and pointed out the target on a map of the Middle East. The route discussed included over flying neutral Turkey. An RAF officer adamantly cautioned the crews to detour around Turkey to avoid violating that country's neutrality.

Clearly, there was a lack of unanimity between the British and HALPRO positions on over flying Turkey's borders.[30]

At lunch, John Payne, HALPRO Operations Officer, announced the crew and plane assignments for HALPRO's first mission.[31]

Meanwhile, Belyaev reported that he had not received a response from his government. He did not know the reason for the delay.[32]

James Sibert and crew were assigned to this mission.

> *Opened my eyes at 07:30* and drank the hot Cup of tea that my batman brought each morning. Ate breakfast and was notified to have my crew and ship in shape for a mission on the following day. Went out to the ship to notify crew and pay

per diem. S/Sgt. Francis J. Laney, my crew chief vouched that it was in best of condition so we talked about conditions in combat and what to expect on our first mission. Came back to Officers Mess and ate some onions, tomatoes, Bully beef, and other mixtures. Went over to operations after dinner and turned in a list of my men on my crew and data on the officers in event we didn't return from the mission. Ate supper went to bed early and slept well.[33]

1942 JUNE 11

BOMBING JAPAN

Arnold responded to the proposal about bombing Japanese volcanoes:

- Bombing volcanoes should be considered due to destructive results of Hawaiian lava flows.
- At this time, however, our opportunities for bombing Japan are very limited.[34]

HALPRO

The Cave crew had damaged *Hellzapoppin* on their landing at Fayid. So, here was a crew without an airplane. Meanwhile, Andrew Moore, pilot of *Blue Goose*, was hospitalized for depression. Moore was left at Khartoum and Col. Feldmann helped ferry the plane to Fayid. HALPRO had a plane without a complete crew. Cave's crew was assigned to take *Blue Goose* on the mission.

In his diary, *The Forgotten Force*, Flight Surgeon George Richardson's recorded his impressions of the take-offs between 10:30 and 11:00 that evening:

> *The take-offs were a thing* of beauty. The big ships trundled to the end of the runway in the darkness, where their searchlight-like landing lights pierced the blackness for seemingly miles. With turbos whining, the ships would thunder past our view point, shaking the ground they left with such effortless ease, running lights and landing lights, together with the exhausts, marking their path upwards. Soon the landing lights were extinguished and the big ships became nothing but four pinpoints, from the exhausts, in the reaches of upper air.
>
> My old ship [Black Mariah II] rolled past the tiny figures on the flight deck vaguely outlined in the half-light of the fluorescent illumination and the landing lights illuminating the edge of the runway. Kalberer, Rhoades, Halverson and Rang fairly pulled their ship into the air. Kal's first takeoff with bombs was a precision set of movements with a four-engine ship. General Whang, Bill Wil-

liams and I watched them all off and bade them Godspeed over target on June 12, 1942.[35]

1942 JUNE 12

HALPRO

After the mission, Halverson sent the following summary to Arnold on June 13:

- Attacked Ploesti refineries and pumping plant at Constanta with 13 planes at dawn June 12. Results not determined. [Bombed] through an overcast at 12,000 feet with intermittent breaks. Part of bombing was under overcast.
- No personnel lost or wounded. Three planes and crews landed at Ankara, Turkey, and one landed at Izmit, Turkey, due to damage. One slightly damaged on landing at Ramadi, Iraq, which will be repaired. One plane damaged on landing 150 miles from Ramadi. Rest in good condition.
- Operation was conducted from Fayid airdrome in Egypt. Distance to target 1269 land miles.
- All planes were to terminate flight at airdrome near Ramadi, Iraq, if possible. My own plane and eight others made this terminus. Total distance 2600 miles.
- Colonel [Demas] Craw has special detail of getting personnel and planes out of Turkey.
- Definite recommendations as to continued use of this force will be sent June 14th.[36]

HALPRO crew members would have many recollections of the first and subsequent mission.(Appendix A)

The approval to use Soviet airbases was tentatively given. A notification was sent to Halverson in Cairo saying the Russians had agreed to permit a landing at Armavir or Krasnodar. (see Figure 42) 95 octane fuel was available at both fields.[37] Unfortunately, the planes had already departed.

The HALPRO personnel who were not executing the Ploesti mission were meeting their British counterparts. Those "meetings" planted the seeds of what would become a rocky relationship between American and British allies. In his diary, George Richardson noted the following:

> *The RAF Liberator Squadron was* equipped with the early LB-30's, that could not get the altitude desired nor the speed required. They had been unable to go [bomb Ploesti themselves], but all the lads [HALPRO] awaited the next target.

We were suspicious. It was later learned that we had assured the RAF that our boys could handle the long distance mission but why, if the British had Palestine and Syria, had we not gone to Habbaniyah, Iraq or to Rayak, Syria, or to Lydda, Palestine for initial take-off? We could have flown to Ploesti much easier and had enough fuel for return. After all, lots of the aviation gasoline came from Mosul to Haifa. Bombs could have been carried to Rayak, armed and gas tanks "topped" for take-off.

The British knew of our coming. We are told by several, after a few drinks, that we were not going to China despite our wishes. And, we had settled down in rooms all prepared – name plates and all, for our coming. These were our Allies – but in some fashion, we were being told nothing of plans.

Nick Craw left for Cairo. Paul Zuckerman remained. When were we going on to China? Whang was worried with the desperation born of frustration. There in his company was still a fair striking force to cheer the Chinese. Our doubts began to assail us. The RAF was making our decisions and the first had not been such a good one. Perhaps we had been over-confident.

The British knew of the decisions made in Washington, but no one told HALPRO. And if that was not depressing enough, George Richardson noted that the Italian prisoners seemed to know what was going on:

> *It was strange how the* "Eyetie" prisoners on the post asked about the raid and knew the target the next day. One ex-Italian dive-bomber's gunner said that he would have liked being along to bomb "dem goddam Germans".[38]

While HALPRO was busy flying its first mission, Allied leadership in Cairo was worried about chain of command. In essence, General Maxwell was concerned about HALPRO operating in his theater as an independent force.[39]

1942 JUNE 13

NORTH AFRICA

On 12 and 13 June, Rommel succeeded in luring Ritchie's numerically superior Eighth Army into a tank trap in the Knightsbridge area, known as the "Cauldron". In the "Cauldron," 230 British tanks were destroyed. Throughout the Eighth Army, June 13 became known as "Black Saturday".[40] (see Figure 47)

HALPRO

Despite its modest results, the Halverson raid was as significant in its way as any the AAF had flown in the six months since Pearl Harbor.

- It was the first American mission in World War II to be leveled against a strategic target, if the Tokyo raid be excepted.
- It struck at an objective, which later would become a favored target for American bombers.
- It was the first blow at a target system whose dislocation contributed mightily to the final German collapse.
- It was the first mission by what later came to be known as the Ninth Air Force.[41]

This is one historian's opinion. Others argue that this raid alerted the Germans to the vulnerability of Ploesti to AAF heavy bombers. Hitler had already expressed concern over the deployment of RAF bombers to Greece in 1941, which he knew could reach Ploesti. But, he did not suspect that an AAF bomber, i.e. the B-24, could reach Ploesti from North Africa. Now, the Germans knew for sure.

The HALPRO crews became aware of the reality of the casualities of war. George Richardson noted:

> *The RAF went out one* mission the night of the thirteenth and returned the morning of the fourteenth. News of the mission circulated rapidly. Fairly successful, but one man killed and two wounded. The dead man had been a top turret gunner. Enemy fire had gone through turret and individual. The deck of the ship was a pool of blood above the bomb bay. The boys began to realize that war was not and could not always be a bowl of big, ripe cherries. They wished for the Servel system of feeding ammunition to the guns, 10,000 rounds of 30 cal. for tail gunner set of four.[42]

1942 JUNE 14

NORTH AFRICA

The British began a headlong retreat towards Cairo, known as the "Gazala Gallop", to avoid being completely cut off.

HALPRO

Something was obviously being planned. HALPRO personnel could only speculate on what was being planned.

There was a constant stream of Group Captains and Air Commodores to see Colonel Halverson. Why?" Whang was infuriated. He was never invited to the conferences. The RAF simply ignored him. Pressure politics was coming to the fore.[43]

The uncertainty about their future was adding to HALPRO's discontent.

1942 JUNE 15

COMMITMENT TO CHINA

Drew Pearson and Robert S. Allen wrote a nationally syndicated article with the following headline:

ARNOLD'S STAFF BLAMED FOR FAILURE TO SEND BOMBERS TO CHINA

The body of the article stated:

- Most tragic neglect of the war probably has been our failure to send bombing planes and munitions to China.
- Without US support, China would be to the USA what Singapore was to the British Empire.
- Despite some US headline's about US supplies to China, actually there was a mere driblet within the past three months.

However, the greatest tragedy was that,

time after time" T. V. Soong would go to the White House and get a promise of a certain number of bombers, "but then the planes are not sent. Someone around Gen Arnold in their Corps holds them up despite White House orders. And little by little, Generalissimo Chiang Kai-shek is beginning to wonder what an American promise amounts to.[44]

This article appeared three days after Halverson's mission, raising the question whether General Whang called Chiang, who in turn called the reporters.

HALPRO

While the Pearson/Allen article was being published, HALPRO flew its second mission - in support of the British. Five HALPRO B-24s[45] and one Liberator of 160

Squadron, RAF were ordered out with torpedo-carrying Beauforts against the Italian fleet, which had now put to sea. The Beauforts sank a cruiser. The HALPRO planes also attacked the fleet, claiming hits on a *Littorio*-class battleship and a *Trento*-class cruiser. According to the RAF, the inflicted damage kept the two ships in dock for the next three months.

Returning to base at minimum altitude, the bomber formation encountered and shot down an Me-110. This was the first aerial victory in which Americans had participated in the Middle East.

On the way to the Italian fleet, the planes passed over the British convoy on its way to Malta. The Italian fleet had been launched to intercept the British convoy. The mission of the HALPRO and RAF planes was to turn the Italian fleet. The British escort vessels fired upon the HALPRO and RAF planes.

> *Colonel Halverson and Kal [Kalberer]* had a piece of British "ack-ack" through the side of the ship. That was their nearest escape and the only opposition encountered on the trip out.[46]

1942 JUNE 16

MIDDLE EAST

On the day after the raid on the Italian fleet, *Völkischer Beobachter* [*People's Observer*], the daily newspaper published by the Nazi Party, ran an article about the mission. In it was the following analysis:

> *Without crushing Malta and impeding* nearly all the movements of British transport by sea it would not have been possible to dispatch to the German and Italian combatants in Africa the weapons and materiel used by them now so successfully in fighting the much vaunted British Eighth Army. One realises how completely have been turned the tables in the Mediterranean.[47]

1942 JUNE 17

NORTH AFRICA

Axis forces now controlled the road to Bardia. Tobruk was surrounded.[48] (see Figure 44)

MIDDLE EAST

Maxwell was named Commanding General of United States Army Forces in the

Middle East (USAFIME) . In the same communiqué, he was informed that HALPRO would report to him. He would formally assume this position on June 19.[49]

WASHINGTON CONFERENCE JUNE 1942

Meanwhile, Churchill departed for conferences in the United States He would later admit that he had not been fully informed of the situation facing the British Eight Army in North Africa at the time of his departure. He was fully apprised when he arrived in Washington. Once there, he made a plea for American aid.[50] At this time, the only available American force was HALPRO.

HALPRO

Halverson informed Arnold that:

- One more cooperative mission in the Mediterranean area would deplete his unit to such an extent that its primary mission could not be accomplished.
- The lack of spare parts for B-24s in this theater would make operational maintenance a serious problem. HALPRO supplies had been shipped to Karachi and were already deep within India, (see Figure 32)
- He would have to resort to cannibalism of existing aircraft, which would prove an expensive and disheartening process,
- He asked permission to proceed on to China.[51]

Arnold ordered him to stay put, and "to employ his force in cooperation with the RAF in its Middle Eastern operations".[52]

COMMITMENTS TO CHINA

HALPRO personnel watched this political intrigue, knowing that their individual futures were being bantered as pawns. George Richardson recorded the feelings of the HALPRO personnel:

> *June seventeenth was a busy* day and an important one. Air Marshals, Commodores, the Air Section of the U.S.N.A.M.M. (Col. Smith, Pennington et al) descended on Colonel Halverson with sheafs of cables from General Marshall "to remain in the Middle East until further orders" and "to cooperate with the British and Sir Arthur Tedder" in every respect.
>
> Forgotten was the secret paper carried by Colonel Halverson giving him clear and unhindered path to China. The boys could see the tight little knot cluster ever tighter around Colonel Halverson. He opened his secret pouch, waved the paper. All were adamant. The final touch was a further order from Major Gen-

eral Russell Maxwell. CG USNAMM assuming command of an air unit and ordering Colonel Hal personnel and materiel to the Mission, the first of the U.S. Middle East Air Force. Just a breath of political but necessary intrigue. General Whang was never invited to any of the conferences threatening his unit. It was the first, as evidenced by his chagrin, of those "Slaps in the face" that Lin Yutanp discusses in his book of Laughter And Tears.[53] The General hid bitter tears.

Well, here it was. We were part and parcel of this hard, desolate land. General Maxwell would never let anything go. Nick Craw was already pulling strings to get away despite the General's wish for him to remain. Whitlock and Medford flew Whang on to India and China. Never can the General's disappointment be forgotten. He said he was glad of one thing, "to know how these American boys felt about China."

Global strategy is hard to understand. I confess that I and many of the boys suffered acute disappointment at being "shanghaied" from the Chinese job and jobs. We were all very fond of General Whang and the feeling was reciprocated. We were specially equipped for a China task. All the medicines and other supplies had been gathered for a specific purpose in a specific location. Our Intelligence Section of Colonel Shumaker, Bill Williams, and Wilfred Smith knew China inside out. Col. Shumaker had been thorough and in China on many occasions. Bill Williams had spent eighteen years in China. Smitty had been born in China, studied China and knew it all. Beyond all of this our entire first team had a score to settle with our "little brown brothers" – the Japs.[54]

1942 JUNE 18

EASTERN FRONT

After 12 days, Germany claimed the northern part of Sevastopol.[55] (see Figure 8)

NORTH AFRICA

Axis forces drove to Gambut, the airfield near Tobruk. Now, Tobruk could not be supplied by air.[56]

With the British lines in North Africa about to collapse, the military commanders should have had a lot on their minds. Instead, June 18 would see an exchange of messages, regarding who was in charge of the day-to-day operation of a group. Maxwell told Marshall:

> *that the experience of the* Halpro Mission in the Middle East had demonstrated the unsoundness of sending small combat groups to that theater without giving orders to the unit commanders to report either to British authorities or to him.

Marshall's response was that:

> *It was not the intention* of the War Department that the planes of the Halverson Detachment should be employed in local tactical operations unsuited to the technical characteristics of heavy bombers.[57]

This debate was nearly identical to the one between Stilwell and Washington.

Then, given the gravity of the situation facing the British Eighth Army, Marshall asked Maxwell to prepare plans to evacuate from Cairo.[58] (see Figure 50)

HALPRO

While the leadership argued about who was in charge, HALPRO had to deal with the realities of working with the British. George Richardson wrote:

> *Our first relations with the* British were not a howling success. My personal travel started when [Kenneth] "Bombsight" Butler, stricken with bacillary dysentery was discharged to the RAF for "boarding" - to pass on his physical fitness to fly. I went to the RAF people saying that the unit had its own medical section, that we had come with the boys, knew them and would pass on them. One pompous Wing Commander, Medical, stated that one man couldn't be a board, and that we were RAF and he would run the medical end. After all, Colonel Kendricks and I had been partially and totally charged with the responsibility of maintaining their health before we ever saw a Wing Commander or the RAF. These individuals, however, were determined to be nasty about it until they "got a ruling from the PMO in Cairo". The PMO is the Principle Medical Officer and quite a "big dog". ... Ah, me, but the tactfulness of these individuals was really furthering amicable relations.
>
> Colonel Halverson and the boys would not have it. Some would not report ill because of fear that the RAF might get them off flying. Beside, their "doc" was not on the board--nor ever 'invited to sit' on such a board. Needless to say, no operations were ever cancelled because of illnesses which lessened support to these nonsensical arguments which went on by the hour -- even at the bar and in the mess. The RAF lost that battle -- they did not "board" our personnel. Gawd! It would be nice to see Colonel Kendricks and have some advice on such procedure. The pressure group was becoming local-the RAF Squadron Leader at Station Sick Quarters constantly telling me that there was no reason not to let my men be boarded, sent to convalescent depots, etc.[59]

OPERATION VELVET

A rather interesting article appeared on the first page of *The New York Times*:

U.S. BOMBERS AIDING SEVASTOPOL FORCE.

The lead sentence of the second paragraph read: "The planes were flown direct to Sevastapol from bases in the Middle East,..."[60]

Since United States bombers had just arrived in the Middle East, it was highly unlikely that any planes had flown to aid Sevastopol, as the article claimed.

COMMITMENTS TO CHINA

> *Gen. Whang left yesterday [June* 18] in one of our planes for Chunking. He has been very patient up till now. We have been fooling around getting tied up with the British, and now we are about to be attached to this Middle East outfit.
>
> So the General is going to Generalissimo Chiang Kai-shek and get him to wire Franklin D. Maybe we'll get action then. Now we are waiting.[61]

NEW GUINEA - SOLOMON ISLANDS

The 435th Bombardment Squadron conducted the first Allied photographic reconnaissance mission over Guadalcanal, Solomon Islands. (see Figure 45)

1942 JUNE 19

OPERATION BOLERO

Stimson sent a six-page memo outlining the reasons why he believed that Operation BOLERO was the most viable military option open to the United States. Any other option, in particular Operation GYMNAST, which jeopardized BOLERO, would be a mistake.[62] Marshall attached a cover note to Stimson's memo indicating that Arnold, McNarney, and Eisenhower had all read the memo and were in complete agreement with it.[63]

Marshall also sent Roosevelt a note stating that Churchill was "pessimistic" about BOLERO and continued to be interested in GYMNAST or a similar venture into Norway.[64]

10TH AIR FORCE

Marshall told Stilwell and Brereton, who were in India, that:

HALPRO would be assigned to the 10th Air Force, but he gave no indication as to when the assignment would be made.[65]

Marshall was telling Maxwell that he was in command of HALPRO at the same time he was telling Stilwell that HALPRO would be part of the 10th Air Force.

REINFORCEMENTS

The B-17, *Topper*, took off on a bombing mission to Rabaul. (see Figure 19) *Topper* was the plane that Paul Eckerly and crew had ferried over as part of Project X. Japanese fighters engaged her in a 35-minute battle. *Topper* made it back to base, but crash-landed in the process, breaking its right wing main spar. Her combat days were over.[66]

1942 JUNE 20

NORTH AFRICA

Axis forces drove into Tobruk.[67] (see Figure 47)
Word leaked out about the tenuousness of the British situation at Tobruk. Brereton, in India, made the following notation in his diary:

General Wavell received a confidential message from Auchinleck stating that Tobruk would fall within two days. The Libyan situation had changed with catastrophic speed from the first of the month, and the British Eighth Army was in full retreat.[68]

Just two weeks earlier, Auchinleck had told Churchill that Tobruk could withstand a three-month siege.

SECOND FRONT

In response to Stimson's memo, Roosevelt asked his military leaders what options the United States had for mounting a mid-September offensive, assuming a German breakthrough on the Russian front.[69]

MIDDLE EAST

At a meeting of the Combined Chiefs, Marshall indicated that he was not enthusiastic about sending more troops to the Middle East other than those absolutely necessary to avert a British collapse. Limited assistance to the British in Egypt could block the German assault in North Africa and thereby keep plans for the invasion of France alive.[70]

HALPRO

The White House *Daily Summary Report* reflected decisions and actions. Regarding the African-Middle Eastern Theater,

> ***Colonel Halverson's heavy bombardment force*** has been ordered to continue to function in the Middle East provisionally, cooperating with the British. It will not be used in local tactical situations.[71]

The decision to keep HALPRO in the Middle East did not go over well with the men. Richard Miller recorded the following:

> ***All the men, including me,*** are hacked at staying here. First, we want to get at the Jap. Second, we think the British pulled a fast one to hold us here. Third, we don't like the prospect of being under General Maxwell, an Ordinance General. Fourth, we think U.S. should keep its promise to China. Here General Whang, the second man of China - successor to Chiang Kai-shek - spent months trying to get us to China to help relieve the pressure and now we are stymied.[72]

1942 JUNE 21

NORTH AFRICA

Rommel captured Tobruk.[73] (see Figure 44)

Churchill received the news of the fall of Tobruk while in Washington. He remembered it as "one of the heaviest blows I can recall during the war . . . a bitter moment."[74] Bonner Fellers was a bit more brutal in his evaluation of the British.

> ***With numerically superior forces, tanks,*** aircraft, artillery, and transports, reserves of all classes the British Army has twice failed to defeat the Axis in Libya. Under their present leadership and with their present casual measures the British cannot be given enough lease-lend equipment to win a victory. The VIIIth Army failed to maintain the morale of its troops. Its tactical conceptions were consistently faulty, it neglected completely the use of combined arms; its reaction to lightning battlefield changes was sluggish; it was without foresight in planning evacuation of supplies. The only remaining certain and effective method of destroying Rommel is to unify Air and Army commands, to reorganize the VIIIth Army under new leadership and new methods, to delay and to contain the Axis forces, at the same time interrupt shipping so as to deny vital supplies to the Axis.

He recommended:

- If the Middle East fell, India would be lost because it would be caught in a vise between two powerful forces; and it would therefore be almost impossible to maintain Allied forces, above all air power, in the theater.
- He recommended that the U.S. Army Air Forces (AAF) in India be immediately transferred to the Middle East along with an air support command consisting mainly of B-24s.[75]

GLOBAL STRATEGY

The press continued to return to the idea that the Germans were pushing towards the Middle Eastern oilfields. *The New York Times* published a map, similar to Figure 49, as their idea of the German objectives.

1942 JUNE 22

NORTH AFRICA

Axis forces occupied Bardia. (see Figure 5) British Eighth Army retreated to Marsa Matruh, Egypt.[76] (see Figure 44)

Maxwell told Marshall that he was "perfecting plans to fall back with their heavy bombers toward the Persian Gulf area, in case the Eighth Army was destroyed."[77]

HALPRO

On the evening of June 21/22, HALPRO flew its third mission, this one against Rommel's supply port at Benghazi. (see Figure 50) They put nine planes in the air. This mission marked the beginning of a series of strikes by HALPRO against Rommel's supply lines.[78]

This was Therman Brown and crew's first mission.

> *I remember my first mission.* This was on the night of June 21-22, 1942. The target was the shipping in the harbor at Benghazi. We were to bomb in the middle of the night and return to our base at Fayid – a ten hour mission.
>
> Spotting the target at night would have been difficult if only the enemy had doused the search lights and held their fire. In some cases, we would never have found the target at all. You could depend on them turning on the lights and start shooting at something.
>
> There was someone ahead of us. Turning onto the target was no problem. All hell was breaking loose up ahead. I don't remember being afraid but I can remember my feet shaking a little on the rudders as we made the turn onto the

inferno above the target. Nothing happened to us. The anti-aircraft fire continued to follow the poor souls ahead of us. We essentially got a free pass. Getting caught in the search lights isn't fun but the lights don't hurt you. They only help the anti-aircraft gunners improve their accuracy.[79]

GLOBAL STRATEGY

As happened during the Arcadia meetings in January, Roosevelt and Churchill made decisions without their military advisors. Stimson recorded the following:

At the War Council this morning . . . Marshall went over the story of his negotiations yesterday with Churchill and the President in a very sketchy way. But after the others had gone, he told me of what had occurred in the evening session that had been called ostensibly for questions of naval cooperation between our navy and the British in the south Pacific. But after that had been handled and when it was very late at night, the President, suddenly suggested that we might throw a large American force into the Middle East and cover the whole front between Alexandria and Teheran. Marshall was of course terribly taken aback and he said to the President that that was such an overthrow of everything they had been planning for, he refused to discuss it at that time of night in any way and he turned and went out of the room.[80]

THE FALL OF TOBRUK

The fall of Tobruk signaled a fundamental change in American strategy. Unlike Churchill, Roosevelt never saw himself as a military strategist. Yet, in the early months of the war, Roosevelt would overrule his military advisors and defer to Churchill. This practice was surprising because, like many of his advisors, Roosevelt was suspicious of British objectives. And United States leadership was not alone in having those reservations. Public opinion polls in the summer of 1942 reflect that about 25% of the American people thought Lend-Lease to the British was a waste, and that Churchill was a schemer, not a heroic leader. While former American isolationists held this Anglophobia, columnist Walter Lippmann admitted that the war had stirred many anti-colonial feelings among Americans. Thus, Roosevelt had to tread lightly. He had to support the British allies, while at the same time, not appearing to be subservient to British objectives.

For one historian, Douglas Porch, the fall of Tobruk was significant for three reasons:

1. Tobruk's fall did not alter the strategic balance in the desert.
2. It signaled to Washington Britain's, and more specifically Churchill's, apparent weakness.
3. Rommel's operational triumph stimulated Hitler's victory fever. Just at the moment that the strategy of the Western Allies found a solid footing, Axis strategy was catapulted into disarray. 'News of the swift capture of a fortress that had been so fiercely contested the previous year sent Hitler and Mussolini into such raptures that they forgot all their existing plans,' concludes the official German history.[81]

As far as Roosevelt was concerned, the Fall of Tobruk had forced his hand. The United States would defend Egypt and the Middle East.

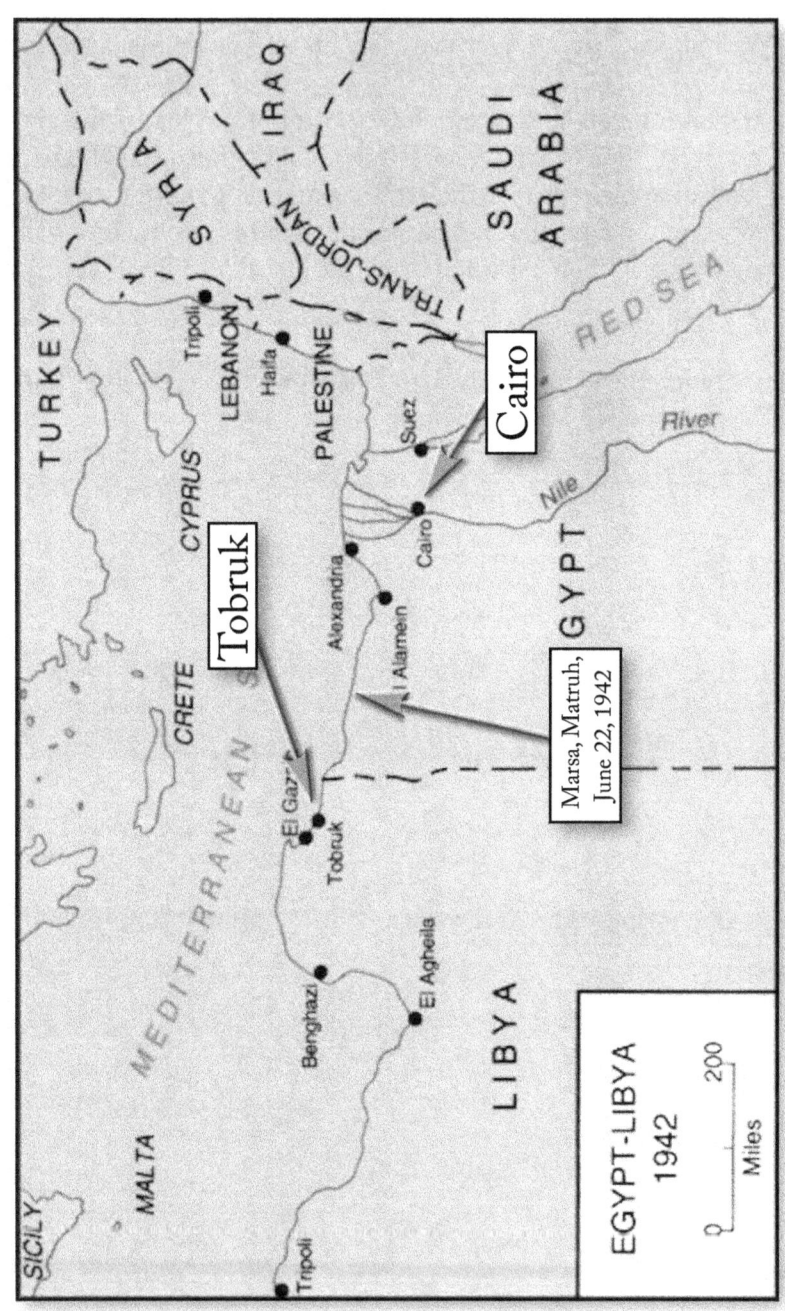

Figure 44– Mediterranean – June 1942

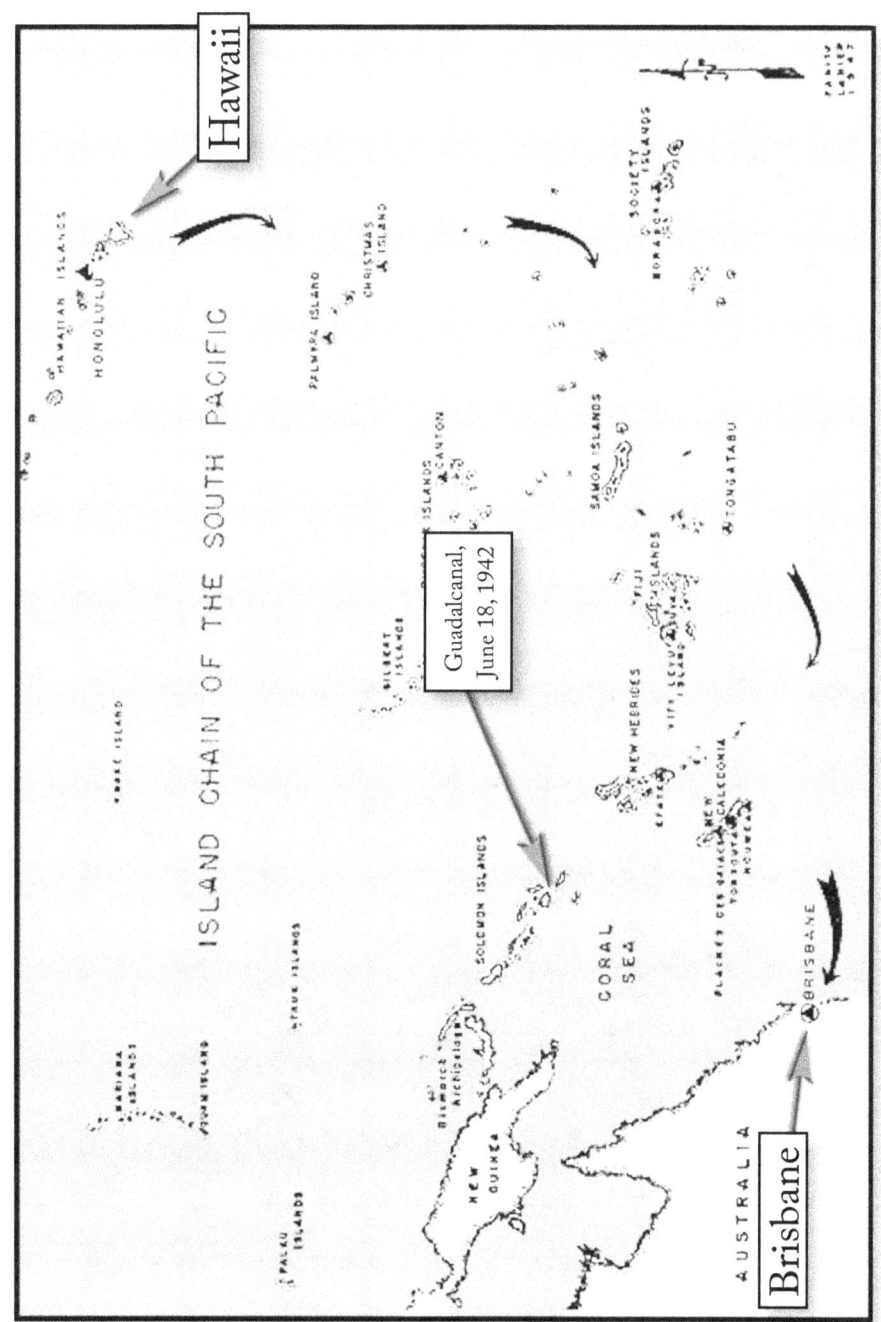

Figure 45 – Supply Chain – June 1942

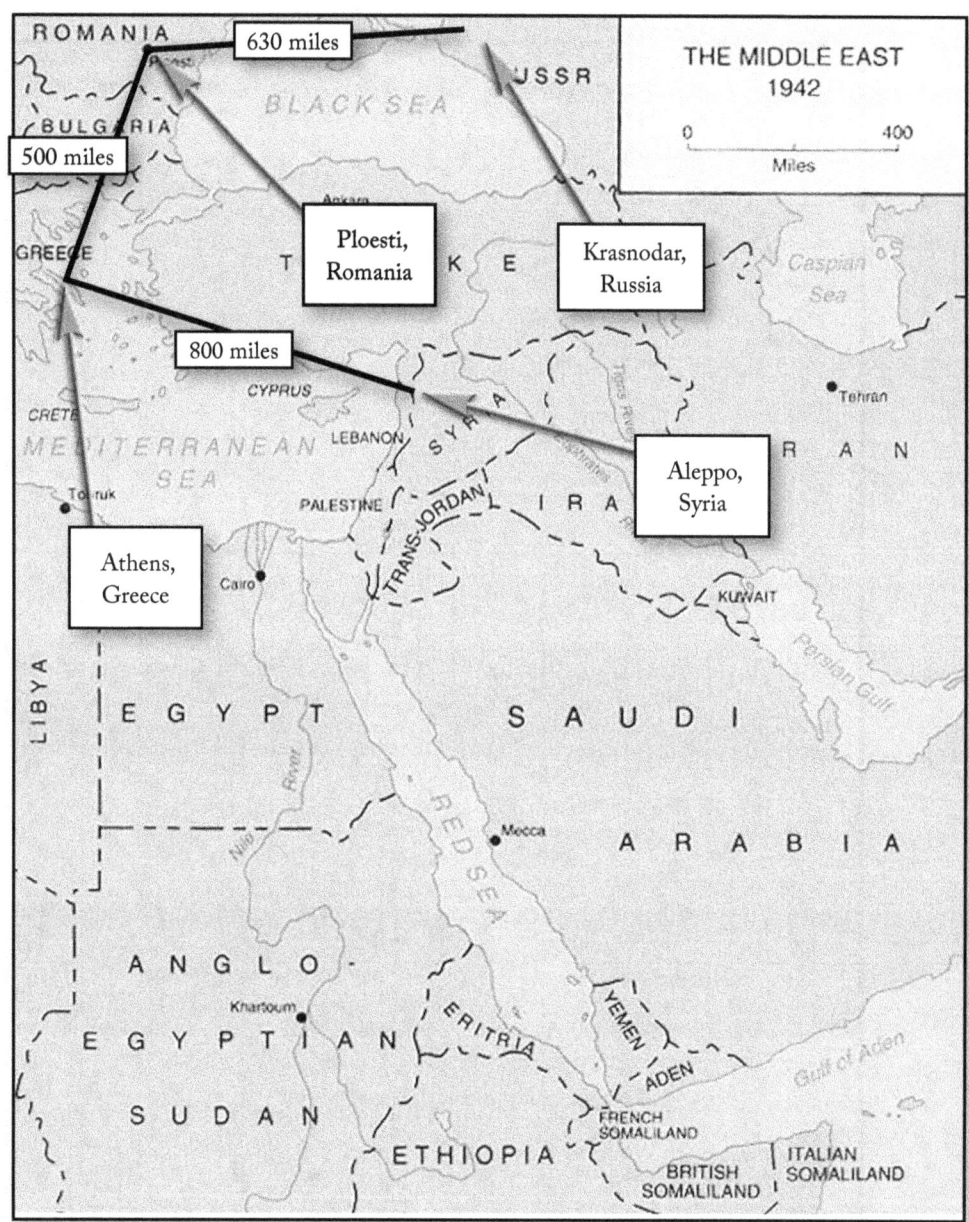

Figure 46– Aleppo, Syria, Ploesti, Romania, and Krasnodar, Russia

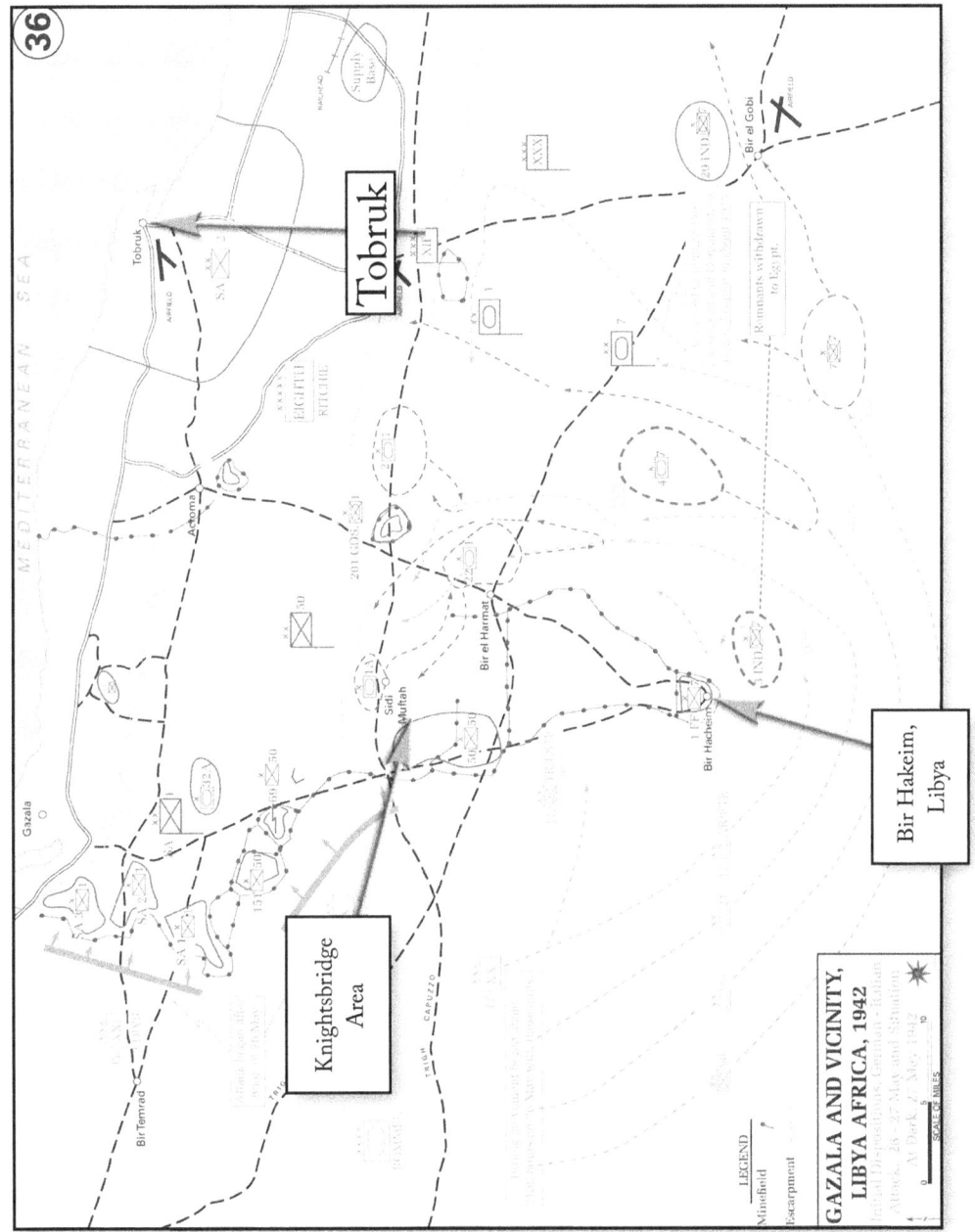

Figure 47 – Knightsbridge Area and Bir Hakeim, Libya

Figure 48– HALPRO Internees in Turkey

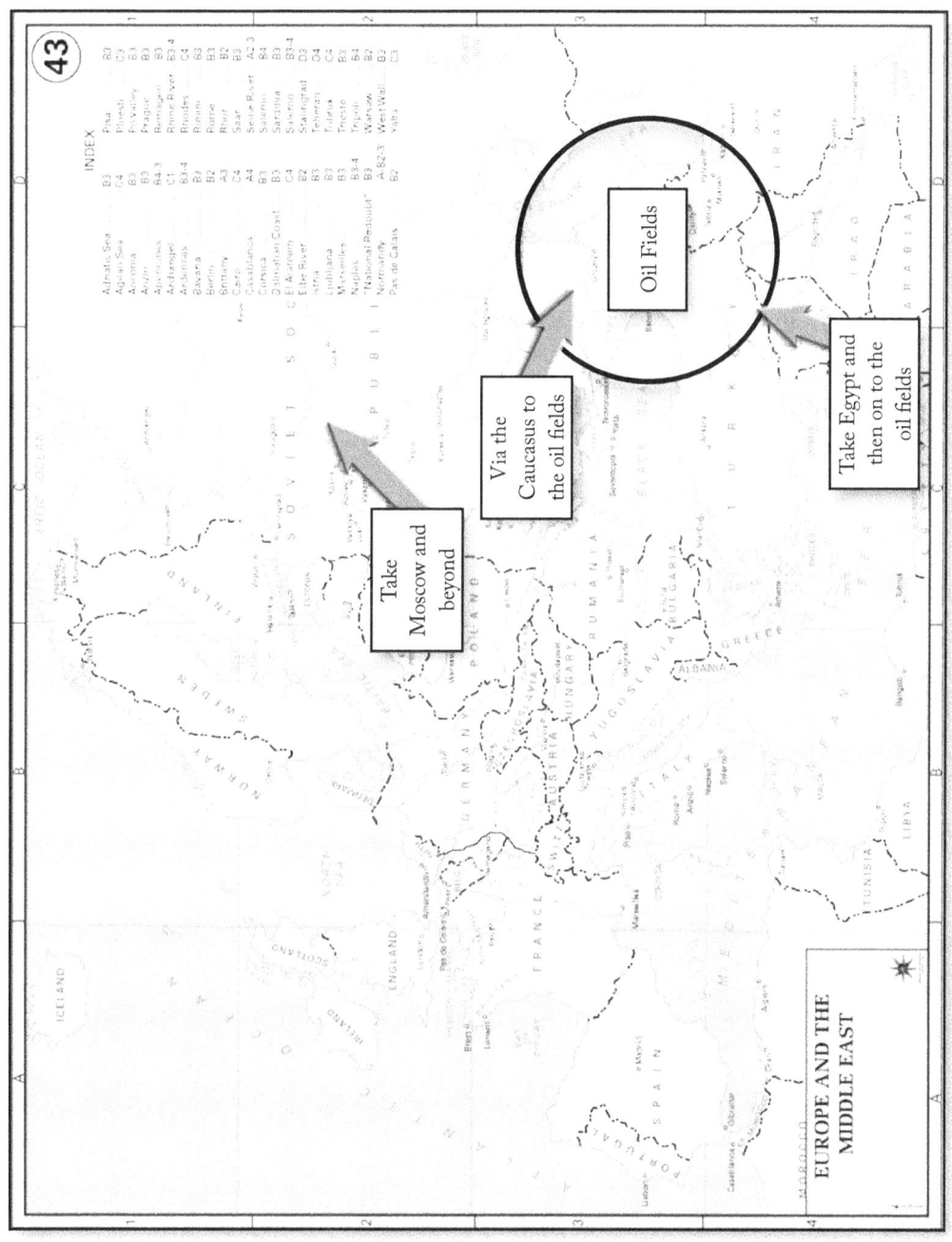

Figure 49 – A Possible Pattern for the Sumer Campaigning – June 21

ENDNOTES

1. Mayhew, *transcript of interview*.
2. Assistant Chief of Air Staff, *The Ploesti Mission*, p. 15.
3. Williams, *Chronology 1941-1945 - 1942*, p. 40.
4. Williams, *Chronology 1941-1945 - 1942*, p. 40.
5. ----, AFHRA, Hap Arnold file, Soong to Arnold, June 2, 1942.
6. Craven, *The Army Air Forces in World War II, Vol. I*, pp. 456-457.
7. Assistant Chief of Air Staff, *The Ploesti Mission*, p. 15.
8. ----, AFHRA, Hap Arnold file, Marshall to Amb. Litvinov, June 4, 1942.
9. ----, FDRL, PSF, Box 82, Arnold Folder, Oil situation, June 4, 1942.
10. Williams, *Chronology 1941-1945 - 1942*, p. 40.
11. Walker, The Liberandos.
12. ----, AFHRA, Hap Arnold file, Harmon to Maj Gen Alexander Belyaev June 5, 1942.
13. Buchanan, *A Friend Indeed?*, p. 279.
14. Williams, *Chronology 1941-1945 - 1942*, p. 40.
15. Assistant Chief of Air Staff, *The Ploesti Mission*, p. 15.
16. Craven, *The Army Air Forces in World War II, Vol. II*, p. 11.
17. ----, *New York Times*, 2, PERHAPS 3, JAPANESE CARRIERS SUNK, June 7, 1942, http://www.freerepublic.com/focus/f-chat/2892593/posts, accessed February 19, 2014.
18. Williams, *Chronology 1941-1945 - 1942*, p. 40.
19. Frank, *Guadalcanal*, p. 32.
20. Frank, Guadalcanal, p. 31.
21. Craven, *The Army Air Forces in World War II, Vol. II*, p. 12.
22. Richardson, *The Forgotten Force*, p. 55.
23. Sibert, *personal diary*.
24. Miller, unpublished diary, p. 16.
25. Therman Brown Diary, 9 June 1942.
26. Richardson, *The Forgotten Force*, p. 55.
27. Miller, unpublished diary, p. 17.
28. Williams, *Chronology 1941-1945 - 1942*, p. 41.
29. ----, AFHRA, Hap Arnold file, Arnold to Soong, June 10, 1942.
30. Walker, The Liberandos.
31. Richardson, *The Forgotten Force*, p. 58.
32. ----, AFHRA, Hap Arnold file, Belyaev to Harmon, June 10, 1942.
33. Sibert, personal diary.
34. ----, FDRL, PSF, Box 82, Arnold Folder, Arnold to FDR, June 11, 1942.
35. Richardson, *The Forgotten Force*, p. 60. General Whang was the Chinese general.

36. ----, FDRL, PSF, Box 136, Halverson to Arnold, June 13, 1942.
37. ----, AFHRA, Hap Arnold file, Harmon to the cable section, June 12, 1942.
38. Richardson, *The Forgotten Force*, p. 62-3.
39. Craven, *The Army Air Forces in World War II, Vol. II*, p. 12.
40. Craven, *The Army Air Forces in World War II, Vol. II*, p. 11.
41. Craven, *The Army Air Forces in World War II, Vol. II*, p. 10.
42. Richardson, *The Forgotten Force*, p. 63.
43. Richardson, *The Forgotten Force*, p. 64.
44. Pearson, Drew, his personal papers.
45. Clendenin, *Mission History of the 376th Bombardment Group*, p. 14.
46. Richardson, *The Forgotten Force*, p. 67.
47. Richardson, *The Forgotten Force*, p. 65.
48. Williams, *Chronology 1941-1945 - 1942*, p. 42.
49. Craven, *The Army Air Forces in World War II, Vol. II*, p. 12.
50. Craven, *The Army Air Forces in World War II, Vol. II*, p. 13.
51. Assistant Chief of Air Staff, *The AAF in the Middle East*, p. 70.
52. Assistant Chief of Air Staff, *The Ploesti Mission of 1 August 1943*, p. 17.
53. Lin Yutang (October 10, 1895 – March 26, 1976) was a Chinese writer, translator, linguist and inventor. His informal but polished style in both Chinese and English made him one of the most influential writers of his generation, and his compilations and translations of classic Chinese texts into English were bestsellers in the West.
http://en.wikipedia.org/wiki/Lin_Yutang
Interestingly, *Between Tears and Laughter* was published in 1943. This indicates that *The Forgotten Force* is not a diary.
54. Richardson, *The Forgotten Force*, p. 68-9.
55. Williams, *Chronology 1941-1945 - 1942*, p. 42.
56. Williams, *Chronology 1941-1945 - 1942*, p. 42.
57. Assistant Chief of Air Staff, *The AAF in the Middle East*, p. 72.
58. Craven, *The Army Air Forces in World War II, Vol. II*, p. 18.
59. Richardson, *The Forgotten Force*, p. 69-70.
60. ----, *New York Times*, U.S. BOMBERS AIDING SEVASTOPOL FORCE, June 18, 1942, http://select.nytimes.com/gst/abstract.html?res=F50615FE3C5A147B93CAA8178DD85F468485F9, accessed February 21, 2014.
61. Miller, personal diary, p. 36.
62. ----, FDRL, PSF, Box 4, Henry Stimson-->FDR-6/19/42, http://docs.fdrlibrary.marist.edu:8000/psf/box4/folo44.html , accessed January 8, 2014.
63. ----, FDRL, PSF, Box 4, Marshall-->FDR-6/19/42, http://docs.fdrlibrary.marist.edu:8000/psf/box4/folo44.html , accessed January 8, 2014.

64. ----, FDRL, PSF, Box 4, -->FDR-6/19/42, http://docs.fdrlibrary.marist.edu:8000/psf/box4/folo44.html , accessed January 8, 2014.
65. Craven, *The Army Air Forces in World War II, Vol. I*, Chapter 14, footnote 66.
66. ----, Army Air Force Forum, 43rd BG, 41-2481 B-17 "TOPPER" thread, http://forum.armyairforces.com/412481-B17-quotTOPPERquot-m152537.aspx, accessed July 12, 2012.
67. Williams, *Chronology 1941-1945 - 1942*, p. 42.
68. Brereton, *Brereton Diaries*, pp. 129-130.
69. Kimball, *Stalingrad*, p. 97.
70. Buchanan, *A Friend Indeed?*, p. 288.
71. ----, FDRL, Map Room, Box 54, Daily Summary report for June 20, 1942.
72. Miller, personal diary, p. 38.
73. Williams, *Chronology 1941-1945 - 1942*, p. 42.
74. Porch, *The Path To Victory*, p. 278.
75. Leiser, *The U.S. Military And Palestine In 1942*,
76. Williams, *Chronology 1941-1945 - 1942*, p. 42.
77. Craven, *The Army Air Forces in World War II, Vol. II*, p. 18.
78. Clendenin, *Mission History of the 376th Bombardment Group*, p. 14.
79. Brown, *War Stories*.
80. Stimson, *Stimson Diaries*.
81. Porch, *The Path To Victory*, p. 278.

WE ARE ALL IN

JUNE 1942

1942 JUNE 23

10TH AIR FORCE

The ramifications from the fall of Tobruk were evidenced within days. Only two days had elapsed when Marshall reassigned Brereton to the Middle East. The accompanying orders stated:

> *In the Middle East the* situation is critical. For Stilwell, Brereton and Maxwell have been sent copies. Brereton will come at once to Middle East with available heavy bombers with mission to assist Auchinleck. Instruct Brereton all our units in Middle East will come under his command. Place at his disposal transport planes and personnel required. Forces taken to Middle East will be for temporary duty only. The War Department will return Brereton's force to your command at the completion of his mission. At destination to be selected by Maxwell, have Brereton arrange with Maxwell for his reception and accommodation. For liaison and coordination with the British, Brereton will make use of Maxwell's headquarters. Because of his wide combat experience, Brereton was selected for this temporary detail. Report size of air force made available for this temporary mission. It is desired Stilwell acknowledge receipt of this message.[1]

Thus, the United States had decided to abandon India, and, indirectly, China. Two reasons seem to justify this decision. The first one is obvious: Rommel's advance posed the most immediate threat to the Suez Canal and then on to the Middle Eastern oilfields. Second, it appeared that Japan had withdrawn from the Indian Ocean and posed no immediate threat to India. The Japanese naval forces were concentrated in the southeast Pacific, where the Allies were beginning to contest Japanese expansion.

1942 JUNE 24

NORTH AFRICA

Axis forces crossed into Egypt, heading for Sidi Barrani.[2] (see Figure 50)

EASTERN FRONT

Germany claimed it had captured the Oskol rail line.[3] (see Figure 51)

HALPRO

Halverson informed his men that General Brereton would be arriving to assume control of the air units in the Middle East; i.e., HALPRO. The general reaction of the men was that at least they would be under the command of an air general rather than an Army man.[4]

The position of HALPRO in the middle of an increasing desperate British position had not gone unnoticed to the HALPRO men. As Richard Miller noted:

> *The position of the British* is quite precarious here in Africa. The Japs landed on Madagascar and at Durban, South Africa. The Germans took Tobruk. More Jerries should start through Syria so that all three can move on the Suez. If the British ever lose North Africa, it will really be rough going, but everything in the Middle East is being drawn back and thrown into this battle and I believe the drive will be stopped. Most of the troops participating in this campaign will be Canadians, South Americans, Aussies and New Zealanders! These men took Tobruk after the toughest sort of going, and the English lost it in six hours. Feeling is high as is contempt for the British fighting ability.[5]

SOUTH PACIFIC

Since it was obvious that the Army was making plans to attack Japanese positions, King was not idle in making Navy plans. He ordered Nimitz to formulate plans to seize "Tulagi and adjacent islands." To cement the idea that the Navy had not been idle, King told Marshall that the United States needed to attack the Japanese on the Santa Cruz Island by August 1.[6] (see Figure 39)

1942 JUNE 25

NORTH AFRICA

General Claude Auchinleck — Commander-in-Chief (C-in-C) Middle East Command — relieved Ritchie and assumed command of Eighth Army. He avoided a direct confrontation with Rommel's forces, choosing to execute a series of delaying actions. British forces would withdraw to El Alamein by the end of the month.[7] (see Figure 50)

HALPRO

On the night of June 24/25, eleven HALPRO planes bombed Benghazi, their 4th mission.[8] (see Figure 50)

Homer Adams, crew, and *Ole Faithful* finally made it to Fayid. And not by a direct route:

> *My crew and I did* not arrive at El Fayid until June 25. Due to an exchange of aircraft, two engine changes et al, the instructions I received sent us all the way to Karachi before we knew we would be going to the Middle East theater instead of China to bomb Tokyo.[9]

98TH HBG

Despite the uncertainty of the availability of Allied air bases in the Middle East, Arnold decided to send the 98th HBG to the Middle East.[10]

NEW GUINEA - SOLOMON ISLANDS

Admiral King issued two communiqués. The first was to Admiral Nimitz, directing him, per orders of the JCS, to begin an immediate offensive action in the Lower Solomons. Allied forces were to seize Santa Cruz Island, Tulagi, and adjacent areas. Operations should begin August 1. (see Figure 39)

The second memo was sent to Marshall. King reminded Marshall that the United States should take the initiative since Japanese focus was away from the Solomons. In fact, the ideal time was early August when the Japanese were focused on Midway.[11]

1942 JUNE 26

COMMITMENT TO CHINA

Stilwell sent the following to Roosevelt:

- Conference with CKS on June 24th and 26th. Gave Chinese news about diversion of HB from India to Egypt and diversion of A-29s "now arriving at Khartoum." Made profound impression on CKS who said he believed FDR sincere. Perhaps orders given without his knowledge and consent
- CKS feels Allies do not regard China as part of war effort. China has done best for 5 years. "If crisis exists in Libya a crisis also exists in China"
- CKS wanted answer: do allies want China Theatre maintained - Mrs. CKS hinted possibility of separate peace with Japan
- Both "bitter" about this matter and they are not mincing words when they asked for an unequivocal answer to the question as to whether or not the Allies were interested in maintaining the Chinese Theater
- This issue had "reached a very serious stage"[12]

This communication was another consequence that cascaded from the Fall of Tobruk. After being promised for months that help was on the way, the military situation had changed all that. China was not pleased with her secondary role.

HALPRO

With the defeat of the British at Tobruk, Rommel's supply line was now much shorter. This made the port at Tobruk an Allied target. So, on the night of June 25/26, four HALPRO planes bombed Tobruk, their 5th mission. HALPRO's next six missions would be against Tobruk. Before the year was out, HALPRO would fly 32 missions against Tobruk.[13] (see Figure 50)

SOUTH PACIFIC

Marshall responded to King's message the previous day. Not only did Marshall disagree with the overall strategy, but he also did not agree with putting a naval officer in the overall command of the operation. Marshall said MacArthur should be in charge.

King was unmoved. The operation would proceed with Nimitz in command. Then, after the landings were completed and the objectives obtained, MacArthur could assume command. King also pointed out that the forces needed to execute the operation would come from the South Pacific zone, not from MacArthur's Southwest Pacific forces.

In addition, King pointed out that, in Europe, where ground forces would be the dominant force, King had already acquiesced to overall Army command. In the Pacific, the reverse would be true. Thus, overall command had to be the hands of the Navy.

On the following day, King told Nimitz to make his plans assuming that he, Nimitz, would only have Navy and Marine assets. Ultimately, no ground troops were ever sent from the Southwest Pacific.[14]

10TH AIR FORCE

Harry Holloway and the Lavin crew joined the 10th Air Force. When Brereton left India, so did all of the heavy bombers and their crews: "On June 26, we left Allahabad for Karachi, to Basra, Iraq, to Hasbbauyia, Iraq, to Fayid, Egypt, to Tel Aviv (Lydda) Airdrome Palestine (Israel)."[15]

Holloway and Lavin would transition to B-24s and complete their tours in April 1943.

1942 JUNE 27

HALPRO

On the night of June 26/27, eight HALPRO planes bombed Tobruk, their 6th mission.[16] (see Figure 50)

CONVOY P.Q. 17

Convoy P.Q. 17 left Hvalfjord, Iceland for the port of Archangel.

1942 JUNE 28

EASTERN FRONT

Germany launched its summer offensive, Operation Case Blue, with a drive towards the Don River from Kursk.[17] (see Figure 18)

GLOBAL STRATEGY

The press continued to return to the idea that the Germans were driving to the Middle Eastern oilfields. *The New York Times* published a map, Figure 52, as their idea of the German objectives.

COMMITMENT TO CHINA

Roosevelt responded to Stilwell's message:

- June 27th cited rapid advance of Axis in ME caused crisis. If not stopped, would sever air routes to India and China. Imperative to hold ME. Therefore, 10th AF bombers sent to ME - this is tem, and as soon as sufficient airpower present to secure lines of ex, planes would be returned to 10th AF
- Re A-29s now departing, US, decision not yet made on theater. Instructions to stop at Khartoum and await Instructions
- Reassured CKS that China regarded as vital to War effort, etc.[18]

USAMEAF

Brereton arrived in Cairo. (see Figure 50) He was placed in command of the newly created United States Army Middle East Air Force (USAMEAF). USAMEAF was comprised of the Halverson Detachment (HALPRO), Brereton's detachment (9th Heavy Bombardment Squadron) and other personnel, which Brereton brought from India, and the Air Section of United States Military North African Mission.[19]

HALPRO

Francis Nestor and crew finally arrived with their plane, *Mona The Lame Duck*.[20]

1942 JUNE 29

NORTH AFRICA

Axis Forces captured Marsa Matruh. (see Figure 44) They were now only 15 miles from El Alamein.[21] (see Figure 50)

NEW GUINEA - SOLOMON ISLANDS

Admiral King proposed that Admiral Robert L. Ghormley be in overall command of the lower Solomon campaign, while General MacArthur would be in command of the New Guinea and New Britain campaigns. This was an attempt to resolve the overarching question of whether the Army or Navy would be in command of Pacific operations.[22]

COMMITMENT TO CHINA

Stilwell met with Chiang, where he delivered Roosevelt's message. In response, Chiang counters with a list of "three minimum requirements" needed to maintain China's participation in the war:

1. Three American army divisions needed to help restore communications between India and China through Burma,
2. The China Air Force needed a minimum of 500 planes, and this level needed to be maintained with adequate replacements,
3. China needed 5000 tons a month of supplies.

The establishment of these supply lines had an August deadline. This list became known as the "Three Demands."[23]

9TH AIR FORCE

Upon his arrival in the Middle East, Brereton found himself in a chain of command dilemma. He was ordered to report to Maxwell, who, in rank, was subordinate to Brereton. Brereton thought this was an unnecessary addition to the chain of command. After all, he, Brereton, was responsible to overall Theater commanders, Auchinleck and Tedder, for the planning and execution of heavy bomber activities. In essence, he had responsibility but limited authority. Therefore, he recommended to Marshall and Arnold the immediate activation of the 9th Air Force and that the two heavy bomber units be given numerical designations.[24]

HALPRO

On the night of June 28/29, ten HALPRO planes flew the group's seventh mission.[25] *Time* printed the following:

U.S. STRIKES A BLOW

The headlines proclaimed that the U.S. Army Air Forces had opened "a new American front," the first on which U.S. crews in U.S. planes could strike directly at Hitler. The fact was that U.S. airmen, flying from R.A.F. bases in North Africa, were in position to peck at the Axis anywhere from the Mediterranean to the Black Sea. But there were not yet enough men and planes in evidence to do much more.

...

Colonel Halverson's squadron is the first of many special, roving units, which the Army Air Forces can send wherever the enemy is in reach. Operations with this one squadron are bound to be limited. But, from North Africa, the U.S. flyers can range over the Near East, Russia's southern front, the Axis fringe of Mediterranean Europe. The bombs on Ploesti, the blows at Italy's shrinking fleet, can well be the heralds of a major U.S. air offensive.[26]

1942 JUNE 30

NORTH AFRICA

The British assigned a portion of their forces to plan the defense of Alexandria and Cairo.[27] (see Figure 50)

HALPRO

On the night of June 29/30, five HALPRO planes flew the group's eighth mission.[28] Sadly, Nestor and crew, who had just arrived from the States, were lost. The circumstances surrounding the loss did not add to the feelings of camaraderie with the RAF.

I remember that one crew was shot down on its first mission (night) by the British because, according to them, no IFF signal was on. I am not at all sure that I remember correctly, but I think this was C. Brown & crew. The British were stupid and mean about shooting first and asking questions later. They shot down a U.S. C-47, Capt. Roth, over the Canal, broad daylight, VFR, with him just out of Port Said.[29]

Author's note: it was Nestor's crew, not C. Brown. The episode about Roth is true.

Two HALPRO planes had been delayed in departing from Florida. On June 30, the last of these planes was approaching Fayid. Scotty Royce was a passenger on board and recounted the landing.

Meanwhile Royce had been shepherding stragglers across the Atlantic and the African continent. The last airplane finally arrived over Fayid on June 30th. As they prepared to land it was discovered that the nose wheel would not come down. A mechanic's tool box had gotten stuck in the mechanism. After all his work Royce was not about to let a new B-24 bite the dust. With crew members holding his legs, hanging half out of the nose wheel opening, he was able to free the gear and allow the airplane to land. The first overseas movement of a B-24 bombardment unit had been completed without losing an airplane.[30]

Meanwhile, plans for the group's evacuation from Fayid were coming together. Missions were suspended for the transition. Orders arrived at noon directing the group to leave by 4 p.m., with the destination of Lydda, Palestine.[31] (see Figure 50)

Therman Brown recounted his departure for Lydda.

By the end of June 1942, Rommel had pushed the British to the Egyptian border. It appeared that it was only a matter of time before he would have all of Egypt. The fighters now need the Egyptian bases and we had to look for a base farther away from the front. It was to be Lydda, the airport for Tel-Aviv.

We had to install our bomb bay fuel tank and baggage racks for the move. While the flight was less then 200 miles, we had to take everything including eighteen men and their baggage. It was just too much weight for the surface of the revetment. Ken DeLong said it best. The wheels were stuck and so were we. We had to stay another night. We had been at Fayid for about twenty days.[32]

One B-17 from Brereton's India 9th Bomb Squadron had already arrived at Fayid. Two more B-17s would arrive on June 30. All three joined HALPRO in the move to Lydda. The rest of Brereton's aircraft flew directly to Lydda. (see Figure 50)

By nightfall, all of HALPRO had relocated. And eight Brereton B-17s were parked on the ramp. Crews descended on the RAF mess hall, who apparently not been informed to expect the two United States units.[33]

The retreat to Lydda did not enhance the HALPRO men's perception of the British:

Well now I have lived; and having lived I am part English; have come to enjoy their tea and now have become part of one of their famous and very prevalent practice of strategic retreats and successful disengagements because of the Nazi approach on Cairo, we have to vacate for Lydda, Palestine.[34] (see Figure 50)

HEAVY BOMBER PRODUCTION

During June, 197 heavy bombers were manufactured. Ninety-four were B-24s.[35]

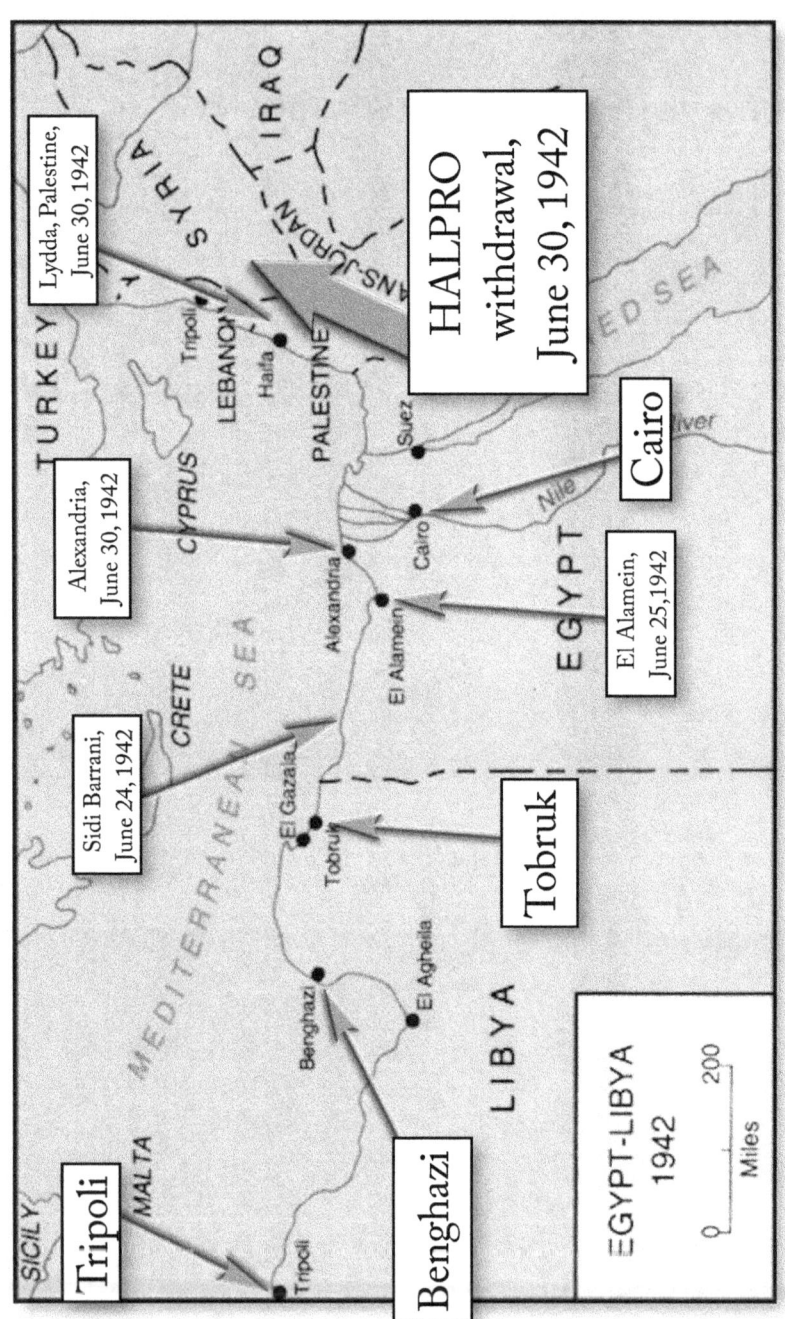

Figure 50– Mediterranean – June 1942

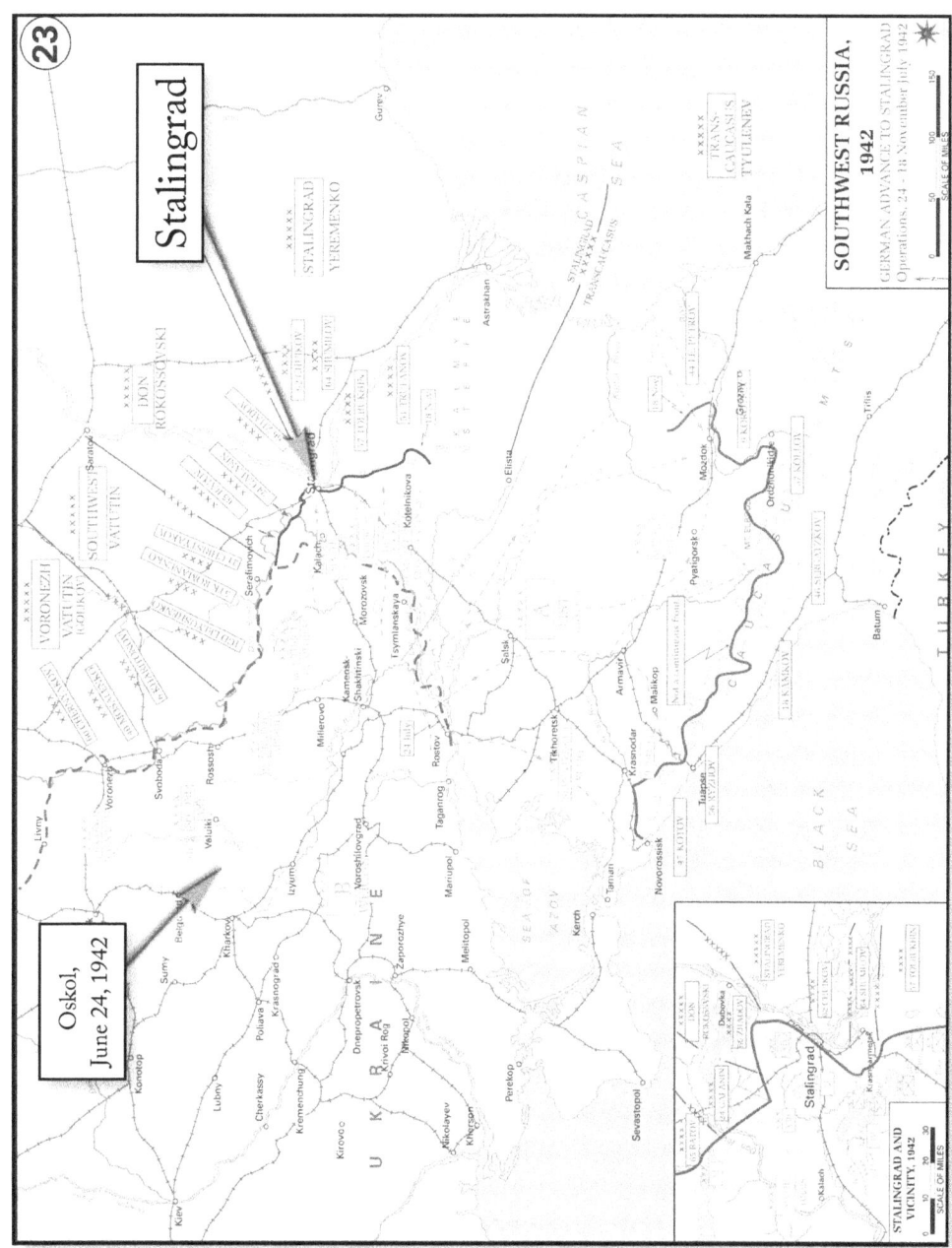

Figure 51 – Stalingrad – June 1942

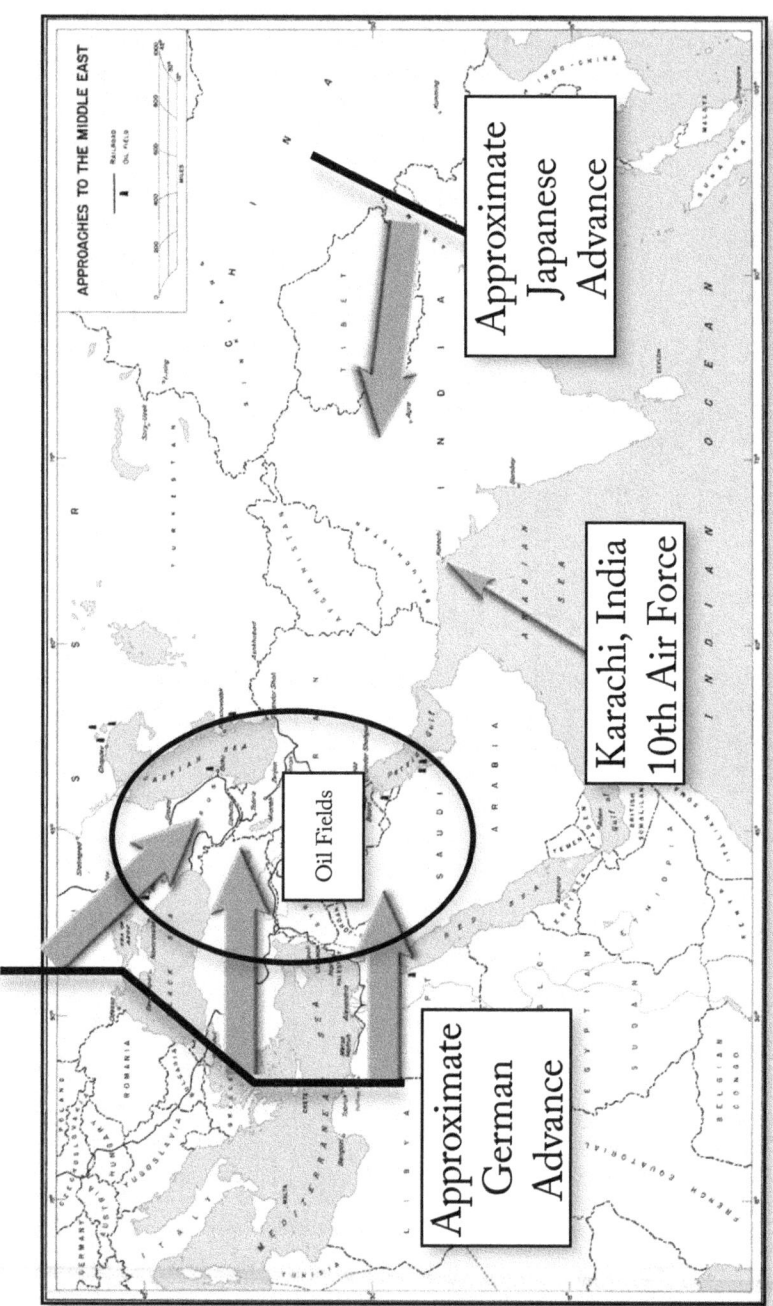

Figure 52– Possible Pattern for the Sumer Campaigning – June 28

ENDNOTES

1. Brereton, *The Brereton Diaries*, p. 130.
2. Williams, *Chronology 1941-1945 - 1942*, p. 43.
3. Williams, *Chronology 1941-1945 - 1942*, p. 43.
4. Richardson, *The Forgotten Force*, p. 74.
5. Miller, Personal diary, pp. 42-43.
6. Frank, *Guadalcanal*, P. 33.
7. Williams, *Chronology 1941-1945 - 1942*, p. 43.
8. Clendenin, *Mission History of the 376th Bombardment Group*, p. 15.
9. Adams, Letter, August 3, 1974.
10. Assistant Chief of Air Staff, *The AAF in the Middle East*, p. 77.
11. ----, *US Marine Corps in World War II, Guadalcanal Campaign*, http://www.ibiblio.org/hyperwar/USMC/Guadalcanal/USMC-M-Guadalcanal-1.html , p. 6.
12. ----, AFHRA, Hap Arnold file, Stilwell to FDR, June 26, 1942. Comments on the meetings between Stilwell and Chiang can be found at ---, *The China-Burma-India Theater*, pp. 168-9, http://www.ibiblio.org/hyperwar/USA/USA-CBI-Mission/USA-CBI-Mission-5.html , accessed January 8, 2014.
13. Clendenin, *Mission History of the 376th Bombardment Group*, p. 15.
14. ----, *US Marine Corps in World War II, Guadalcanal Campaign*, http://www.ibiblio.org/hyperwar/USMC/Guadalcanal/USMC-M-Guadalcanal-1.html , pp. 6-7; Frank, *Guadalcanal*, p. 33.
15. Holloway, letters to the author.
16. Clendenin, *Mission History of the 376th Bombardment Group*, p. 15.
17. Williams, *Chronology 1941-1945 - 1942*, p. 43.
18. ----, AFHRA, Hap Arnold file, FDR to Stilwell, June 28, 1942.
19. http://en.wikipedia.org/wiki/United_States_Air_Forces_Central , access December 6, 2013.
20. Richardson, *The Forgotten Force*, p. 74.
21. Williams, *Chronology 1941-1945 - 1942*, p. 43.
22. Williams, *Chronology 1941-1945 - 1942*, p. 43.
23. ---, *The China-Burma-India Theater*, p. 172.
24. Assistant Chief of Air Staff, *The AAF in the Middle East*, pp. 68-9.
25. Clendenin, *Mission History of the 376th Bombardment Group*, p. 16.
26. ----, *Time* archive, *U.S. Strikes a Blow*, June 29, 1942, http://www.time.com/time/magazine/article/0,9171,795914,00.html , accessed November 28, 2008.
27. Williams, *Chronology 1941-1945 - 1942*, p. 43.
28. Clendenin, *Mission History of the 376th Bombardment Group*, p. 16.
29. Adams, Letter, August 3, 1974.
30. Cilio, John, *Two Plans-One Target*,
31. Richardson, *The Forgotten Force*, p. 80.

32. Brown, *War Stories*.
33. Richardson, *The Forgotten Force*, p. 83.
34. Miller, personal diary, p. 48.
35. ----, Army Air Force Statistical Digest, http://www.ibiblio.org/hyperwar/AAF/StatDigest/aafsd-3.html , accessed January 22, 2014, p. 112; B-24 production from Author's personal papers.

JULY 1942

Rommel began his attack on El Alamein, less than 100 miles from Alexandria and the Nile River.

Organizationally, HALPRO was merged with the Brereton's Indian forces to form the 1st Provisional Group. They began a concerted effort to attack Rommel's supply lines, primarily the port of Benghazi. The 98th Group began to arrive.

The fall of Tobruk had another consequence. Roosevelt ordered his military chiefs to visit London and finalize a plan with their British counterparts. Their orders were to arrive at a military action for 1942, preferably before the November elections. The result would be approval of Operation TORCH, necessitated the postponement of a "second front."

Meanwhile, German forces in Russia seemed to be repeating their 1941 unstoppable performance. Churchill suggested to Roosevelt that they send air units to aid the Russians.

In the Pacific, plans were underway to stop the Japanese drive through the Solomon Islands. Guadalcanal was as good a place as any to start. (see Figure 45)

1942 JULY 1

NORTH AFRICA

The First Battle of El Alamein began. (see Figure 50) Axis forces made their deepest advance into Egypt. They captured Deir el Shein, south of El Alamein.[1]

EASTERN FRONT

Germany captured Sevastopol.[2] (see Figure 8)

HALPRO

Therman Brown still had to get his plane, crew, and baggage from Fayid to Lydda. (see Figure 50)

> *The next morning, July 1*, 1942, we were able to get the plane out of the revetment and on the way to Tel-Aviv. Lydda is just a little inland from the Mediterranean

Sea. It was a bright, clear day. From six thousand feet, I could not see the airfield. There was a reason. All the buildings and the runway had been camouflaged with paint so you could see no straight lines. It was quite effective. Once we had figured it out, we had no trouble but I took an extra pass over the airstrip to be sure of what I was doing. As we became familiar with the field, we were not conscious of the camouflage. We landed without any problem but we were a day late. Most of the other planes were already there.[3]

TASK FORCE BR

The first B-17 landed at Prestwick, Scotland. It had just completed the 3,000-mile journey from Maine via Greenland and Iceland.

CONVOY P.Q. 17

A German U-boat spotted Convoy P.Q. 17 as it made its way to Archangel.

1942 JULY 2

HALPRO

On the night of July 1/2, six HALPRO planes resumed their mission against Rommel's supply base at Tobruk, their 9th mission.[4] (see Figure 50)

GLOBAL STRATEGY

Meanwhile, the blame game for the Tobruk debacle continued. Churchill's political position continued to be tenuous. *The New York Times* ran a front-page article exposing Churchill's vulnerability:

BRITAIN'S FORCE IS INFERIOR IN DESERT, MINISTER ASSERTS.[5]

The front page also contained two other articles about the situation in the Middle East – *British Ready to Move Fleet* and *Egyptian Line Hit*. The situation did not look promising.

1942 JULY 3

NORTH AFRICA

Rommel ordered the attack on British positions to resume. By days end, the Afrika Korps only had 26 operational tanks.
Meanwhile, capital ships of the Royal Navy had withdrawn through the Suez Canal to Red Sea. An evacuation of civilian and military personnel from Egypt began.[6]

HALPRO

Four of the Brereton B-17s joined six HALPRO B-24s for another raid on Tobruk, HALPRO's 10th mission and the eighth against Rommel's supply lines.[7] When the crews returned, they reported that German anti-aircraft fire (ack-ack) was becoming more accurate and, for the first time over North Africa, German fighters were "looking them over."[8] (see Figure 50)

Brereton had been asked to prepare plans for the withdrawal of his air units if the Afrika Korps had continued success. He argued that his units should stay put to continue aiding the British Eighth Army. If the British forces were defeated, then he proposed to withdraw his bombers to either Khartoum, to defend the trans-Atlantic ferry route, or Iran/Iraq to defend the oil fields.[9]

OPERATION VELVET

Arnold's staff submitted its response to the question of air support for the Russian front. In their opinion, once the conflict in the Middle East was resolved, the two bombardment groups should be sent to Russian air bases. From these bases, the bombers could operate against German targets in southeastern Europe. In particular, they could attack the Ploesti oil fields.[10]

10TH AIR FORCE

Brereton had already left India, and so had most of the crews. Now it was Thomas' crew's turn. He recorded his last days in Karachi (see Figure 32):

> *We were alerted to move* out at a moment's notice. Major Wade took off for Cairo. General Brereton had already gone. The inference was we were to be shifted to Egypt to help stop Rommel. We paid farewell visits to our friends in Karachi and visited the palatial home of the famous Baluch regiment, where we were toasted in silver goblets with crystal bottoms. Hawkins had another of his frequent birthdays. Finally we were put on a six-hour alert and told to be ready to take off for Egypt at 6 A.M. The Fourth of July eve Schilling, Murphree, Stewart, and I held our last celebration in India.[11]

1942 JULY 4

HALPRO

Two of the Brereton B-17s attacked German troop concentrations during the night of July 3/4.

At the same time, four HALPRO B-24s raided a German convoy. These were HALPRO's eleventh and twelfth missions.[12]

CAUCASUS

Russian forces surrendered Sevastopol. (see Figure 8) Additional German forces were now free to pursue Case Blue.

REINFORCEMENTS

Bill Mayhew's crew joined Brereton in the Middle East.

In late June and early July 1942 Major General Lewis Brereton took all the remaining 17 heavy bombers (12 B-17Es, 4 B-24Ds and 1 LB-30) of the 7th Bombardment Group (all were assigned to the 9th Bombardment Squadron at that time) to the Middle East to try to help stop Field Marshal Erwin Rommel and his Afrika Korps. At that time Field Marshal Rommel was pushing the British 8th Army out of Libya and threatening all of Egypt. As there were few heavy bombers in that area to help stop him, we were sent there to help. Twenty crews were sent to the Middle East. We became part of the 9th Air Force.

All the planes of our Group were assembled at Karachi before departing for the Middle East. The morning of 4 July 1942 four of our B-17Es left Karachi for the Middle East. About 45 minutes after take-off, Major Fennell had to turn back to Karachi because of engine trouble. The other three planes continued on. When engine repairs were completed, we took off again at 11:45 (2345) that night.

....

Our movement to the Middle East was typical for the squadron. Our first stop was at Sheiba, Iraq , which we reached after a flight of nearly eight hours. That town is a short distance from Basra, Iraq. All you could see in any direction was sand, sand, sand. We were there only one day, but that was enough for us. Our next stop was Habbaniya, Iraq, which was about 20 miles from Baghdad, Iraq. Some of the fellows went into Baghdad that evening, but I had seen enough of the place while flying over it to suit me. When the boys returned from their trip, I was glad I'd stayed at the airdrome. They certainly did not like it. The airdrome at Habbaniya has a high plateau overlooking it. Early in World War II the Axis troops had gotten as far as the plateau, and the airdrome still showed the effects of the shelling it received. All the hangers were riddled with holes, and the barracks looked pock marked.[13]

Meanwhile, Max Fennel noted:

At last I've learned what to name my Fortress. We're going out to whip Rommel and I'm going to name my ship 'Fennell vs. Rommel.'

"I'll bet Rommel is scared to death," I said, grinning.
He doesn't know the best fliers in the world are coming after him![14]

10TH AIR FORCE
The Flying Tigers flew their last mission as an independent air force.[15]

1942 JULY 5

EASTERN FRONT
German forces reached the Don River at Voronezh. The Russian defenders surrendered the next day.[16] (see Figure 54)

GLOBAL STRATEGY
Continued German success on the southern portion of the Eastern Front seemed to confirm previous speculation about German strategy. They were headed for the Caucasus oilfields. Rommel's North African campaign could end in the Iraqi oilfields. If both campaigns succeed, then they could drive onto India, as shown in Figure 56, a map published in *The New York Times*.

HALPRO
Six HALPRO B-24s bombed Rommel's supply port of Benghazi during the evening of July 4/5, their 13th mission.[17]

Halverson had complained to his leadership that the British were running the show. General Elmer Adler had been sent to evaluate the situation. He reported that the approved procedure was for the HALPRO planes to depart Lydda to their former base at Fayid, Egypt. There, they received a briefing and final instructions from RAF personnel. As a consequence, Halverson believed that he was not being given proper or complete information. In essence, Halverson had lost control of his men and planes once they departed Lydda. (see Figure 50)

As a result, Kalberer was ordered to report RAF Group 205 at Ismailia, where the mission planning for the heavy bomber forces was done. He was to act as the liaison officer between HALPRO and 205 Group. Several other officers were assigned to Fayid.[18]

1942 JULY 6

HALPRO
Nine HALPRO B-24s bombed Rommel's supply port of Benghazi during the evening of July 5/6, their 14th mission.[19] (see Figure 50)

SOUTH PACIFIC

A Japanese convoy arrived at Guadalcanal and proceeded to land men and material. Allied coastal watchers deduced that the Japanese were intent on building an airfield.[20] (see Figure 45)

GLOBAL STRATEGY

Time published a piece that succinctly stated what was at stake in the Middle East.

IF EGYPT FALLS

Within reach of his sand-scarred hands Field Marshal Erwin Rommel had the greatest war prize since the Japanese took the Dutch East Indies. If he lays his hands on Alexandria, already within easy bombing range of his forward bases, the Mediterranean will belong to the Axis. Give him Suez, and he will have opened the gate of the Near East, to the Axis.

More than that, the fall of Egypt would be a threat to the entire Russian front, for the eastern European front extends in fact from Murmansk down through the Caucasus to the wreckage of Matrûh.

By the mere threat of an offensive based on Suez, the Axis might well force Turkey out of her stubborn neutrality into collaboration with Germany. Then Russia would have the foe on her left flank, within reach of the Caucasian oilfields. But the threat would be greater than that. Once in the Near East, the German would be near British oil—the great wealth of the Iraq fields. These fields (with Russia's) are the last big oil source for a vast strategic area in which the Japanese have already snatched the rest of the wells. The sub-harried tankers of the United Nations could not make up the deficit for many exhausting months to come.

From Egypt, the Nazis could also swing south, strike at the great U.S. concentration base in Eritrea. From the Near East, they could swing farther east into India, striking at the communications lines of the United Nations to China. Wherever they went, south, east, or both, they could well expect to get some help from the native populations, for the loss of Egypt would strike British prestige a severe blow in all Mohammedan countries.

Yet, except for the taking of the Mediterranean (which could be done at a stroke by depriving the British Fleet of its Egyptian bases) all this would take time—time to organize and employ greater armies than Germany now has on the south side of the Mediterranean. And time would give the United Nations opportunity to organize new resistance at the southern end of the Red Sea and the head of the Persian Gulf. At those two strategic gateways a fight can still be made, which, if successful, will still keep the Nazis and their allies, the Japanese, from joining hands.

But the words of Churchill and Roosevelt last week intimated that the United Nations would stake their forces not so much upon defense in the Near East as on offense in Europe (see below). If it succeeds, Germany will be defeated soon. If it fails, the war may last for many years and the democratic world will bleed.[21]

1942 JULY 7

1ST PROVISIONAL GROUP

Two Brereton B-17s bombed Rommel's supply port of Tobruk during the evening of July 6/7, their 15th mission.[22] (see Figure 50)

10TH AIR FORCE

Bill Mayhew described their flight from Karachi to Lydda.

The next day (7 July 1942) we took off again, landing four hours later at Lydda, Palestine which is about 5 miles from Tel Aviv, Palestine. We were now 3,000 miles west of Karachi. The flight across most of Iraq and Trans Jordan was over scrub desert. However, when we reached the Jordan River between Trans Jordan and Palestine, the appearance of the land changed spectacularly. Now we were flying over grain fields, vineyards, orchards, and citrus groves. I could imagine we were flying over southern California. After nearly four months in the Sind Desert of western India, then Iraq and Trans Jordan, we could hardly believe such lush places still existed in the world.

When we left India our new base was to have been on the banks of the Suez Canal at Fayid, Egypt, approximately 250 miles west of Lydda. However, by the time we reached Lydda, we received orders to set up our base there. By this time the Germans were so close to our original base that it was unsafe to base heavy bombers there. Naturally, this situation did not help our morale any. We expected to see "Jerry" come marching down the road from Cairo, Egypt or drop parachutists from the sky any day. This move to the Middle East meant that at this time we were the only American combat group to fight the Japanese, Italians and Germans (all the Axis partners). Our B-17 crews began flying combat missions immediately to the harbor of Tobruk, Libya and after convoys in the Mediterranean Sea.[23]

1942 JULY 8

COMMITMENT TO CHINA

The thirteenth meeting of the Pacific War Council occurred in the Executive offices of the White House. Joining Roosevelt were nine representatives of various governments having an interest in the war with the Japanese. The President began by summarizing what had happened since their last meeting. He then proceeded to inform the group of his latest decision:

> *The President continued his remarks* by stating that he had recently been in correspondence with Chiang Kai-shek and that the Generalissimo appeared to be greatly depressed. This downheartedness was no doubt due to the fact that when the Libyan situation became acute and something had to be done in a hurry to save Egypt, "I ordered the Tenth Air Force to proceed to the Middle East and help Egypt. Time for consultation with the Generalissimo simply did not exist, and something had to be done very quickly. I am sure if the Generalissimo had been here and had all the facts before him, he would not only have heartily approved, but would have urged that I take the action I did. If the Middle East falls, the problem of the supply of China will be made immeasurably harder. The stand that the British are now making on the El Alamein line is a last ditch stand. If successful, we can (1) reinforce the British, and (2) seize the initiative."[24]

1942 JULY 9

HALPRO

Six HALPRO planes attacked Benghazi during the evening of July 8/9. At the same time, three Brereton B-17s bombed Rommel's supply port of Tobruk. (see Figure 50) It was their 16th and 17th missions.[25]

Then during the daylight hours of July 9, five HALPRO planes attacked a convoy.[26] The plane piloted by Paul Davis, *Yank*, was shot up so badly that he had to land at an alternative field. Scotty Royce remembered the required repairs to Davis' plane.

> *We did have one rather* serious incident happen to Paul Davis. He caught a large shell of undetermined size in the wing - left wing, just outboard of the gas tank - which blew up in such a manner that it expanded the wing like a balloon. All the wing stringers were torn loose from the ribs, but essentially the strength of the wing was still there, although it made it considerably more limber than the wing should have been. Paul's description of flying it back and watching the wing wave up and down as he moved the aileron and as the airplane hit bumps was quite

interesting, but anyway he got the airplane back in one piece, got it landed, and I was able to repair it by putting the wing from Sturkey's crashed airplane on it.[27]

It would take a month for *Yank* to see action again.

But worse yet, Kenneth Butler and crew, flying *Ole Rock*, never returned from the mission.

The next day, Davis recounted what happened:

Paul's story the next day was sad for us all. He had been flying with Crouchley as co-pilot and Butler had been flying one of the wing positions in this element. They were jumped by Me 109's before they reached the target and shot one down. The fanatic German pilots kept boring in and after making a pass at the formation would proceed out of range and slow roll and do acrobatics. The noticeable feature of the attack was the newness of the airplanes the Germans were supplied with, and how they used it. One German pilot came from the front directly at the formation. Just as the lead ship thought this bastard would ram head on, the Nazi pulled his ship over in a vertical wing up and went right through the formation, laughing as he went.

Another ship had hopped them from above and his shooting had caught both Paul and "Bombsight" getting some cannon bursts into both. Butler's engines and wing tanks had caught fire on one side. Crouchley had seen Butler and Kysar lift a hand in salute and leave the formation. Another burst of cannon fire had gone in Paul's left wing and rather than going straight through had gone the full length, ballooning a furrow through the central portion of the wing. Both Paul and Ted had worked like the devil to keep up with the formation while an element of four was made up and Uhrich had gone down to give protection to Bombsight, because he was unable to keep up.

Ferdinand Schmidt had not seen anyone leave the injured ship, and during another determined attack lost sight of Butler. He was never seen again. Nobody was injured in Paul's ship.[28]

Homer Adams also remembered Butler and crew.

I do not remember the date for sure – maybe July 4 1942 – and I am not sure which crew it was, probably Butler; but I sure remember the terrible feeling in the pit of my stomach when ME 109s shot down the B-24 flying on my right wing and the B-24 crew talked to us on the radio as they went down.[29]

One of the most important lessons from this recounting was the description of the

German fighter attack – they attacked from the front. Arnold had opposed sending United States heavy bombers piecemeal into the war. He wanted to hold them back until a sufficient force could be deployed. He was concerned that the Germans would develop a tactic to use against the bombers. Both the B-17 and B-24 did not have a nose turret in their original design. The Germans had discovered this vulnerability and would now use it to their advantage. In response, future models of both bombers would have nose turrets.

EASTERN FRONT

Hitler divided German Army Group South into German Army Group A and German Army Group B. Hitler and Field Marshal Fedor von Bock (commander of Army Group South) had heated discussions about Case Blue. Ultimately, Hitler dismissed von Bock on July 11.[30]

1942 JULY 10

NEW GUINEA – SOLOMON ISLANDS

Nimitz issued operational orders for the capture of Tulagi and Guadalcanal. The operation was given the code name WATCHTOWER. The target date for beginning operations was August 1.[31] (see Figure 55)

CONVOY P.Q. 17

The remnants of Convoy P.Q. 17 arrived at Archangel. Only 11 of the original 35 ships made it through the gauntlet of German forces. The next convoy would not depart for Archangel until September. Despite the severe losses, the Russian press "hailed" their victory over the Germans in providing escort duty for the convoy.

EASTERN FRONT

German Army Group A headed for Rostov, while German Army Group B drove along the Don River toward Stalingrad.[32] (see Figure 54)

GLOBAL STRATEGY

At a meeting of the Joint Chiefs, Marshall expressed a willingness to abandon the "Germany First" strategy and throw his support behind proposed operations against Japan. He was frustrated by British intransigence about operations in Europe. This change in position and its adoption by the Joint Chiefs was the first formal recognition of the Solomon Island campaign, Operation WATCHTOWER.[33]

1942 JULY 12

HALPRO

A single Brereton B-17 bombed Rommel's supply port of Tobruk during the evening of July 11/12, their 19th missions.[34] (see Figure 50)

REPLACEMENTS

The first group of nine additional B-24 crews and bombers left Morrison Field in Palm Beach Florida. One of those crewmembers was Howard Conlee, the assistant engineer on Alexander E. Munsell's crew. This was his recollection of the trip overseas:

> *We left Morrison Field West* Palm Beach, Fla. on July 12, 1942. We flew to Africa by way of Trinidad, Belem Brazil, Natal Brazil, then to Roberta Field in Liberia. Then down to Accra and on over to Khartoum. We were part of the Halverson detachment headed for China. We were delayed in Khartoum and received new orders. The orders sent us to Gura Eritrea, where our plane was dismantled for spare parts. We made the rest of our journey to Lydda Palestine in a DC-3. Having no plane our crew was split up and from that time on I felt like a spare part.[35]

On June 17, Halverson had sent a message to Marshall that he did not have sufficient spare parts to keep all of his planes in the air. All the spare parts had been sent to India because that was where HALPRO was headed. The only option HALPRO had was to use the new planes for parts. It was interesting that the decision was to strip a perfectly good plane for parts and not remoe them from veteran plane for parts.

Conlee was not the only crewmember to recall a similar fate for their replacement plane. Harold Christiansen was on Norman Appold's plane. This was his recollection of their trans-Atlantic trip:

> *The next morning we took* off for Morrison Field, West Palm Beach, Fla. where we were outfitted with high altitude flying gear, .45 caliber pistols, oxygen masks and machines.
>
> Two days later, with an ominous sky overhead, we took off, ostensibly for South America. However, we encountered a fierce Atlantic storm and as skillful and resourceful a pilot as Norman Appold was, we just couldn't make headway through the rain, sleet, and hurricane winds generated by the storm. So reluctantly, Lt. Appold turned and headed back for Morrison Field.
>
> The next morning the weather was not quite so inclement and we got off without incident and headed for our first overseas stop, Belem in Brazil. After landing in Belem we welcomed a hot meal after the long flight from Florida.

Our stopover was brief and after refueling we headed for our Atlantic jumping off spot: Natal. Brazil. On that leg of our journey the pilot flew at a lower altitude and we could clearly see how dense and seemingly impenetrable was that part of the Brazilian jungle as we skimmed the treetops at less than 500 feet. I then realized why we had been issued machetes because if we managed to survive a forced emergency landing there, we would have had to hack our way through the thick underbrush.

As soon as we landed in Natal we headed for the showers with the anticipation of another welcome hot meal after we refreshed ourselves. From a geographic standpoint, Natal was the shortest distance from the two continents: South America and Africa to our destination in Accra, Ghana.

After a two day rest in Natal, we refueled, put aboard food supplies and headed out across the broad expanse of the South Atlantic. Knowing what a long flight lay ahead of us, I remember saying silently to myself: "Christ I wonder if we'll make it.

I must have killed quite a few of the fish indigenous to that part of the Atlantic for I then received my first instructions on how to operate the tail turret from Lt. Appold. After showing me how to turn on the turret and manipulate both the turret and fire the twin .50 caliber machine guns, the pilot said I could practice firing at the waves below as soon as I received an all clear signal from Lt. Gerry, who was now flying our bomber. The pilot stressed the importance of firing short bursts so as not to burn out the barrels of the guns. So during our 11 1/2-hour flight across the Atlantic I had plenty of time to practice and familiarize myself with the controls of my turret.

...

I still didn't know our final destination. The closest battle front was in the Middle East so we had guessed correctly that was where we were going. The pilot didn't brief us on the overall flight plan. He merely told us of our next stop just prior to takeoff.

When the pilot announced on the intercom that we were approaching the coast of Africa, I rejoined the two waist gunners, Frank and Ray, amidships. We discussed how great it would fell to again get off the bomber and stretch our legs on good old terra firma. It had been a long and arduous flight and I'm sure that they were as relieved as I was that the most hazardous part of our journey would soon be behind us.

Our stay in Accra, Ghana, was of short duration and early the next morning we headed for Khartoum, Sudan. After a lengthy flight across the heart of the African continent we landed, and as we stepped down from the plane. It was as though we had stepped inside a blast furnace. The temperature that day in Khartoum was 124 degrees! We couldn't get into the showers fast enough.

When the officers returned from preflight briefing, the pilot informed us that we were headed for our first base Haifa, Palestine, (Israel). Now the excitement we had all felt the past few days as we neared the end of our journey mounted as it meant that we would soon be engaged in combat. I looked forward to our first combat mission with a mixed feeling of awe and in trepidation. I now realized that the next time I faced my guns someone would undoubtedly fire back and try to blow me right out of my tail turret!

As we headed north two days later we gazed down upon the Nile River as it meandered northward in the same direction that we were going. After a long flight we finally saw the Mediterranean stretched out below, and followed the coastline northeast to our destination.

When we sighted Tel Aviv the crew that the next town of any consequence would be Haifa.

Shortly after landing at the Royal Air Force Airdrome, several miles inland, we were assigned quarters in a Quonset hut with a corrugated tin roof. We were then treated to a hot meal, British style. The English don't have much imagination in the preparation of food. At least not at that base. Our menu seldom varied from two staple items: bully beef (vaguely similar to corned beef) and mutton stew. But famished as we were, the Piece de resistance of that day, mutton stew, soon disappeared from our mess kits.

Airmen, like sailors, develop a kinship for their particular mode of transport. So it was disheartening news that Andy (Charles Anderson) and I learned the next morning when we asked the skipper (Andy and I had already applied that sobriquet to Lt. Appold and later Andy was to refer to our pilot as "Johnny Appleseed," albeit respectfully) where our plane was parked.

It was then that we suffered one of the many disillusionments in store for those who set out to fight a war. The pilot informed us that the brand-new bomber, fresh out of the Consolidated plant at San Diego, Calif., was already being dismantled and the four Wright Cyclone engines were to be used for spare parts. It was considered simply the most feasible way to obtain spare parts: fly the planes overseas and just take what parts were needed. We all had had visions of giving our bomber a name and perhaps even painting a picture on her nose.[36]

57TH FIGHTER GROUP

The first units of the 57th Fighter Group arrived at an airfield near Lydda. Now, the bombers might have fighter escort. (see Figure 50)

1942 JULY 13

HALPRO

Five HALPRO B-24s attacked Benghazi.[37] (see Figure 50) Unfortunately, HALPRO lost another plane and crew - Charles Brown and *Eager Beaver*. This was George Richardson's recounting of what happened:

> *It was known that ack-ack* and pursuit were plenty hot because the boys had hit the place at noon. Ferdinand Schmidt had said that apparently an eighty-eight millimeter anti-aircraft shell had gone right through the flight deck, the ship going out of control. Some of the crew bailed out and were picked up by launch below. We later learned that Charlie had died in a German hospital in Benghazi. Red Cross stated that he had received the very best medical attention.[38]

1942 JULY 14

1ST PROVISIONAL GROUP

Five Brereton B-17s attacked Tobruk during the evening of July 13/14.[39] (see Figure 50)

1942 JULY 15

LONDON JCS MEETING

In anticipation of the London JCS Meeting, British Field Marshall Dill, who was stationed in Washington, prepared a memo for Churchill. It outlined what he, Dill, thought the American position would be at the upcoming JCS meetings. His analysis was that the Americans would oppose Operation GYMNAST because:

1. it would require withdrawing forces from the Pacific
2. it would require new lines of communication
3. a strike at Casablanca would not result in any relief to the Russians, while attacking on the Mediterranean was too hazardous, and
4. a commitment of forces to GYMNAST would result in the postponement of ROUNDUP

Dill's perception was that Marshall was committed to the cross channel invasion. Despite continual discussions, Marshall believed that most of the German military was occupied on the Russian front. So, German defenses in Western Europe would never be weaker than they were now. A German success in Russia could preclude any future invasion. Dill's analogy was that a business facing either a risky venture or going bankrupt would always opt for the venture.

Dill also pointed out that Admiral King favored a campaign in the Pacific. And, Dill believed that many Americans thought that a stalemate in Europe was the most likely outcome. Dill's conclusion:

> *May I suggest with all* respect that you must convince your visitors that you are determined to beat the Germans, that you will strike them on the continent of Europe at the earliest possible moment even on a limited scale, and that anything which detracts from this main effort will receive no support from you at all? Marshall believes that your first love is GYMNAST, just as his is BOLERO, and that with the smallest provocation you always revert to your old love. Unless you can convince him of your unswerving devotion to BOLERO everything points to a complete reversal of our present agreed strategy and the withdrawal of America to a war of her own in the Pacific, leaving us with limited American assistance to make out as best we can against Germany.[40]

HALPRO

Hannah and crew finally got their orders.

> *After 30 days to Basra* then to what is now Lod Airport Israel & the HALPRO boys. 9th Sqdn 7 BG (B-17s) & other odds & ends from India as well. Now nobody knew what to do with our radar ship so they kept sending us to 221 Sqn RAF. They were a radar outfit (Wellington Bombers I think). Obsolete & no range.
>
> We flew long range – no set schedule – depended on intelligence reports. We did find & sink one freighter.
>
> The mission of HALPRO, the 98th BG & 9th Sqdn was deny fuel to Rommel & we were very successful. I remember one hit in Benghazi harbor – the ship was almost to the dock. A moment later & the smoke was over 30,000 feet & rising.[41]

1942 JULY 16

HALPRO

Six HALPRO B-24s attacked Benghazi during the evening of July 15/16. At the same time, four Brereton B-17s attacked Tobruk. It was the group's 22nd and 23rd missions.[42] (see Figure 50)

98TH BG

The men of the 98th Bombardment Group left New York Harbor for points unknown.[43]

LONDON JCS MEETING

Hopkins, Marshall, and King were about to depart for consultations with their British counterparts. Roosevelt outlined the objectives for the meetings – definitive plans for 1942 and tentative plans for 1943. There were five criteria that plans had to meet:

1. defeat of the Axis
2. concentration of forces
3. coordination of British and American forces
4. action as quick as possible
5. United States ground troops must see action in 1942

Roosevelt supported Operation SLEDGEHAMMER as he viewed this action as critical to saving Russia. If, and only if, it was the consensus that SLEDGEHAMMER was not feasible should they consider other options for 1942. Roosevelt pointed out that if Germany prevails on the Russian front, German forces could be freed up to face Allied forces in Western Europe. This would make Operation ROUNDUP even more difficult.

Roosevelt had finally come around to see the importance of the Middle East, regardless of what happened on the Russian front. Thus, he wanted Hopkins and company to remember what a defeat would mean:

1. loss of Egypt and the Suez Canal
2. loss of Syria
3. loss of the oil fields
4. possible link-up between German and Japanese forces
5. possible loss of the rest of North Africa, and
6. threat to shipping lanes across the southern Atlantic

Therefore, he asked the planners to determine what could be done to hold the Middle East.

Roosevelt also opposed any plan that emphasized the defeat of Japan. He believed that the defeat of Japan would not result in the defeat of Germany. However, the reverse would happen. Any diversion of American forces to the Pacific enhanced the prospect of a German victory in all of Europe.

He gave them one week to reach an agreement with the British.[44]

1942 JULY 17

HALPRO

Two missions were conducted on July 17. Six HALPRO B-24s attacked Tobruk. Later, three Brereton B-17s attacked Tobruk. They were the group's 24th and 25th missions.[45] (see Figure 50)

OPERATION WATCHTOWER

New orders were issued, changing the start date to August 7.[46]
A B-17 of the 435th Bombardment Squadron conducted a photographic reconnaissance mission over Guadalcanal. Marine Corps observers onboard noted the progress the Japanese were making on construction of an airfield. (see Figure 55)

OPERATION VELVET

The United States and Great Britain had been sending supplies to the Russians via naval convoy. Military realities forced both countries to suspend those shipments. To Stalin, the more relevant decision was the decision to postpone "The Second Front." To assuage Stalin, Churchill wired a message to him stating:

> *We are studying how to* help on your southern flank. We might be able to send powerful air forces in the autumn to operate on the left of your line.[47]

The "left of your line" would be the Caucasus.

COMMITMENT TO CHINA

In an effort to examine all possible avenues of getting supplies to China, the United States had asked Russia for permission to fly over Siberia. On July 17, the Russians refused the request.[48]

1942 JULY 18

LONDON JCS MEETING

The American Chiefs of Staff arrived in London for strategy discussions with their British counterparts. Harry Hopkins accompanied them and had directions from Roosevelt to reach an accord with the British. Roosevelt gave the American team a list of acceptable strategic objectives. While the Americans continued their push for the cross-channel invasion of Europe, the British refused to even consider the idea. From the British perspective, the only viable option was an operation in the Mediterranean.

Ultimately, given the intransigence of the British, Marshall and King faced the choice of action in either the Middle East or North Africa. Operation GYMNAST (which would evolve into Operation TORCH) was approved.[49]

1942 JULY 19

1ST PROVISIONAL GROUP

Seven HALPRO B-24s attacked Tobruk, the group's 26th mission.[50] (see Figure 50)

1942 JULY 20

1ST PROVISIONAL GROUP

Three Brereton B-17s attacked Tobruk, the group's 27th mission.[51] (see Figure 50)

COMMITMENT TO CHINA

Roosevelt had sent Currie to China to deal first hand with Chiang. Upon his arrival and meeting with Chiang, Currie discovered that the lines of communication between Roosevelt and Chiang via Soong and Stilwell were less then totally forthright.[52]

GLOBAL STRATEGY

As the second year of the Russian-German conflict began, the Allied speculation about German strategy was confirmed. *Time* printed the following analysis of the 1942 campaign:

TIME WILL NOT WAIT

The Russians and the Germans, in these July days of 1942, are fighting the battle that may decide the world's fate.

On the plains beside the Don the battle has only begun. Already it is erupting and spreading along the vast Russian front. No Russian loss or retreat in any one sector will be a final loss. But if the Germans win this battle—and in the Don sector they were still winning this week—the war will be indefinitely lengthened; the 1939-42 phase of it will be definitely lost. The U.S. and Great Britain, invading Hitler's Europe and fighting him on his own fronts, will then have an infinitely harder task than Hitler had in Russia. And it will be a task that they must take on while in the Far East Japan is still winning, and growing stronger, and becoming as hard to defeat in her area of conquest as Hitler will be in his—if Russia falls.[53]

OPERATION VELVET

Having promised Allied Air support on Stalin's left flank, Churchill now had to convince the Americans. At the Joint American-British military conference, he outlined why the potential battlefront between the Middle East and the Caspian Sea was, for all intents and purposes, void of any Allied military units. If the British could not stop Rommel, Stalin could not be expected to stop the southern German Armies from reaching the Caucasus oil fields. Allied Air power was the only military force offering a valid deterrent.[54] Churchill could have said American Air Power as they were the only ones present.

That argument convinced Hopkins to send the following to Roosevelt:

> *It is also my hope* that you will consider putting some of our air squadrons into Russia.[55]

1ST PROVISIONAL GROUP

On 20 July 1942 the 9th Squadron was combined with Halpro (a B-24 outfit that had been stopped in Egypt while on its way to China) to form two squadrons of the 1st Provisional Bombardment Group. We referred to these two squadrons as Halpro (B-24Ds) and "Brereton's Bastards", "Orphans", " Bengal Bombers", or "Toomey's Flying Circus" (B-17Es).

Our facilities at Lydda (today Ben Gurion International Airport) were the best we had experienced since leaving the States. We were housed in stone barracks shaped like Quonset huts. As we had arrived with little equipment, we had to sleep on the cement floors for several nights until cots could be located for us. We each had only two blankets, but the temperature at that time of year was mild, so it was not bad, considering what we had come from. The airfield had been built by the Germans before the war. In our early days there, our crews had difficulty locating the field from the air, because it was so well camouflaged.

We had brought a limited number of ground personnel from India. Sufficient maintenance personnel [were] always lacking. A normal ground crew for a B-17 was 32 men. The [entire] 9th Squadron in the Middle East had 19.[56]

1942 JULY 21

1ST PROVISIONAL GROUP

Nine B-24s attacked a new target, Suda Bay on the Island of Crete. Crete had been captured by Germany in 1941 and, according to Allied Intelligence, was a departure point for Rommel's supply convoys.[57] Flying distance from Crete to Palestine was about 650 miles.

The condition of the B-17s was beginning to manifest itself in operations. These planes had been flying combat missions since December 1941. They were in need of a major overhaul. It was only a matter of time before the planes would become operationally useless. It was time to transition the crews to the B-24.

1942 JULY 22

EASTERN FRONT

The Germans mounted an all-out assault for Rostov.[58] (see Figure 54)

1ST PROVISIONAL GROUP

Three B-17s attacked Tobruk, the group's 29th mission.[59] (see Figure 50)

OPERATION WATCHTOWER

The 11th Bomb Group moved to New Caledonia. This base would allow them to strike Japanese targets in the Solomon Islands.[60] (see Figure 55)

LONDON JCS MEETING

As originally proposed, Operation SLEDGEHAMMER was an emergency landing along the French coast. Eisenhower and his staff had made plans for the Cherbourg and the Cotentin Peninsula. This operation was to be executed in the event of a collapse of the Russian army on the eastern front. Marshall had slowly evolved the mission into the first phase of a major invasion in 1943, and was independent of events on the Russian front. His major concern was that if an invasion of Europe were delayed, resources would be diverted to other objectives.

At the Joint Chiefs of Staff meetings in London, the participants finally agreed that a cross channel invasion could not occur in 1942. They then turned to evaluating other alternatives proposed by the staff planners.[61]

1942 JULY 23

1ST PROVISIONAL GROUP

Nine HALPRO B-24s attacked Benghazi. Meanwhile, four Brereton B-17s attacked Tobruk. They were the 30th and 31st missions.[62] (see Figure 50)

One of the B-17 crews transitioning to the B-24 was the crew of Willard (Chick) Fountain, flying *Mona the Lame Duck*:

On return that evening the wheels could not be lowered by the hydraulics and not one of the old B-17 crew knew how to operate the manual control. Chick circled the field but nobody could get the method of lowering the wheels across the radio. The crew gradually dropped by parachute and Chick took the ship over the hills and bailed out himself.

That trip ended disastrously. Sergeant George Wrigley, enlisted bombardier of the old 19th Bomb Group and veteran campaigner was found the next day. He had used the nose wheel escape hatch. His chute had fouled on the plane and dropped Wrigley to earth. Sergeant La Londe had fractured his femur in the jump, was put in a Thomas splint and sent to the hospital. And, a quarter million dollars worth of airplane was gone, a valuable Sergeant with thirteen years of experience, and another had broken his leg because transition had failed to cover such an emergency. The depletion of our airplanes was becoming serious.

Mengel had been Fountain's co-pilot. As he started to jump he remembered his Leica camera and retraced his steps to get it. He had to take his parachute off, sling the camera beneath the harness and jump--"but one hundred and eighty dollars is still quite a lot of dough". He was not bothered by the jump, but was mentally kicking his pants for almost forgetting the camera. Rodriguez, another member of the crew had taken his girl's picture, which he had planed on the windshield for a long stay and stuffed it in his pocket before he jumped. All the crew-felt that it was "like leaving home", but it could not have been a better home to leave. Scott Royce took two rolls of film of the many pieces of wreckage.[63]

Scotty Royce added an interesting commentary about the local Arab population.

Spare parts arrived when Fountain bailed out of Mona the Lame Duck over Lydda airport. I forget exactly why he bailed out. The aircraft crashed in the desert, but unfortunately with such an impact that very little was recoverable. We went out and salvaged what we could and I noticed that our operations were watched with interest by about fifty Arabs. After we picked over the parts and filled a truck up with what we thought we could use and took the guard off the airplane, the Arabs moved in. I'm sure they used everything but the squeal. I remember looking back as we topped the hill about a mile away, and I could see the fuselage moving off on the shoulders of maybe thirty Arabs. It looked like a large caterpillar.[64]

OPERATION WATCHTOWER

Due to Japanese activities in the South Pacific, the JCS agreed that the Allies needed to maintain the southern ferry route to Australia. Thus, the Japanese advance had to be stopped. Operation WATCHTOWER, the seizure of Guadalcanal and Tulagi, was approved.[65] (see Figure 55)

LONDON JCS MEETING

Monitoring the negotiations in London, Stimson took great exception to the actions of the British. He was troubled by the British tendency of agreeing to something and then reneging on the deal.[66]

1942 JULY 24

EASTERN FRONT

German Army Group A captured Rostov.[67] (see Figure 54)

1942 JULY 25

1ST PROVISIONAL GROUP

Six B-24s and four B-17s flew a mission to Tobruk.[68] (see Figure 50) Therman Brown was on the mission, but he did not fly his plane *Draggin Lady*.

> *Ever since April 28, 1942,* we regarded the B-24s that we had personally sign for at the Brookley Field Depot in Mobile Alabama as our very own. We didn't want to fly another pilot's plane and we didn't like them flying ours. This is one of the reasons we did not go on the first Ploesti mission or the raid against the Italian Fleet. Our plane did not have the range because of the fuel cell leaks.
>
> This had to change. Too many ships had been lost or were out of commission. We only put five B-24s against the Italian fleet although the sixth one was ready. It was a very early takeoff and the pilots overslept. After that we all had wake up service. The system of being able to claim our own plane had to break down. We flew number 17, our plane, for eleven missions but on Sunday July 25, on a mission to Tobruk, we flew another B-24. Once in awhile, we would get our beloved number 17 back. Actually, if number 17 were available, we would fly it.[69]

98TH BG

Air elements of the 98th began to arrive at their bases near Haifa. The Middle East Theater now had two heavy bombardment groups. While management may have considered locating the 98th BG near the 1st Provisional Group a necessary step, the men had their reservations:

> *During this latter part of* July rumor had it that another heavy bombardment

group was coming into the theater. It was our old 98th Bomb Group. Nothing could be more ill-considered than to send the 98th. After all, we had all been more or less "stolen" from that group. Colonel Rush had his best enlisted men "hired" from his group. These old cankers would fester anew. This rumor was confirmed with the arrival of Colonel Rush in Cairo and a large group of his airplanes at Khartoum.

...

On July 25th the first squadron of the 98th, the 344th arrived in the theater, to go to fields previously selected by General Brereton and Colonel Halverson. Colonel Rush [CO of the 98th] and his men were asked over to see us, but the entire group evidently had something bothering them and would not.[70]

In mid August, George Richardson wrote the following:

The original rupture never healed, for not one time did the 98th call on us for any of our experience. Though the men and crews were on the best of terms, the two group commanders had nothing to do with one another. The 98th would be scheduled on a mission and both groups would rendezvous in the vicinity. There was always rivalry to see who could be in best formation and at the rendezvous first. The 98th always followed Nathan Bedford Forrest's admonition about "the mostest" but were not always "the firstest."[71]

The servicemen of HALPRO were dealing with the RAF and the merger of the remnants of MacArthur's Philippine Air Force. Now, they were facing the resentment of the 98th. And unbeknown to them, they would soon face another challenge, as if fighting the Germans was not enough.

LONDON JCS MEETING
Admiral King continued his campaign against any "agreement" with the British. His position was that, while the American contingent supported Roosevelt's criteria of defeating Hitler and keeping Russia in the war, the British only gave lip service to those objectives. In his opinion, the British were only interested in "preserv(ing) its empire in the Middle East."[72]

1942 JULY 26

LONDON JCS MEETING
Roosevelt had received two telegrams, the first on July 24 and the second on July 25, which outlined the agreement the American and British planners had reached during

their meetings in London. Roosevelt then cabled his approval to Hopkins. He agreed in principle, but insisted that Operation GYMNAST had to occur before October 30, 1942. Planning for the operation should begin immediately. Hopkins was to tell "That Naval Person" that he, Roosevelt, was "delighted" they had reached an agreement.[73]

While there is no written evidence to support the belief that Roosevelt insisted on the October 30 date as it preceded the November 3, 1942 general elections. There are ample public opinion surveys from the period, and newspaper articles reporting on the surveys, that reported that the American people

1. did not understand "the Germany First" strategy
2. did not understand America's apparent unwillingness to send material to China, and
3. favored a negotiated peace with Hitler.[74]

Within the White House, there were staffers who monitored these surveys. Roosevelt could have been concerned that the electorate would send more dovish representatives to Washington.

1942 JULY 27

EASTERN FRONT

German Army Group B cleared the Don bend near Stalingrad of Russian resistance.[75] (see Figure 54)

Stalin issued Order 227: No retreat. Any commander who disagreed with this order would be dealt with.

1ST PROVISIONAL GROUP

Four B-17s flew a mission to Tobruk, the group's 33rd mission.[76] (see Figure 50)

1942 JULY 28

OPERATION VELVET

With the return of the American military chiefs to Washington and the approval of Operation TORCH, Roosevelt dealt with Churchill's request for air support in the Caucasus.

I will do my best to get the air squadrons on the Russian southern flank.[77]

1ST PROVISIONAL GROUP

Twelve B-24s left on a mission to find a convoy.[78] They used one of the new B-24s, which was equipped with submarine searching equipment. This type of plane was commonly referred to as a "sub-sniffer."[79]

1942 JULY 29

GLOBAL STRATEGY

Roosevelt finally replied to Stimson's July 23 memo about the British. His first sentence said it all: "This memorandum from the Secretary of War is not worth replying to in detail."

And the lead phrase of his second paragraph did not lighten up: "The Secretary of War fails to realize . . ."[80]

OPERATION VELVET

Roosevelt followed up his previous message to Churchill concerning American air support to Russia. He explained his rationale as follows:

> *Russia's need is urgent and* immediate. I have a feeling it would mean a great deal to the Russian Army and the Russian people if they knew some of our air force was fighting with them in a very direct manner.[81]

1942 JULY 30

NORTH AFRICA

During the month of July, there had been numerous attempts by both the Allies and the Axis to overrun the other's position. All efforts by both sides failed to achieve significant territory. Auchinleck decided to cease offensive actions until he received reinforcements.[82]

Roosevelt had asked Marshall to estimate when he, Marshall, thought Rommel might reach Cairo, and what the United States could do to aid the British. Marshall's guess for the first part was one to two weeks. As to the second, he was waiting to see if Auchinleck could stop the German advance. And there was nothing more the United States could do.[83] (see Figure 50)

1ST PROVISIONAL GROUP

The Group flew two missions. First, five Brereton B-17s attacked Tobruk. (see Figure 50) Then nine B-24s left on a mission to bomb Suda Bay.[84]

OPERATION VELVET

Churchill was worried about Stalin's reaction to the western Allies' decision to postpone the cross channel invasion of Europe. As a consolation, Churchill informed Stalin that plans were being made to send a supply convoy, P. Q. 18, to the Russians via Archangel. The expected delivery date would be early September. He also offered to meet with Stalin at a time and place of his convenience.[85]

EAST INDIES

Japanese troops made multiple landings on Aru, Kei and Tanimbar Islands.[86] (see Figure 53)

REPLACEMENTS

Norm Appold was one of the pilots of the first replacement group.

After accumulating approximately 60 total B-24 flying hours, all training, during mid 1942, I was sent to Morrison Field, West Palm Beach Florida, in late July to be given an overseas assignment involving ferry/combat crew duty in a B-24D. Initially, I was not aware of where I was being sent, with whom, nor to what unit I would be attached. Within a few days after arrival at Morrison Field, orders were cut listing me as pilot, identifying the other crew members by name, and assigning a specific B-24 to us by tail number . . . [July 30, 1942] Not until we were airborne several days later, were we authorized to open our sealed duty orders. Those orders directed us to report to the 1st Provisional Bomb Group (Halpro) at Lydda via the south Atlantic ferry route, ASAP.

Our B-24D was fully equipped and heavily loaded with many spare parts, ammunition, and full bomb bay tanks, and was identified as a combat ready aircraft. My crew was as inexperienced a group of young airmen as any checking out of Morrison Field, but we were eager and enthused, and frankly innocently unaware of what was in store for us overseas. Howard Sturkie and his crew had left a few days earlier, and did indeed arrive in Lydda as the first Halpro replacement crew, ahead of us. Due to a minor mix-up in routing instructions we first landed in Haifa on the 1st of August. That was where the 98th Bomb Group was located, and since they were not expecting replacements, we were told to proceed to Lydda. There, on the 2nd of August we reported in to Lt. Col. McGuire as the second replacement crew and airplane for Halpro. We were given quarters and late that night we watched from our bunks as the B-24s returned from their raid of that day.[87]

1942 JULY 31

EASTERN FRONT

By mid-July, the Russians realized that the Germans were intent on capturing Stalingrad. Therefore they started to remove foodstuffs from the city and into military zones across the Volga River. The citizens of Stalingrad were left to fend for themselves. (see Figure 54)

The Germans, seeing what was happening, attacked the Russian ferry ships. By the end of July, they had sunk thirty-two of them and damaged nine more.

OPERATION WATCHTOWER

B-17s of the 11th Bomb Group began flying "softening-up" sorties against the Japanese position on Tulagi and Guadalcanal. Due to the range of the B-17s, the initial operations were conducted from the Island of Efate. These type missions would be flown daily.[88] (see Figure 55)

The Allied Guadalcanal invasion force departed from Fiji.[89]

1ST PROVISIONAL GROUP

With all of the discussions between United States and British leaders about who was in charge, Richard Miller recorded one small event:

> *Today we all took part* in a ceremony — a very simple one, but it was probably the combination of the two that made it impressive. We had a flag raising and was the first time an American flag has been raised over Palestine, and as far as I know, any part of the Middle East. At this stage of the War, most of us are beyond the false thrill of jumping on a band wagon, but yet this didn't seem to fall under the ordinary type of flag waving that was so prevalent in the States. It gave a sort of settled, more secure, feeling — one which we were proud to recognize. Can't figure why the hauling down of the Union Jack and pulling up of our Stars and Stripes should make any difference, but I believe nearly all of us are the better off under "Old Glory".[90]

HEAVY BOMBER PRODUCTION

During July, 213 heavy bombers were manufactured. One hundred were B-24s.[91]

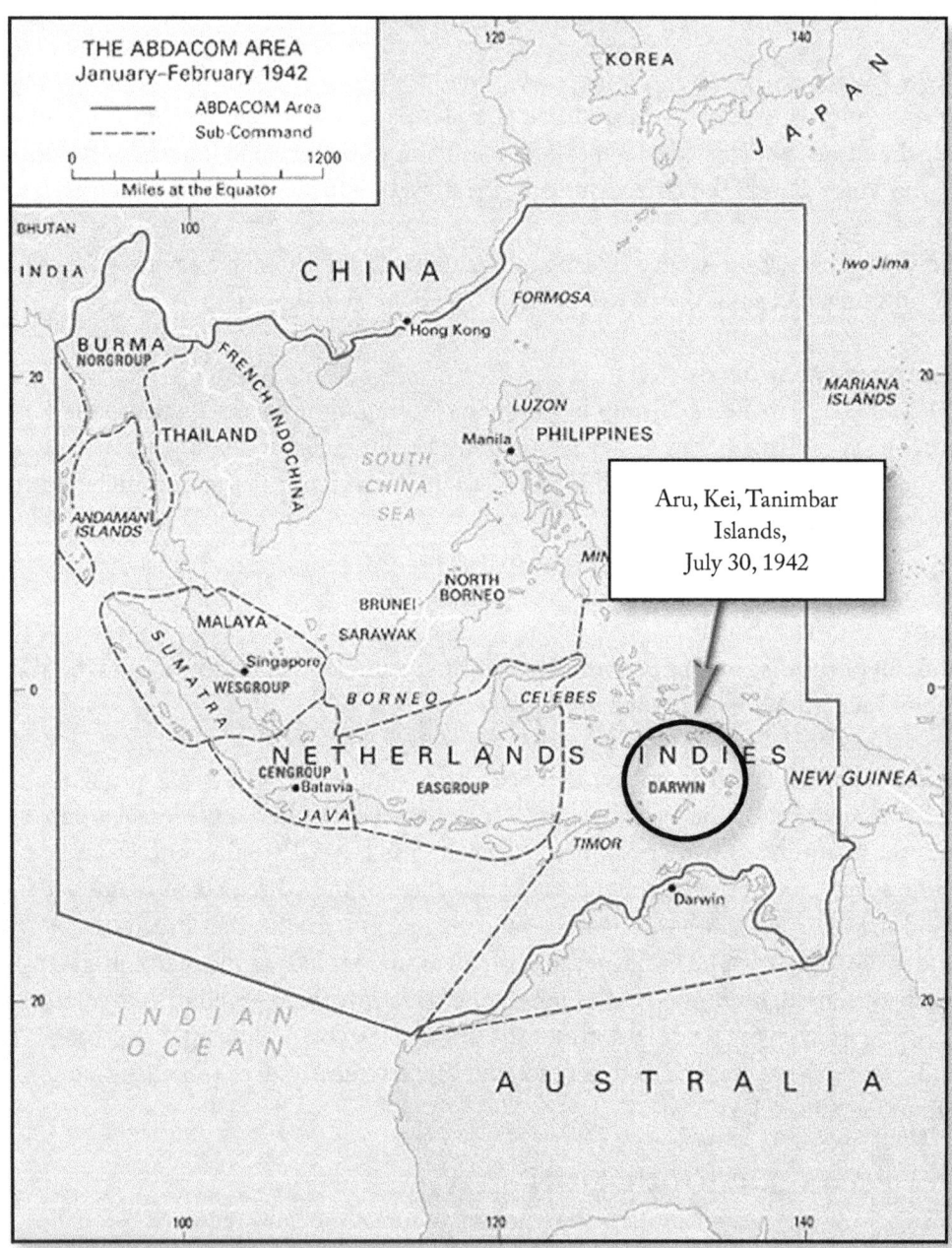

Figure 53 – ABDA – July 1942

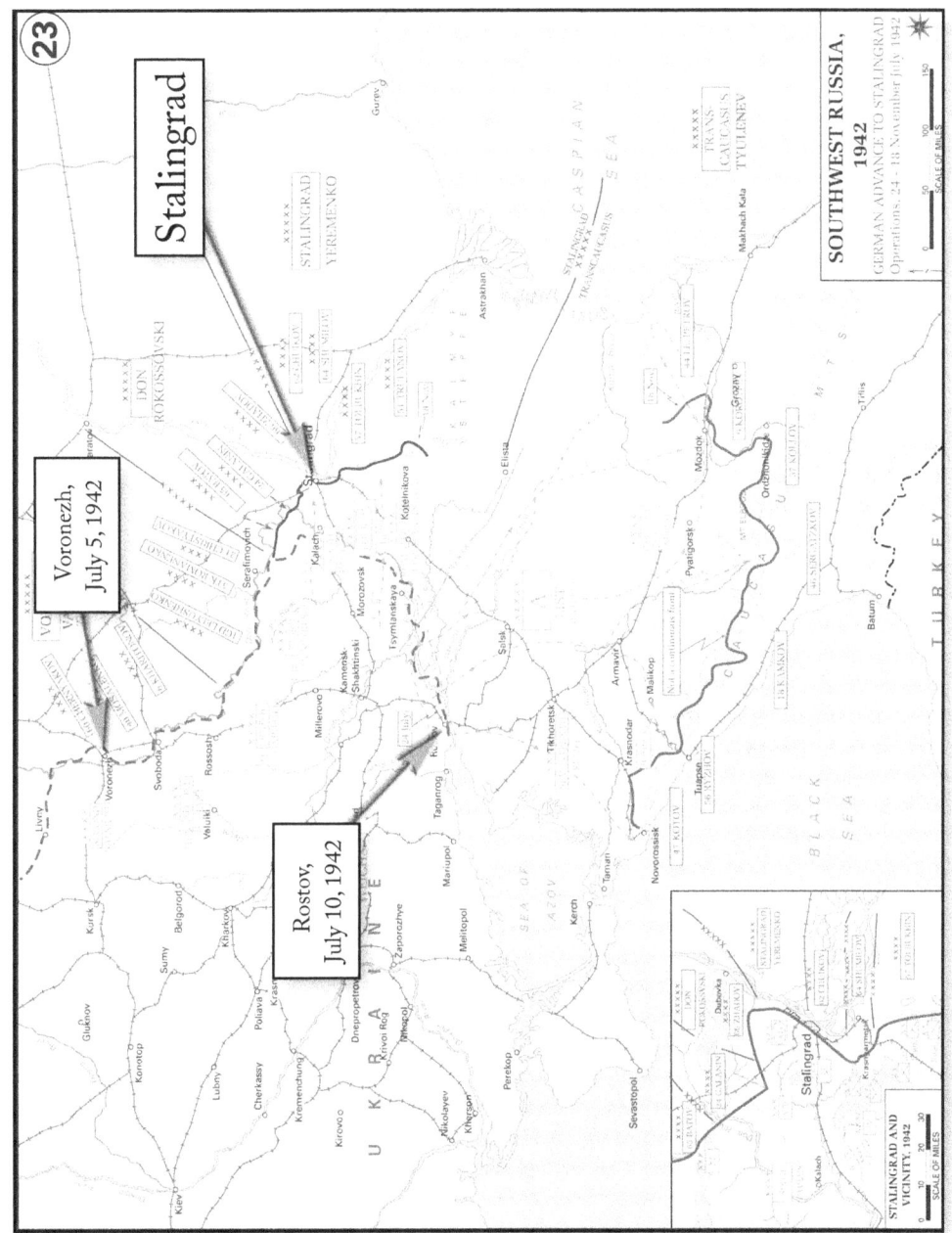

Figure 54– Stalingrad – July 1942

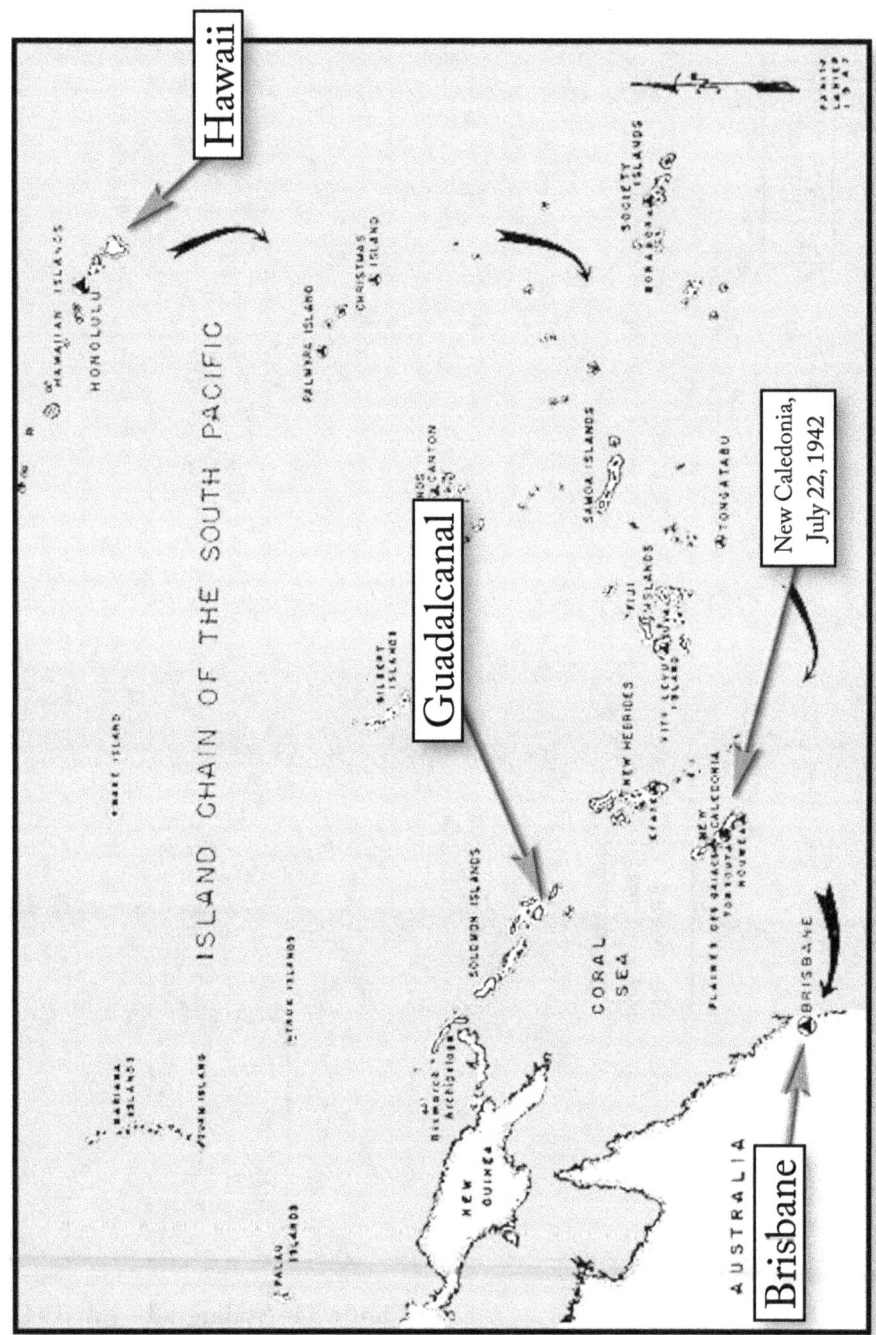

Figure 55 – Pacific Supply Chain – July 1942

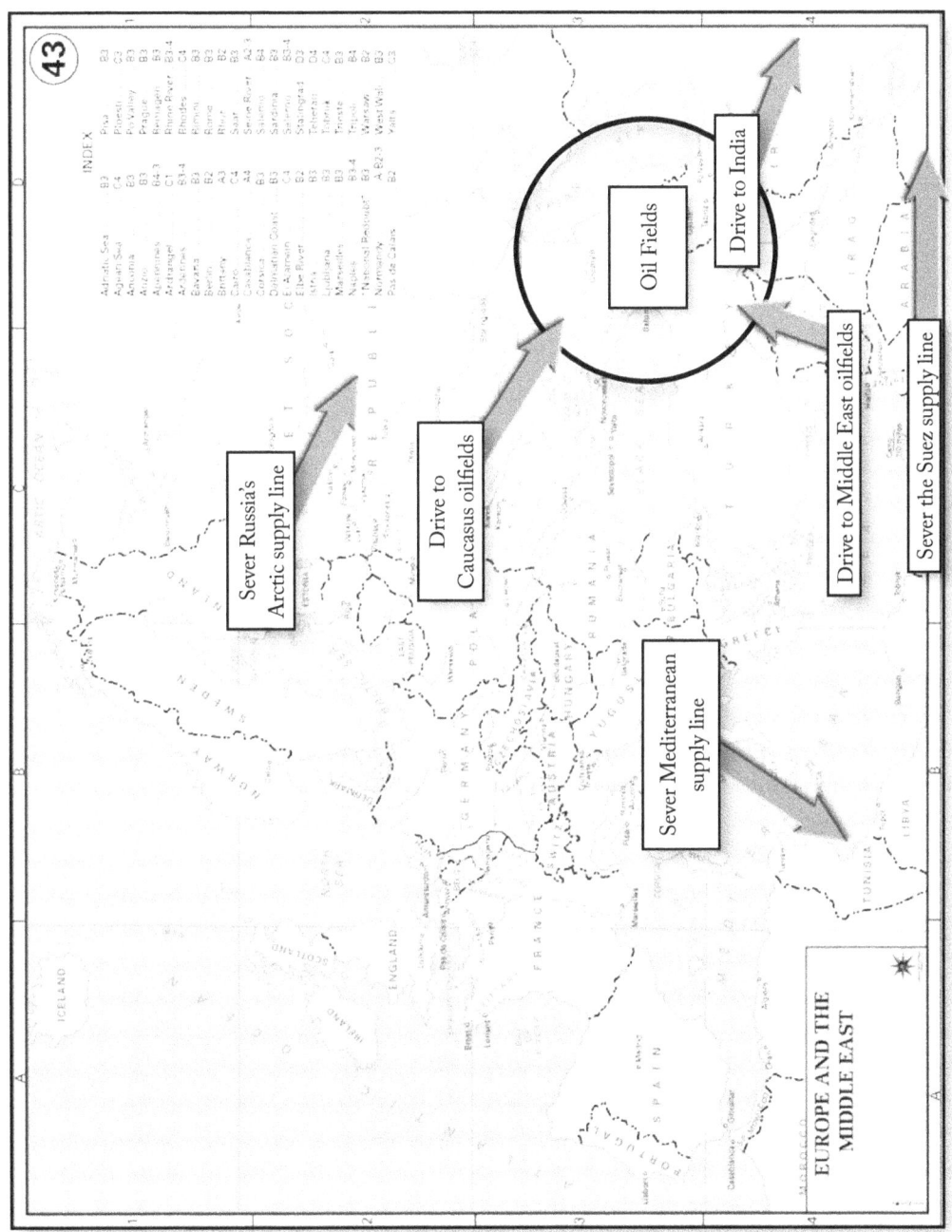

Figure 56 – Egypt Holds Fate of a Vital Area

ENDNOTES

1. Williams, *Chronology 1941-1945 - 1942*, p. 44.
2. Williams, *Chronology 1941-1945 - 1942*, p. 44.
3. Brown, *War Stories*.
4. Clendenin, *Mission History of the 376th Bombardment Group*, p. 16.
5. Daniell, Raymond, *New York Times*, BRITAIN'S FORCE IS INFERIOR IN DESERT, July 2, 1942, http://select.nytimes.com/gst/abstract.html?res=F10916F73858167B93C0A9178CD85F468485F9, accessed February 22, 2014.
6. Craven, *The Army Air Forces in World War II, Vol. II*, p. 18.
7. Clendenin, *Mission History of the 376th Bombardment Group*, p. 17.
8. Richardson, *The Forgotten Force*, p. 86.
9. Assistant Chief of Air Staff, *The AAF in the Middle East*, p. 73.
10. Lukas, *The Velvet Project*, p. 147.
11. Thomas, *Born in Battle*, p. 150.
12. Clendenin, *Mission History of the 376th Bombardment Group*, p. 17.
13. Mayhew, *transcript of interview*.
14. Thomas, *Born in Battle*, p. 151.
15. Williams, *Chronology 1941-1945 - 1942*, p. 44.
16. Williams, *Chronology 1941-1945 - 1942*, p. 44.
17. Clendenin, *Mission History of the 376th Bombardment Group*, p. 17.
18. Assistant Chief of Air Staff, *The AAF in the Middle East*, p. 75.
19. Clendenin, *Mission History of the 376th Bombardment Group*, p. 18.
20. Frank, Guadalcanal, p. 31.
21. ----, *Time*, Strategy: If Egypt Falls . . ., July. 6, 1942, http://www.time.com/time/magazine/article/0,9171,932052,00.html, accessed August 1, 2012.
22. Clendenin, *Mission History of the 376th Bombardment Group*, p. 18.
23. Mayhew, *transcript of interview*.
24. ----. FDRL, Map Room, Box 168, 7/8/42 Pacific War Council minutes.
25. Clendenin, *Mission History of the 376th Bombardment Group*, p. 18.
26. Clendenin, *Mission History of the 376th Bombardment Group*, p. 19.
27. Royce, Transcript of an interview. Author's note: Sturkie would be involved in a landing accident on August 1st.
28. Richardson, *The Forgotten Force*, p. 92.
29. Adams, Letter, August 3, 1974.
30. Williams, *Chronology 1941-1945 - 1942*, p. 45.
31. ----, *US Marine Corps in World War II, Guadalcanal Campaign*, http://www.ibiblio.org/hyperwar/USMC/Guadalcanal/USMC-M-Guadalcanal-1.html , p. 10.

32. Williams, *Chronology 1941-1945 - 1942*, p. 45.
33. Frank, *Guadalcanal*, p. 43.
34. Clendenin, *Mission History of the 376th Bombardment Group*, p. 19.
35. Conlee, *transcript of interview*.
36. Christensen, *transcript of interview*.
37. Clendenin, *Mission History of the 376th Bombardment Group*, p. 19.
38. Richardson, *The Forgotten Force*, p. 93.
39. Clendenin, *Mission History of the 376th Bombardment Group*, p. 19.
40. Wedemeyer, *Wedemeyer Reports!*; Ben-Moshe, Tuvia, *Winston Churchill and the "Second Front*, p. 513.
41. Hannah, letters to the author.
42. Clendenin, *Mission History of the 376th Bombardment Group*, p. 19-20.
43. Baroni, *The Pyramidiers*, p. 10.
44. ----. FDRL, PSF, Box 4, Commander in Chief -> Hopkins, Marshall, King 7/16/1942, http://docs.fdrlibrary.marist.edu/PSF/box4/folo44.html , accessed January 7, 2014.
45. Clendenin, *Mission History of the 376th Bombardment Group*, p. 20.
46. ----, *US Marine Corps in World War II, Guadalcanal Campaign*, http://www.ibiblio.org/hyperwar/USMC/Guadalcanal/USMC-M-Guadalcanal-1.html , p. 13.
47. Churchill, *The Hinge of Fate*, pp. 267-69.
48. Anderson, Col. Orville A., Notes from Library of Congress Box 142, Hap Arnold File, AFHRA.
49. Wedemeyer, *Wedemeyer Reports!*
50. Clendenin, *Mission History of the 376th Bombardment Group*, p. 20.
51. Clendenin, *Mission History of the 376th Bombardment Group*, p. 20.
52. ---, *The China-Burma-India Theater*, p. 180.
53. ----, *Time, BATTLE OF RUSSIA: Time Will Not Wait*, July 20, 1942, http://www.time.com/time/magazine/article/0,9171,795999,00.html , accessed August 1, 2012.
54. Churchill, *The Hinge of Fate*, p. 446.
55. Lukas, *The Velvet Project*, footnote 15.
56. Mayhew, *transcript of interview*.
57. Clendenin, *Mission History of the 376th Bombardment Group*, p. 21.
58. Williams, *Chronology 1941-1945 - 1942*, p. 46.
59. Clendenin, *Mission History of the 376th Bombardment Group*, p. 21.
60. Williams, *Chronology 1941-1945 - 1942*, p. 46.
61. Craven, *The Army Air Forces in World War II, Vol. I*, p. 572.
62. Clendenin, *Mission History of the 376th Bombardment Group*, p. 21 and 22.
63. Richardson, *The Forgotten Force*, p. 99-100.
64. Royce, transcript of interview.
65. ----, *US Marine Corps in World War II*, http://www.ibiblio.org/hyperwar/USMC/ , *First Offensive: The Marine Campaign for Guadalcanal*, p. 1.

66. ----. FDRL, PSF, Box 4, Stimson-->FDR-7/23/42, http://docs.fdrlibrary.marist.edu/PSF/box4/folo44.html , accessed January 7, 2014.
67. Williams, *Chronology 1941-1945 - 1942*, p. 46.
68. Clendenin, *Mission History of the 376th Bombardment Group*, p. 22.
69. Brown, *War Stories*.
70. Richardson, *The Forgotten Force*, p. 101-102.
71. Richardson, *The Forgotten Force*, p. 116.
72. ----. FDRL, PSF, Box 4, ?-> FDR, 7/25/1942, http://docs.fdrlibrary.marist.edu/PSF/box4/folo44.html , accessed January 7, 2014.
73. ----. FDRL, PSF, Box 4, FDR-->Hopkins, Marshall, King, http://docs.fdrlibrary.marist.edu/PSF/box4/folo44.html , accessed January 7, 2014.
74. See for example, the above reporting on April 21 and 29.
75. Williams, *Chronology 1941-1945 - 1942*, p. 47.
76. Clendenin, *Mission History of the 376th Bombardment Group*, p. 22.
77. Churchill, *The Hinge of Fate*, p. 449.
78. Clendenin, *Mission History of the 376th Bombardment Group*, p. 23.
79. Miller, personal diary, p. 70.
80. ----. FDRL, PSF, Box 4, FDR-->Sec of War-7/29/42, http://docs.fdrlibrary.marist.edu/PSF/box4/folo44.html , accessed January 7, 2014.
81. Churchill, *The Hinge of Fate*, p. 272.
82. Williams, *Chronology 1941-1945 - 1942*, pp. 44 - 47.
83. ----. FDRL, PSF, Box 4, Chief of Staff-->FDR-7/30/42, http://docs.fdrlibrary.marist.edu/PSF/box4/folo44.html , accessed January 6, 2014.
84. Clendenin, *Mission History of the 376th Bombardment Group*, p. 23.
85. Churchill, *The Hinge of Fate*, p. 453-4.
86. Williams, *Chronology 1941-1945 - 1942*, p. 47.
87. Appold, letter to Robert Brooks, dated December 28, 1973.
88. Frank, *Guadalcanal*, p. 45; ----, *The 11th Bomb Group*, p 14.
89. Williams, *Chronology 1941-1945 - 1942*, p. 47.
90. Miller, Personal diary, pp. 70-1.
91. ----, Army Air Force Statistical Digest, http://www.ibiblio.org/hyperwar/AAF/StatDigest/aafsd-3.html , accessed January 22, 2014, p. 112; B-24 production from Author's personal papers.

AUGUST 1942

By the end of the month, the unstoppable Germans were at the outskirts of Stalingrad and were threatening to capture the first Russian oilfield.

Churchill continued pressing Roosevelt for a commitment to sending air units to Russia. Marshall opposed the move as he, Marshall, needed all the assets he could gather to support TORCH.

Dissatisfied with his military leadership in Egypt, Churchill replaced General Auchinleck with Harold Alexander. Lieutenant-General Bernard Montgomery was placed in command of the Eighth Army.

Meanwhile, the 97th BG, the initial AAF Task Force BR bomb group, finally flew its first mission against the Germans. It was thirty-six weeks after Pearl Harbor and nearly two months after the first HALPRO raid.

In the Pacific, United States Marines landed on Guadalcanal.

1942 AUGUST 1

EASTERN FRONT

Stalin appointed Marshal Andrey Yeryomenko the commander of the Southeastern Front, following the success of the German forces driving towards the Caucasus. The task given to him and Commissar Nikita Khrushchev was simple: stop the Germans.

98TH BG

Air elements of the 98th BG flew their first mission.

1942 AUGUST 2

1ST PROVISIONAL GROUP

Twelve B-24s attacked a convoy, during the evening of August 1/2.[1] All went well until the group returned from the mission. Eleven of the planes landed. However, the eleventh plane was tardy in clearing the runway, thus forcing the twelfth plane to abort its landing and go around. Howard Sturkie piloted that plane, *Hellzapoppin*. For reasons unknown, on the second attempt, the plane slammed into a nearby blockhouse and burst into flames. All of the crew were killed except Sturkie. Five locals manning the blockhouse were also killed.[2]

Queen B was damaged in early July. It was repaired and being prepped when the following happened:

> ***Yesterday we got a chance*** to see more results of our brilliant Operations Officers' thinking. The ship I came in, No. 9, has been out getting repaired because an enlisted man let us land without a nose wheel - this repair job took 3 weeks. The plane had been test hopped, and was waiting to go on a mission, and some officers from the 17 Outfit [Brereton] taxied it into a ditch - knocked nose wheel and left landing gear back off. After losing one B-24 by the 17 crew, I don't figure how the boys that make all the rules could trust more 24s to them, but then I guess that is why I wear only one Silver Bar. The accident came as the result of taxiing with Nos. 1 & 4 engines, and not using hydraulic booster pump - therefore they had no brakes.[3]

Max Fennell was another of the B-17 pilots that was transitioning to the B-24. Bill Mayhew was the waist gunner on this day's transition crew.

> ***Our crew's first combat mission*** in a B-24 was on 1 August 1942 when we were sent after a convoy in the Mediterranean Sea. Ships bound for Tobruk or Benghazi came from Greece via Crete, and Naples or other Italian ports. Major Fennell was the only member of the crew that had flown in a B-24 before. These B-24s did not have a ball turret, so I flew the right waist window gun. The waist positions were about six feet apart and about 20 feet from the tail guns. The window openings were three feet by four feet. The enlisted men of the crew were not at all impressed with this aircraft. The back of the plane vibrated a great deal and there was the smell of gasoline permeating the interior. We wondered why anyone would fly in such a "coffin". However, we eventually learned to love to fly them (especially our "Pink Lady"). Nevertheless, the mission was successful — we got a ship. On our night return we hit the coast of Egypt instead of Palestine. We were lost for 1 hour. We unintentially flew over the Suez Canal, but we were not shot at. The return to Lydda, however, was very traumatic. Our crew was next to last to land. The jeep that led us to our parking spot led us down the runway, so the last plane had to abort its landing and go around the field again. We had just parked our aircraft and I reached down to pick up my parachute when an enormous red light flooded the interior of our plane (our plane was parked nearest to the accident). Lieutenant Sturkie had undershot the field on his final approach and

hit the top of a two-story steel-reinforced concrete police building. As I looked up, I saw the plane coming to earth in three pieces, followed by a parachute that survived the explosion. Lt. Sturkie was thrown free of the plane and survived (I do not know how). The eight other members of his crew and seven policemen in the building were killed. Twenty people had left the building a few minutes before the accident.[4]

OPERATION WATCHTOWER
The airfield on Espiritu, New Hebrides was complete. The 11th Bomb Group used this base to refuel for the flight to the Solomons. (see Figure 39)

1942 AUGUST 4

1ST PROVISIONAL GROUP
This day would be forever remembered by the HALPRO servicemen.

The day began with a ten-ship attack on a German convoy, returning the next day. The sortie reports claimed two direct hits.

Brereton, realizing that the crews were exhausted, authorized a ten-day leave for the crews. Seven crews would fly a mission on August 6, but the Group would not fly another mission until August 24.[5]

It is unclear if the issuance of the leaves was done to soften the blow of what came next. Regardless, Halverson was relieved of command.

The reasons for this action were never made clear. Surveys of various HALPRO servicemen after the war were numerous. Others claimed that the British had demanded his removal. What is known was that he had not made friends with the leadership on either side of the pond. His position was always that his men came first.

Regardless of why, the seeds for his removal were planted when HALPRO was ordered to stay in the Middle East following the June 12 Ploesti raid. Other similar tirades followed. George Richardson made note of one of these episodes:

> *Colonel Halverson had a big* argument with the RAF ACC (Levant) who had come down to complain about our putting our men on cots, fixing showers (heated) and lighting their barracks. The RAF argued that since their boys had none of these and slept on floors, it would create resentment. Funny war — the RAF wanted our standards to come down, they never wanted to elevate theirs. Air Commodore Brown parted in a huff over this. Colonel Kendricks and I could see no medical way to think that our men should undergo this for the simple reason that the 'RAF so desired. The RAF had a standard argument — "now down in the 'blue' the boys don't have these things". The "blue" was the Western Desert

and no argument could make them use the logic of improving conditions when all facilities were available. One would think that such a radical departure would cause a breakdown in the Empire.[6]

Homer Adams echoed George Richardson's reasoning.

Roosevelt relieved Col. Halverson to please Winston Churchill because Col. Halverson wanted screens on the mess hall At Lydda airport. The British said, "You silly Americans, you will learn that you can't keep flies out of a mess hall." Col. Halverson put up the screens anyway. The British sent us rotten tomatoes to eat. Col. Halverson rejected them. The British said, "That's alright, our troops up in the desert will be glad to get them." Col. Halverson bought cots for us to sleep on; the British said he would have to turn them back because their troops had none. Col. Halverson said, "Go to Hell."[7]

George Richardson devoted one paragraph to Halverson's departure, but his last two sentences said it all:

Some of the boys actually cried when Halverson walked out for the United States. McGuire [George McGuire replaced Halverson] was not as well liked as Colonel Halverson who had been father confessor and father to these boys.[8]

1942 AUGUST 5

EASTERN FRONT
German Army Group A established a beachhead across the Kuban River, threatening the Maikop oilfields.[9] (see Figure 57)

1942 AUGUST 7

OPERATION WATCHTOWER
The United States Marine Corps landed on Guadalcanal Island and Tulagi.[10] (see Figure 58)

NORTH AFRICA
Churchill visited the British Eighth Army in Egypt.[11]

1942 AUGUST 8

NORTH AFRICA

Churchill relieved Auchinleck of his command and replaced him with Harold Alexander. Churchill also intended to place Lt. Gen. William H. E. Gott, Commanding General of British 13 Corps, in command of the Eighth Army. Unfortunately, Gott was killed when his plane was shot down.[12]

OPERATION TORCH

Brereton was informed that he would not be receiving any new aircraft other than those currently deployed. Operation TORCH had priority claim on future assets.

EASTERN FRONT

German Army Group B captured Surovikino, west of Stalingrad.[13] (see Figure 54)

OPERATION WATCHTOWER

A Japanese naval force intercepted the Allied naval forces covering the Guadalcanal landings. The Japanese attacked at night, catching the Allies by surprise. This engagement, which would become known as the Battle of Savo Island, resulted in the sinking of three United States cruisers, one Australian cruiser, and one United States destroyer. The United States lost a total of 1,077 men.

On Guadalcanal, the Marines captured the unfinished Japanese airfield at Guadalcanal. (see Figure 58) It would later be renamed to Henderson Field.

1942 AUGUST 9

EASTERN FRONT

The First Panzer Army, attached to German Army Group A, reached Maikop, at the foothills of the Caucasus range. (see Figure 57) The oil fields near Maikop had been captured the previous day. Unfortunately, from the German viewpoint, the Russians had destroyed the refineries. The Germans also captured Krasnodar, northwest of Maikop.[14]

OPERATION WATCHTOWER

Due to Japanese air attacks, the Allied naval force defending the Guadalcanal landing force withdrew before it could unload all of the supplies. (see Figure 58)

1942 AUGUST 10

NORTH AFRICA
Alexander received a simple order — destroy Rommel's forces.[15]

COMMITMENT TO CHINA
Two months after the decision to use HALPRO in the Middle East, China was still waging its public relations campaign for aid. Under the title

SO NICE, YES?

It was not words China needed; it was planes. China needed no promises; she needed tanks. China needed no more goodwill ambassadors; she needed guns.

All this was well and bitterly known to Lauchlin Currie when he faced Mme. Chiang Kai-shek and many another Chinese personage last week on Vice President H. H. Kung's lawn in Chungking. President Roosevelt had sent Dr. Currie to Chungking once before, in 1941, when Lend-Lease was a glowing promise. Now he was back, in the sixth and darkest year of China's war.

The Chinese listened for words of action. They were grateful for what Lieut. General Stilwell and Brigadier General Chennault were doing with so little (see p. 37). They hoped that would be proof at last of how much more a little more would do. Had Dr. Currie come to talk real business—like Hopkins & Harriman on their trips to London and Moscow?

Lauchlin Currie denounced Japan's "tawdry mask." He restated "our pledge to deliver to China's veteran armies and experienced generals a striking power that will turn a long and glorious war of resistance into offensive campaigns." He assured the Chinese that President Roosevelt and his advisers "have a full conception of this as a global, worldwide war." He said that China's war is a United States war.

The Chinese faces on the lawn did not change. Chinese pulses did not quicken. The applause was polite.[16]

1942 AUGUST 11

AAF TASK FORCE BR
Thirty-six weeks after Pearl Harbor (and nearly two months after the HALPRO raid on Ploesti), the United States finally sent a group of heavy bombers against a Nazi target from a base in England. Twelve B-17Es of the 97th HBG bombed the railroad mar-

shalling yards at Rouen-Sotteville in France. Six other B-17s flew diversionary missions along the French coast.

Despite "The Germany First" philosophy of the ABC planners in the spring of 1941, military realities had forced the diversion of men and material to other theaters.[17]

The never-ending conflict between the ABC plan and the military realities continued. General Spaatz opposed the diversion of heavy bombers, originally destined to England, to the Middle East. His argument was that neither the B-17 nor the B-24 could reach targets in Germany from the Middle East.

> *Regardless of what operations are* conducted in any other theater, in my opinion, this England still remains the only base area from which to launch aerial operations to obtain air supremacy over Germany, and until such air supremacy is established there can be no successful outcome of the war.[18]

1942 AUGUST 12

NORTH AFRICA

Churchill had announced that Lieutenant-General Bernard Montgomery would command the Eighth Army. Montgomery arrived in Cairo. He assumed command the next day.[19] (see Figure 50)

EASTERN FRONT

German mountain troops hoisted the Nazi flag on the highest mountain of the Caucasus, Mount Elbrus.

However, the more the German forces advanced, the longer the supply lines became. These advanced units were desperately short of gasoline. Armored units often were at a standstill. Even oil tankers were short of fuel. At one point, camels were transporting containers of fuel.

The Germans began shipping fuel from Rumanian ports across the Black Sea in early September.

OPERATION VELVET

Churchill flew to Moscow to have a face-to-face meeting with Stalin. Stalin was pressing for the Second Front. Stalin was not pleased when Churchill told him that any planned invasion of Europe must be postponed. But Stalin was somewhat appeased when informed of Operation TORCH. Then:

> *...the Prime Minister brought* the discussion back to the Russian front stating that you [Roosevelt] and he were exploring the possibility of lending an air force

to the South Russia [sic] front but only after Rommel was defeated. He asked how such a suggestion would be received by Stalin. Stalin's answer was brief and simple, 'I would gratefully accept it.'[20]

1942 AUGUST 13

NEW GUINEA

Japanese troops landed at Basabua, New Guinea. They engaged and pushed back the Allied Maroubra Force at Deniki.[21] (see Figure 58)

1942 AUGUST 14

OPERATION TORCH

Eisenhower was told that he would be the Commanding General of Operation TORCH.[22]

1942 AUGUST 15

OPERATION WATCHTOWER

Four United States destroyers executed a supply run to Guadalcanal. (see Figure 58)

1942 AUGUST 16

98TH BG

The ship carrying personnel of the 98th docked at the port of Suez. Unloading of the ship began.[23]

OPERATION TORCH

Near the end of his Moscow visit, Churchill cabled Roosevelt that he had managed to convince Stalin that the invasion of North Africa was the only viable operation for 1942.[24]

1942 AUGUST 17

OPERATION VELVET

Maj. Gen. Follett Bradley and Maj. Gen. Russell L. Maxwell, representing the United States during Churchill's Moscow visit, met with Russian air staffs to discuss the offer of Anglo-American air force operations from Russian bases. The two staffs agreed to undertake studies once the respective governments had approved the concept.[25]

1942 AUGUST 18

1ST PROVISIONAL GROUP

Brereton argued strongly that the 1st Provisional Group should not be deployed to their original destinations: China for the former HALPRO planes and the 10th Air Force for the B-17s. His basic argument was that this plan would reduce the combat effectiveness of his Middle Eastern heavy bomber force by half; only the 98th BG would remain. And the RAF had no equivalent force.[26]

NEW GUINEA

A group of Japanese troops landed at Basabua, New Guinea.[27] (see Figure 58)

OPERATION WATCHTOWER

The Japanese landed a force on Guadalcanal in an attempt to retake the island. (see Figure 58)

1942 AUGUST 19

DIEPPE

At 5:00 a.m., Canadian forces, with the support of some British troops, mounted an invasion of the French coast town of Dieppe. Within five hours, the invasion was deemed a failure and the commanders issued orders for a withdrawal. By 3 p.m., the invasion troops had been evacuated, killed, or left behind. Of the 6,000 plus men who made it to shore, 3623 had become causalities.[28]

1ST PROVISIONAL GROUP

Brereton repeated his August 18 plea not to remove half of his heavy bomber force.[29]

AAF TASK FORCE BR

B-17Es of the 97th HBG bomb the German airfield at Abbeville, France in support of the Dieppe raid.[30]

OPERATION WATCHTOWER

The initial group of Marine aircraft, 19 F4F fighters and 12 SBD-3 dive-bombers, arrived at Henderson Field.

1942 AUGUST 20

1ST PROVISIONAL GROUP
Marshall decided to keep the 1st Provisional Group in the Middle East, rather than send them to the CBI.[31]

AAF TASK FORCE BR
United States Flying Fortresses bombed the marshalling yards at Amiens, France.[32]

OPERATION VELVET
Averill Harriman had accompanied Churchill during the Churchill's visit to Moscow. After the meetings, Harriman visited the Allied headquarters in the Middle East. There, he briefed the leadership on the situation in the Caucasus. He also told Brereton about the proposal to send his heavy bombers to the Caucasus, but without their American crews. That idea did not sit well with Brereton.

> *Mr. Churchill proposed to the* President an offer of immediate air support to help the Russians in the defense of the Caucasus barrier. The plan called for us to furnish planes to the RAF to man. I pointed out to Mr. Harriman that I would object to this program of furnishing additional American equipment to the RAF. I am convinced that American pilots are at least the equal of the RAF pilots, and any further equipment furnished this theater must in my opinion, be manned by American personnel.[33]

1942 AUGUST 21

AAF TASK FORCE BR
United States Flying Fortresses bombed the shipyards at Schiedam.[34]

OPERATION VELVET
Roosevelt was concerned about British participation in the project. Churchill had proposed that the United States provide the planes and the British provide the men. Brereton took offense saying the American airmen were more than capable of manning the planes.[35] Roosevelt then asked Marshall, "If such an enterprise could be accomplished ... would it be advisable to have British air [sic] also represented?"[36]

1942 AUGUST 22

OPERATION WATCHTOWER

The first group of United States Army aircraft arrived at Henderson Field, flying from New Hebrides. They would be used to provide fighter cover for off-loading supply ships.[37]

1942 AUGUST 23

EASTERN FRONT

The German 6th Army reached the Volga, on the outskirts of Stalingrad.[38] (see Figure 54)

Generaloberst Wolfram von Richthofen, in command of Luftflotte 4, began a massive bombing campaign against the city. The air campaign also targeted the attempts by the Russians to ship reinforcements across the Volga. Ultimately, ninety percent of the city would be left as a massive pile of rubble.

1942 AUGUST 24

AAF TASK FORCE BR

In response to Spaatz's plea for more bombers, Maj. Gen. George E. Stratemeyer, Arnold's chief of staff, attempted to console Spaatz with the promise that Operation TORCH would not be executed "at the expense of the bombing offensive from the UK but in addition to it, and therefore at the expense of anything but the UK."

These words were in direct conflict with a statement made by Admiral Leahy four days earlier. Leahy had told the Combined Chiefs that Operation TORCH might require the diversion of currently planned deployments.[39]

Meanwhile, United States Flying Fortresses bombed the airdromes at Le Trait, France.[40]

1942 AUGUST 25

EASTERN FRONT

Brereton, who opposed sending his planes to the Russians without his crews, realized that the situation was ominous:

> *The situation in Russia is* extremely grave. The Axis High Command may try for a quick decision in Egypt which would put more pressure on Russia. If the German armies in Russia and the Middle East could bring about the capture of the

Caucasian oil fields and the oil resources of Persia, the war would be won for Hitler.[41]

1942 AUGUST 27

AAF TASK FORCE BR
United States Flying Fortresses bombed the Rotterdam shipyards.[42]

1ST PROVISIONAL GROUP
A ten B-24 formation attacked a German convoy, the group's 41st mission.[43]

EASTERN FRONT
German Army Group A closed in on the oil fields at Grozny.[44] (see Figure 57)

1942 AUGUST 29

OPERATION VELVET
Marshall responded to Roosevelt's query about sending American air units to Russia and about British participation in the project. Marshall opposed sending any air units until January 1943 for three reasons:

1. Any resources would require diverting from other theaters
2. The weather in the Caucasus would limit continuous air operations
3. All supplies for such operations would divert the use of the Iranian supply lines from current Lend-Lease projects

Regarding the British, if the project went forward, the British would have to participate.[45]

1942 AUGUST 30

1ST PROVISIONAL GROUP
Twelve B-24s attacked Tobruk, the group's 42nd mission.[46] (see Figure 50)

OPERATION VELVET
Following his recent visit, Churchill was apparently disturbed by the impression he had left with Stalin. Churchil then pushed Roosevelt to communicate to Stalin Roosevelt's commitment to support the Russians. Churchill followed these messages with another message to Roosevelt outlining the political fallout for not helping Russia. One

of his arguments was the "moral effect of comradeship with the Russians, which will be out of all proportion to the forces employed."

Therefore, "It is urgently important that this be put in hand without delay."[47]

1942 AUGUST 31

NORTH AFRICA

During the night of August 30/31, Rommel resumed his drive to the Nile. Initially, his assault was successful.[48]

1ST PROVISIONAL GROUP

Rumors were rampant within the 1st Provisional Group about what was happening. As George Richardson reported:

> *Mussolini was reported at Benghazi* prepared to make a triumphal entry before those citizens of Italian origin. Hitler and Mussolini were reported to have picked out their rooms at the Mena House Hotel. The staff of the Germans would take over Shepheard's and undoubtedly the Italians would take over the Continental Savoy. Egypt, "land of mystery" — in reality "land of misery" — was threatened by unveiling.
>
> Things were hot in Cairo for all concerned. The Italians there were all ready for Musso and the same in Alexandria. Arrests were frequent and extraordinary sessions of the cabinet of King Farouk were always meeting.[49]

But it was not just the threat from the Germans that was affecting morale. The Americans were also fighting the press:

> *[T]he constant newspaper distortions in* the PALESTINE POST and EGYPTIAN DAILY MAIL as regards activity of "RAF heavy bombers" and almost a total lack of credit for the boys who were doing the work for the heavy bombardment. This and the constant raids on Benghazi and Tobruk did little for our morale. Benghazi was as distant from our base as was the Italian Turin from England; ack-ack was perfect, missions were by day over long stretches of water, and enemy territory.
>
> YANK, our own magazine, came out with a description of the preliminary battle of this period and forgot the 1st Provisional Group in its new paper accounts of air activity. We were forgotten all around, and the bastard child of the Middle East Air Force (or 9th as it was soon to become), with nothing to indicate our presence or the fact that we had dropped roughly more bombs than

many units and certainly more than any other American Unit of similar size in the entire American Air Force. We did not, as Colliers and Frank Gervasi [his article in COLLIERS entitled "Rommel Meets the AFF"] put it, "consider ourselves the elite of the gang" — just the best but all on the same side.[50]

OPERATION VELVET

Given Marshall's objection to the timing of VELVET, Roosevelt told Churchill that he understood the "desirability" of the project and he would get back to him. Roosevelt seemed to be postponing any decision.[51]

HEAVY BOMBER PRODUCTION

During August, 234 heavy bombers were manufactured. One hundred and ten were B-24s.[52]

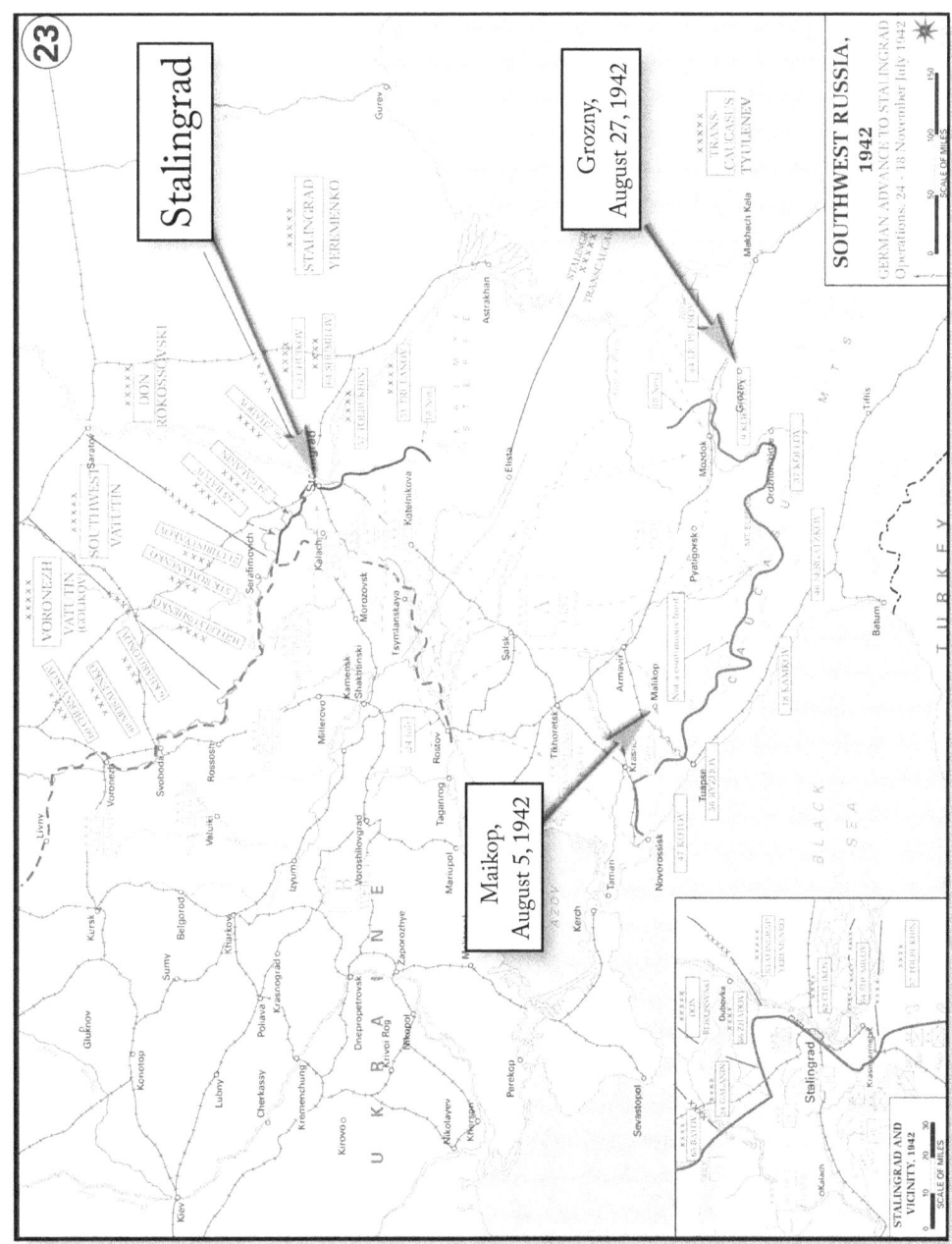

Figure 57 – Stalingrad – August 1942

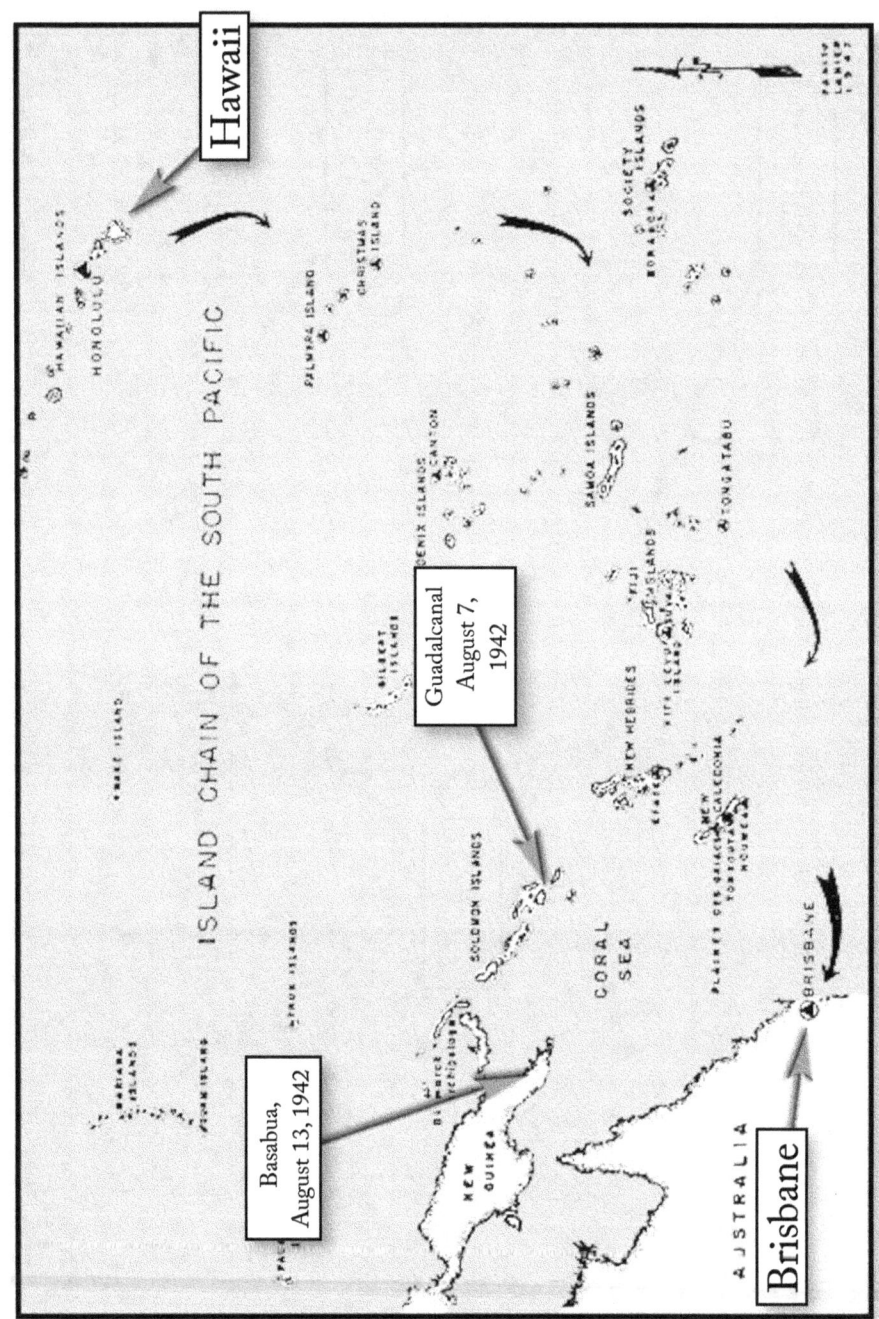

Figure 58 – Pacific Supply Chain – August 1942

ENDNOTES

1. Clendenin, *Mission History of the 376th Bombardment Group*, p. 24.
2. Richardson, *The Forgotten Force*, p. 105-110.
3. Miller, personal diary, pp. 71-2.
4. Mayhew, *transcript of interview*.
5. Clendenin, *Mission History of the 376th Bombardment Group*, pp. 24 and 25.
6. Richardson, *The Forgotten Force*, p. 96.
7. Adams, Contact Report, p.2.
8. Richardson, *The Forgotten Force*, p. 112.
9. Williams, *Chronology 1941-1945 - 1942*, p. 48.
10. Williams, *Chronology 1941-1945 - 1942*, p. 48.
11. Williams, *Chronology 1941-1945 - 1942*, p. 48.
12. Williams, *Chronology 1941-1945 - 1942*, p. 49.
13. Williams, *Chronology 1941-1945 - 1942*, p. 49.
14. Williams, *Chronology 1941-1945 - 1942*, p. 49.
15. Williams, *Chronology 1941-1945 - 1942*, p. 49.
16. ----, *Time, So Nice, Yes?*, August 10, 1942, http://www.time.com/time/magazine/article/0,9171,773413,00.html, accessed August 1, 2013.
17. ----, http://www.reddog1944.com/Missions_Targets_Combat_Action.htm , accessed November 23, 2015.
18. Craven, *The Army Air Forces in World War II, Vol. I*, p. 574.
19. Williams, *Chronology 1941-1945 - 1942*, pp. 49 - 50.
20. Lukas, *The Velvet Project*, p. 148. Message from Harriman to Roosevelt, August 13 and 14, 1942. Harriman had accompanied Churchill to Moscow; Churchill, *The Hinge of Fate*, pp. 477-83.
21. Williams, *Chronology 1941-1945 - 1942*, p. 49.
22. Williams, *Chronology 1941-1945 - 1942*, p. 50.
23. Baroni, *The Pyramidiers*, p. 10.
24. Churchill, *The Hinge of Fate*, pp. 501-2.
25. Lukas, *The Velvet Project*, p. 148-9. Message from Maj. Gen. Maxwell to Marshall, August 31, 1942; notes on Moscow Conference by Maxwell, August 17, 1942.
26. Assistant Chief of Air Staff, *The AAF in the Middle East*, p. 85.
27. Williams, *Chronology 1941-1945 - 1942*, p. 50.
28. Williams, *Chronology 1941-1945 - 1942*, p. 50.
29. Assistant Chief of Air Staff, *The AAF in the Middle East*, p. 85.
30. ----, http://www.historyofwar.org/air/units/USAAF/97th_Bombardment_Group.html , accessed November 23, 2015.
31. Assistant Chief of Air Staff, *The AAF in the Middle East*, p. 85.

32. ----, http://www.reddog1944.com/Missions_Targets_Combat_Action.htm , accessed November 23, 2015.
33. Brereton, *The Brereton Diaries*, p. 44-5.
34. ----, http://www.reddog1944.com/Missions_Targets_Combat_Action.htm , accessed November 23, 2015.
35. Brereton, *The Brereton Diaries*, p. 47.
36. Lukas, *The Velvet Project*, p. 149.
37. Williams, *Chronology 1941-1945 - 1942*, p. 51.
38. Williams, *Chronology 1941-1945 - 1942*, p. 51.
39. Craven, *The Army Air Forces in World War II, Vol. I*, p. 574.
40. ----, http://www.reddog1944.com/Missions_Targets_Combat_Action.htm , accessed November 23, 2015.
41. Brereton, *The Brereton Diaries*, p. 46
42. ----, http://www.reddog1944.com/Missions_Targets_Combat_Action.htm , accessed November 23, 2015.
43. Clendenin, *Mission History of the 376th Bombardment Group*, p. 25.
44. Williams, *Chronology 1941-1945 - 1942*, p. 52.
45. Lukas, *The Velvet Project*, p. 150.
46. Clendenin, *Mission History of the 376th Bombardment Group*, p. 26.
47. Churchill, *The Hinge of Fate*, pp. 564-6.
48. Williams, *Chronology 1941-1945 - 1942*, p. 53.
49. Richardson, *The Forgotten Force*, p. 124.
50. Richardson, *The Forgotten Force*, p. 125-6.
51. Churchill, *The Hinge of Fate*, pp. 566.
52. ----, Army Air Force Statistical Digest, http://www.ibiblio.org/hyperwar/AAF/StatDigest/aafsd-3.html , accessed January 22, 2014, p. 112; B-24 production from Author's personal papers.

SEPTEMBER 1942

While the German forces in Russia appeared to be unstoppable, Rommel may have reached the limits of his supply line. He renewed his drive towards Alexandria, but was driven back.

Churchill and Roosevelt faced another political-military dilemma. Operation TORCH was placing demands on all available Allied resources. The second front had already been postponed. Now, the supply convoys to Russia would have to stop. Stalin's reaction to that news was unknown. Russia had to be kept in the fight. With the German successes, military planners feared that Stalin and Hitler might reach another 1939-like accord. Even worse, Germany might defeat Russia.

Meanwhile, air operations continued their attacks on Rommel's supply lines.

1942 SEPTEMBER 1

NORTH AFRICA
The British stopped Rommel's advance.[1] Still, there was frustration with the British. Brereton noted the following:

> *The war in the desert* has started in full force again. Damnit!! The British promised us if nothing happened until Aug. 27 that we'd take the initiative and the 8th Army would do the unprecedented attack instead of retreat. Here the Allies have had a full month to prepare and then Hitler hits first! [2]

1ST PROVISIONAL GROUP
Nine B-24s attacked Canea, Crete, another departure port for Rommel's supply convoys, the group's 43rd mission.[3]

EASTERN FRONT
German Army Group A captured Anapa, a port on the Black Sea.[4] (see Figure 59)
The first shipment of fuel from Rumania arrived for the German forces.
The Russian Air Force had lost 201 aircraft by the end of August. Only 192 aircraft were operational.

OPERATION WATCHTOWER

The United States 6th Naval Construction Battalion arrived on Guadalcanal to improve and expand Henderson Field. (see Figure 58)

1942 SEPTEMBER 2

NORTH AFRICA

Axis forces ceased their advance, yet the British remained on the defensive.[5] Their apparent inaction seemed to justify Brereton's comments the previous day.

CONVOY P.Q. 18

Convoy P.Q. 18 departed Loch Ewe, Scotland enroute to Archangel. Churchill was keeping his July 30 promise to Stalin.

1942 SEPTEMBER 3

NORTH AFRICA

Axis forces began to withdraw. The official record acknowledged the contribution of Allied aircraft attacks on Rommel's supply lines.[6]

1ST PROVISIONAL GROUP

Nine B-24s attacked a convoy, the group's 44th mission.[7]

1942 SEPTEMBER 5

EASTERN FRONT

The Russians launched an attack against the Germans in Stalingrad. Within hours, the Germans stopped the attack. The Russians lost 30 out of 120 attacking tanks. (see Figure 54)

USAMEAF

A discussion occurred between Tedder's senior air staff officer, Air Vice Marshal H. E. P. Wigglesworth, and G-3 officers of USAMEAF. Wigglesworth asserted that he had control, delegated by Tedder, over the target selection for the United States heavy bombers. Col. Patrick W. Timberlake, G-3 of Brereton's staff, argued that this position violated the Arnold-Portal-Towers agreement. According to Timberlake, American combat units assigned to theaters of British strategic responsibility were to be organized in "homogeneous American formations" under the "strategic control" of the appropriate British commander in chief.[8]

AAF TASK FORCE BR

The B-24s of the 93rd Heavy Bomb Group began their overseas deployment. They would be assigned to the RAF base at Alconbury, England, from which they would conduct air operations.[9]

NEW GUINEA

The Japanese began an evacuation of Milne Bay, New Guinea.[10] (see Figure 60)

1942 SEPTEMBER 6

EASTERN FRONT

German Army Group A captured Novorossiysk, a port on the Black Sea.[11] (see Figure 59)

1942 SEPTEMBER 7

NORTH AFRICA

Montgomery ordered the cessation of offensive actions. This left the Axis with some of its ground gains. The British continued the build-up for the up-coming offensive.[12]

1ST PROVISIONAL GROUP

Twelve B-24s attacked a convoy, the group's 45th mission.[13]

USAMEAF

Following the September 5 meeting, Timberlake admitted that the Arnold-Portal-Towers agreement might not apply to the 12th Medium Bombardment and 57th Fighter Groups. However, he could see no reason why operational control of the 1st Provisional and 98th Groups should not remain in American hands.

NEW GUINEA

The Japanese effort to seize Port Moresby, New Guinea was at an end.[14] (see Figure 25)

1942 SEPTEMBER 9

1ST PROVISIONAL GROUP

Nine B-24s attacked Tobruk, the group's 46th mission.[15] (see Figure 50)

1942 SEPTEMBER 11

1ST PROVISIONAL GROUP

James Sibert attended a planning meeting.

> *Ate dinner and read in* the Pilot's lounge until 15:00 when I went into Maj. Payne's office for a meeting with them. The Major told us of General Brereton's visit and the British Plan for future operations in the desert. Learned that we were only allowing 20% of Axis shipping to reach its destination. Also learned that we will get no relief and plans are for us to operate three more months with the present planes we have. Pilot fatigue disregarded! Learned that a big mission was on for tomorrow. 6 B-17s and 9 B-24s scheduled - me included.[16]

EASTERN FRONT

The Russian commander Andrey Yeryomenko created the 62nd Army and placed Lt. Gen. Vasiliy Chuikov in command. The 62nd Army promptly retreated into the city of Stalingrad the next day. The 62nd was now comprised of a mere force of 90 tanks and 20,000 men. (see Figure 54)

1942 SEPTEMBER 13

NORTH AFRICA

British commandos raided Tobruk and Benghazi.[17] (see Figure 50)

1ST PROVISIONAL GROUP

Five B-17s attacked Tobruk while nine B-24s attacked Benghazi, their 47th and 48th missions.[18] (see Figure 50) They were a diversion for the above British commando raids.

Therman Brown gave an excellent rationale for the mission.

> *My seventeenth mission was to* Benghazi. Again it was at night. It was a special mission in support of a British commando unit that was to attack Benghazi from the ground while we dropped bombs, one at a time, to keep Jerry in his foxhole. There was a similar commando raid on Tobruk at the same time. This bombing raid took place from two to four o'clock in the morning of September 14, 1942. The idea of dropping one bomb at a time was to prolong the raid and give the impression that the attacking force was larger than it actually was. The impression worked. The Italian radio said that one hundred heavy bombers had attacked Benghazi that night.
>
> This was like several combat missions all wrapped as one. We made the usual

bomb run taking the search lights and the flak from a well defended target. We would then circle around and come in again from the same direction for another run. We were doing more than trying to keep the Germans in their bomb shelters. We were also trying to do some damage with each bomb we dropped.

Lt. Norman Davis was the navigator-bombardier. When he finished the first run, he announced that all our bombs had dropped on the first pass over the target. Norman thought we should go home. I saw it differently. The commandos were depending on us to keep the enemy pinned down for a critical two hours while they got in place. We were obligated to keep making the bomb runs to give that illusion of a large attacking force. Ken DeLong's records show that we made four passes at the target.

The plane was much lighter without the bomb load. On each pass we would come in at a higher altitude. Our chances of picking up some flak damage was that much less. We were still doing our job. Everyone complained of the bitter cold.

This was an eleven hour and forty minute mission. With so much time in the target area, there wasn't a lot of extra gas at the end of the flight but everything worked out well. We had no trouble finding our little beds that night.

We later found out that the commando unit did not make it to Benghazi that night. They were still forty miles away. It appears that no commando attack was made at all. The commando raid on Tobruk that night was more successful.[19]

James Sibert was also on the mission.

Ate dinner and was briefed at 13:00 for night mission to Benghazi. Learned just before take-off that we were to be cover and help attract ack-ack fire and cause an air raid while a ground and sea commando raid took place. We were to be over the target from 21:00 to 24:30. I was in second flight of three and we took off at 16:00 and were over target from 22:00 to 23:30. Took off with 2700 gal gas and 9 - 500 lb bombs.

Flew out over desert in a three ship formation until dark and then proceeded individually. Met no opposition on trip and arrived at target at 21:55. Made 8 runs over target while searchlights raked the sky and AA guns belched below 22,000 ft. Was caught in searchlight once but eluded them. Counted 12 searchlights and as many A.A. batteries (4 guns to a battery). Ack-Ack came very close but we were not hit by any. Left target at 23:30 and took up a course for home. Passed within 35 miles of Tobruk where our B-17s were carrying out a similar mission and 40,000 commandos were to land. We could see 4,000 lb bombs bursting and the heavens full of ack-ack and searchlights.

> Arrived back at field at 04:00 after 12 hours of continuous flying. Had checked gas and had 300 gal left. Was interrogated ate breakfast and went to bed at 05:30.[20]

CONVOY P.Q. 18
German aircraft attacked and sank numerous ships.[21]

1942 SEPTEMBER 16

1ST PROVISIONAL GROUP
Nine B-24s attacked Benghazi, their 49th missions.[22] (see Figure 50)

EASTERN FRONT
German Army Group B entered the northwestern suburbs of Stalingrad.[23] (see Figure 54)

1942 SEPTEMBER 18

EASTERN FRONT
The Russians launched a second attack at the German forces within Stalingrad. And again, the Germans stopped the attack. Forty-one Russian tanks were destroyed while the Luftwaffe shot down 77 Soviet aircraft. The Germans effectively had control of the airspace over the city. (see Figure 54)

OPERATION VELVET
Marshall and his staff had time to further evaluate the proposal to send planes to the Russian front. He listed several reasons for his opposition to the idea. However, his concluding argument was the most realistic:

> ***Whereas the British, for political*** reasons, may feel obligated to send combat forces to fight beside the Russians for 'moral effect of comradeship,' our past and present policy of extending lend-lease aid to Russia to the fullest extent possible would seem to obviate the necessity for our sending combat troops for political or ideological reasons. So far as possible we should endeavor to deal with the military realities.[24]

Once again, Churchill had been arguing that military decisions reflect political concerns. The United States military was arguing for military necessity.

1942 SEPTEMBER 21

COMMITMENT TO CHINA

With the collapse of Burma Road and Allied bases in the Far East, the only viable means of supplying China was through India. But India was in the midst of its own independence movement, in particular those of Gandhi's Indian National Congress party. Every attempt by the United States to work with the native population was met with resistance from Churchill. China, who was leading the major opposition to the Japanese, was well aware that her existence was dependent upon getting Allied war material via India. China had a vested interest in this squabble as it played out. American reporters were also aware of the issue. The following appeared in a *Time* article titled,

SALT IN THE SORES OF INDIA

U.S. on India. The reasons for Chinese anxiety and that of all United Nations, including Russia and the U.S., are at least fivefold: 1) only through India can fighting China be supplied; 2) only from India can the United Nations launch a campaign to recover Burma; 3) control of India means control of shipping in the Bay of Bengal and the Indian Ocean; 4) India's position enables United Nations' air power to strike east or west; 5) Indian supply routes to Iran and Russia will be increasingly important if Russia's southern armies are defeated.

Returning from India, U.S. Industrialist Henry Francis Grady revealed last week that U.S. experts had thoroughly canvassed India's possibilities as a strategic military bastion and a source of materiel. Wherever he went Grady heard the old complaint that war production was hampered by Britain's protection of her own interests and by the apathy of labor and industry. The full Grady report was not released, but its recommendation that U.S. engineers and technicians be sent to speed up Indian war production was evidence of the U.S. stake in India and the need for internal stability.[25]

In July, Brereton's remaining heavy bombers had been removed from India to help the British stop Rommel.

1942 SEPTEMBER 22

1ST PROVISIONAL GROUP

Nine B-24s attacked Benghazi, their 50th missions.[26] (see Figure 50)

OPERATION VELVET

At the Churchill-Stalin meetings in Moscow, Stalin was concerned with the decision to postpone the invasion of Europe, known as the Second Front. Now, Churchill and Roosevelt were about to suspend Lend-Lease shipments via convoy to north Russia. Churchill reminded Roosevelt that the success of upcoming Operation TORCH would be due in no small part to this diversion. And if TORCH was successful, this would apply pressure on the Germans to divert resources from the Eastern Front.[27]

Apparently, Churchill wanted Roosevelt to break the news to Stalin. Harry Hopkins agreed, and added that if the shipments were suspended, then the United States had to offer to place military units along side the Russians.[28]

1942 SEPTEMBER 26

OPERATION VELVET

Wendell Wilkie, the 1940 Republican candidate for president, was sent by Roosevelt on an informal mission to Moscow. He was scheduled to leave on September 27. Stalin gave him a farewell banquet. Stalin began his opening remarks complimenting "airmen of the United Nations." He paused, then continued with a blast at the United States and Great Britain for not sending better planes to the Russians.

> *The American Government has furnished* the Soviets P-40 fighters, not Airacobras; the British have supplied Hurricanes, not Spitfires. Both of these aircraft are inferior to German fighters they have to meet in combat.

Stalin failed to mention that he had previously agreed to the diversion of the P-39 Airacobras to Britain.[29]

1942 SEPTEMBER 27

1ST PROVISIONAL GROUP

The group continued attacking Benghazi with nine B-24s bombing the port, their 51st missions.[30] (see Figure 50)

NEW GUINEA

The Japanese were in retreat from Ioribaiwa Ridge, New Guinea.[31] (see Figure 60)

OPERATION VELVET

Roosevelt advised Churchill to postpone telling Stalin about the cancellation of the Lend-Lease shipment. Roosevelt wanted to combine this notice with a firm commitment of an offer to send air units to the Eastern Front. Roosevelt was ready to make that commitment without further study by his military advisors.[32]

HEAVY BOMBER PRODUCTION

During September, 263 heavy bombers were manufactured. One hundred and twenty were B-24s.[33]

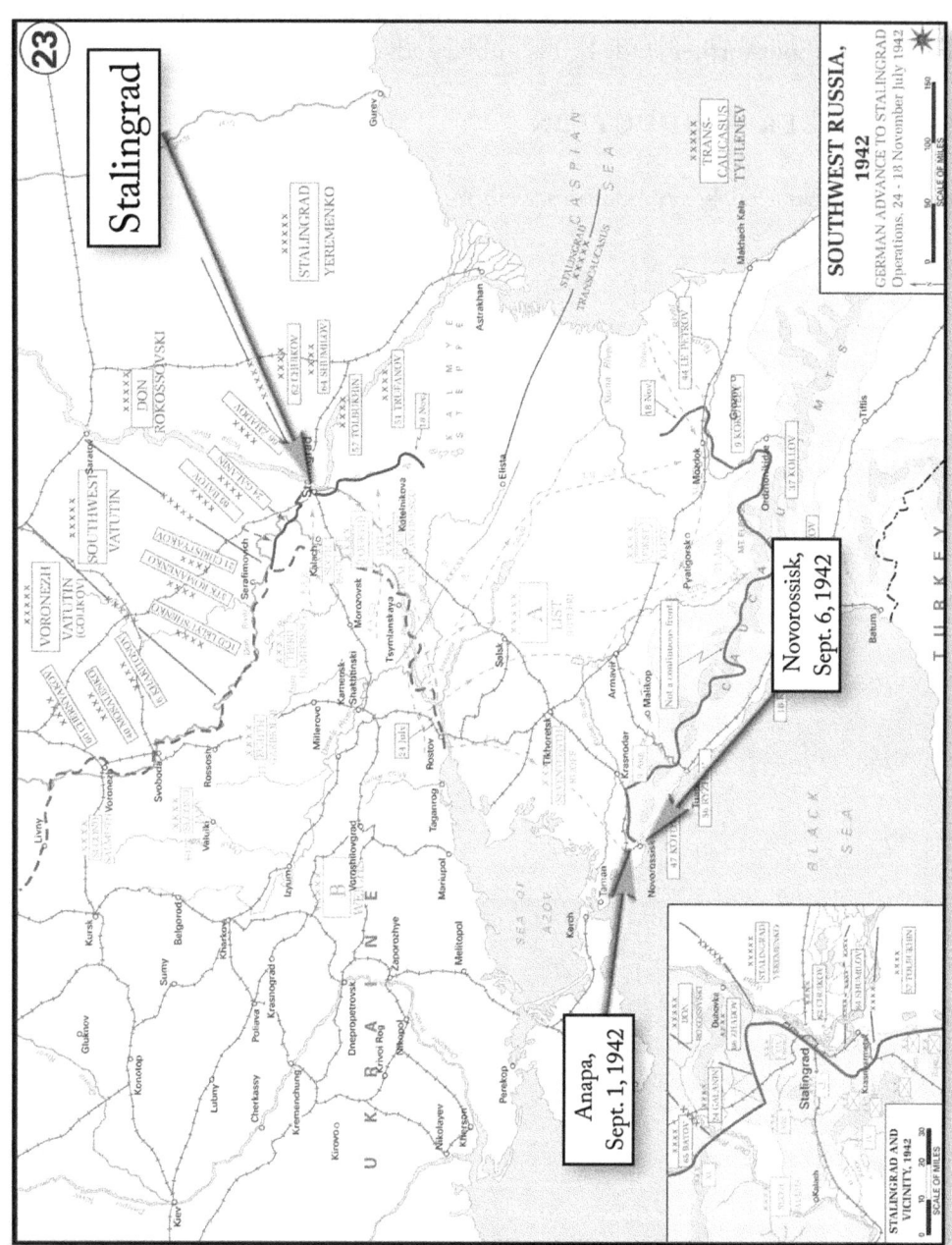

Figure 59 – Stalingrad – September 1942

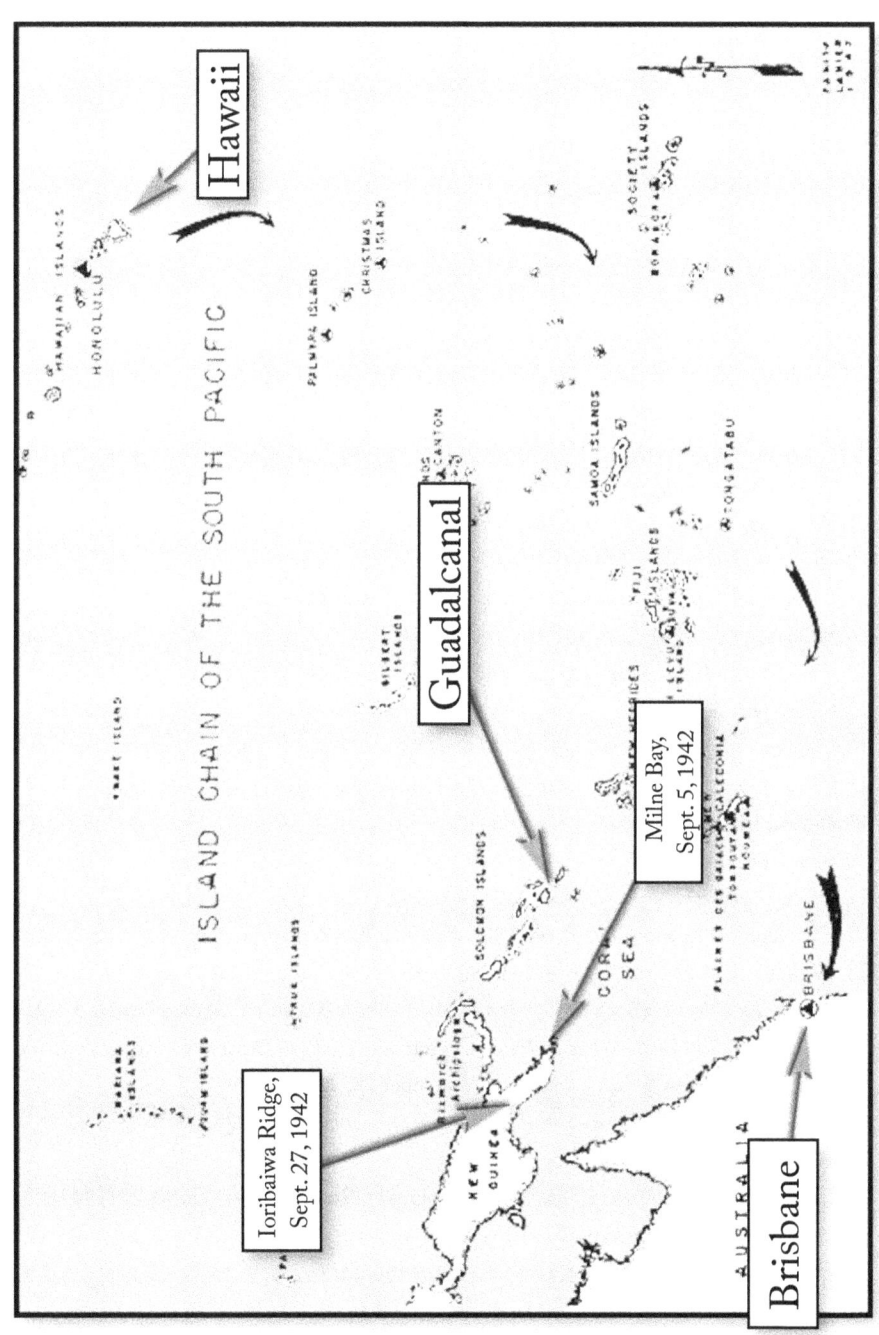

Figure 60 – Supply Chain – September 1942

ENDNOTES

1. Williams, *Chronology 1941-1945 - 1942*, p. 53.
2. Miller, personal diary, p. 81.
3. Clendenin, *Mission History of the 376th Bombardment Group*, p. 26.
4. Williams, *Chronology 1941-1945 - 1942*, p. 53.
5. Williams, *Chronology 1941-1945 - 1942*, p. 53.
6. Williams, *Chronology 1941-1945 - 1942*, p. 53.
7. Clendenin, *Mission History of the 376th Bombardment Group*, p. 26.
8. http://en.wikipedia.org/wiki/United_States_Air_Forces_Central , accessed December 6, 2013.
9. ----, 93rd Bomb Group, http://www.93rdbombardmentgroup.com/ , accessed January 6, 2014.
10. Williams, *Chronology 1941-1945 - 1942*, p. 53.
11. Williams, *Chronology 1941-1945 - 1942*, p. 54.
12. Williams, *Chronology 1941-1945 - 1942*, p. 54.
13. Clendenin, *Mission History of the 376th Bombardment Group*, p. 27.
14. Williams, *Chronology 1941-1945 - 1942*, p. 54.
15. Clendenin, *Mission History of the 376th Bombardment Group*, p. 27.
16. Sibert, personal diary.
17. Williams, *Chronology 1941-1945 - 1942*, p. 55.
18. Clendenin, *Mission History of the 376th Bombardment Group*, pp. 27 and 28.
19. Brown, *War Stories*.
20. Sibert, personal diary.
21. Williams, *Chronology 1941-1945 - 1942*, p. 55.
22. Clendenin, *Mission History of the 376th Bombardment Group*, p. 28.
23. Williams, *Chronology 1941-1945 - 1942*, p. 55.
24. Lukas, *The Velvet Project*, p. 150, Memo, Marshall for Roosevelt, September 18, 1942.
25. ----, *Time, Salt in the Sores of India*, September 21, 1942, http://www.time.com/time/magazine/article/0,9171,773584,00.html , accessed November 28, 2008.
26. Clendenin, *Mission History of the 376th Bombardment Group*, p. 28.
27. Churchill, *The Hinge of Fate*, pp. 572.
28. Lukas, *The Velvet Project*, p. 151, Msg., Hopkins to Roosevelt, September 22, 1942.
29. Lukas, *The Velvet Project*, p. 151.
30. Clendenin, *Mission History of the 376th Bombardment Group*, p. 29.
31. Williams, *Chronology 1941-1945 - 1942*, p. 57.
32. Churchill, *The Hinge of Fate*, pp. 573.
33. ----, Army Air Force Statistical Digest, http://www.ibiblio.org/hyperwar/AAF/StatDigest/aafsd-3.html , accessed January 22, 2014, p. 112; B-24 production from Author's personal papers.

OCTOBER 1942

The September pattern continued through October. The German offensive in Russia was still proceeding, albeit at a reduced pace. In North Africa, all remained quiet until the British began their offensive near the end of the month.

Roosevelt finally committed to sending air units to aid the Russians. However, while Stalin had agreed, logistical issues never seemed to be resolved. Stalin had always wanted the equipment, but never the troops.

1942 OCTOBER 1

1ST PROVISIONAL GROUP

The group attacked a German convoy with nine B-24s, their 52nd mission.[1]

1942 OCTOBER 2

AAF TASK FORCE BR

United States Flying Fortresses bombed the German airdromes at Meaulte and St. Omer, France.[2]

1942 OCTOBER 3

EASTERN FRONT

German Army Group A captured Elkhotovo, west of Grozny.[3] (see Figure 57)

GLOBAL STRATEGY

The American Joint Chiefs, who were interested in husbanding their resources for decisive air and amphibious actions in western Europe in 1943, were thus presented with a dilemma. To lose the Middle East meant to lose the southern supply routes to the U.S.S.R. and the main air ferry route to India. India itself would be rendered difficult, if not impossible, to defend, and the life line to China would be correspondingly endangered. Loss of the oil wells in Iraq and Iran would be a most severe blow, tantamount to cessation of Allied air and naval activity in the Indian Ocean. The economic gain to the Axis, although admittedly substantial, would not be so great as the economic and strategic loss to the Allies. And the key to the Middle East was Egypt: the best hostile avenue to the Persian Gulf, the Allied base most convenient for reinforcing any threatened part of the Middle Eastern area.

Joint Intelligence Comm.,
Memo for Information 29, Axis Capabilities and Intentions in the ME [4]

OPERATION VELVET

The situation at Stalingrad had worsened for the Russians. (see Figure 57) The German Luftwaffe had sent additional planes to provide more air cover. Consequently, Russian troops were now fighting with limited air support. Stalin proposed that the shipments of land equipment be suspended in exchange for increased aircraft. He wanted Britain to send 300 Spitfires and the United States to send 500 P-39s.[5]

1942 OCTOBER 5

OPERATION VELVET

Stalin's October 3 "status report" had its desired effect. Roosevelt was now ready to commit American air units to the Caucasus without regard to any other issue. He was ready to divert the American air units now operating in the Middle East: "We shall on our part undertake to replace in the Middle East all of our own planes which are transferred, and assist you in every way possible with your own air problems in the Middle East."

Roosevelt also suggested that the Western Allies not tell Stalin that P.Q. 19 would not be arriving. Instead, Roosevelt opted for a proposal by Admiral King to send P.Q. 19 in smaller units instead of one large convoy.[6] What effect this might have had on Operation TORCH was unclear as King's proposal never materialized. P.Q. 19 never sailed.

Roosevelt summarized his recommendations with: "I think it is better that we take this risk than endanger our whole relations with Russia at this time."[7]

1942 OCTOBER 6

NORTH AFRICA

Montgomery issued instructions for the El Alamein offensive.[8] (see Figure 50)

1ST PROVISIONAL GROUP

Nine B-24s flew to Benghazi.[9] (see Figure 50) Therman Brown described what happened after their bombs were dropped.

On still another occasion, October 6, 1942, returning from a raid on Benghazi, we were jumped by two ME 109's. The main difference this time was that one ME 109 was making head-on attacks. I think the reason that I remember this one after all these years, is that on one frontal attack, the string of tracers looked as if they were going to hit my hand that was on the throttles. I instinctively snatched my hand back. That burst of fire did not come very close to our ship and I was a little chagrined.[10]

The Germans had learned how to attack the American heavy bombers.

EASTERN FRONT

German Army Group A captured Malgobek, near Grozny.[11] (see Figure 57)

OPERATION VELVET

As a result of Stalin's appeal, a modification to the Moscow Protocol was signed on October 6. The United States had agreed to supply 212 planes in October. Additional planes would be provided through the end of the year, but at a rate that would be determined by "the progress of the war".[12]

General Bradley was in Moscow on another mission. Stalin told Bradley that he, Stalin, was now supportive of Anglo-American air operations in the Caucasus and authorized Bradley to conduct a survey of the potential airfields in the area.[13]

1942 OCTOBER 7

376TH HBG

In support of Roosevelt's desire to send American air units to the Caucasus, Arnold's Air Staff proposed the creation of the 376th Heavy Bombardment Group. It would operate with B-24s and would be formed from the 1st Provisional Group. It would be supplied with additional men and planes sent from the United States. While awaiting transfer to the Caucasus, it would operate in the Middle East.[14]

1ST PROVISIONAL GROUP

The crews of the 1st Provisional Group continued to harass Rommel's supply lines. However, there were no heavy bombers left in the CBI. So if a plane was needed, it had to come from the Middle East. Bill Mayhew's crew was selected for one such mission.

We continued to bomb convoys in the Mediterranean and harbor installations along the coast of North Africa until 7 October 1942. Then our crew received orders to proceed at once back to India for a special assignment. There was a reason for us to be selected. Our pilot, Major Fennell, made the first flight in a 4-engined plane (B-24D) over "The Hump" from India to China. In fact, he made three trips into China from India in the spring of 1942. The first of these trips took Colonel Caleb Haynes to China. The other two trips were to retrieve some of Colonel Jimmie Doolittle's crews after their Tokyo raid from the aircraft carrier "Hornet".

The rumor was that we were going on to China. So, about 4:30 P.M. (1630) on the afternoon of 7 October 1942 our crew took off from Lydda, Palestine in a B-24D Liberator, serial number 41-23660 (Group battle number 53) for India. This was a brand new plane fresh from the States. As it was a bright pink, we named it the "Pink Lady", even though "Poison Ivy" was painted on the nose. We never removed "Poison Ivy" from the nose. There was no ball turret in that B-24D, so I flew at the right waist gun position. Here I had a single .50 caliber machine gun protruding through an open window that was approximately 3 feet by 4 feet in size. Needless to say, with a wind blowing in at 180 miles an hour, it became extremely cold at times (on one mission the temperature dropped to minus 80 degrees below zero F.).[15]

The crew would not return to the Middle East until early November.

OPERATION VELVET

At some point in the communication between Roosevelt and Churchill, Roosevelt must have conveyed the impression that he, Roosevelt, was concerned that Stalin would negotiate a separate peace with Hitler. After all, there was a precedent in the August 1939 Non-Aggression Treaty.

Churchill responded to Roosevelt's October 5 communiqué, advising Roosevelt that he was treading on political thin ice.

1. P.Q. 19, if it sailed, could not be broken up into smaller convoys
2. P.Q. 19 would not sail
3. The truth about P.Q. 19 could not be withheld from Stalin. Anyway, he probably already knew via Maisky
4. Stripping air units from the Middle East was not an option. (Unstated was the reality that protecting Russia's front line was one thing, but not if it meant exposing the British theater.)
5. I don't know what your emissary, Admiral Standley,[16] was telling you, but there was no possibly of a separate Russia-German peace treaty.[17]

1942 OCTOBER 9

1ST PROVISIONAL GROUP

Nine B-24s flew to Benghazi on the group's 54th mission.[18] (see Figure 50) Flying missions was getting trickier, as remembered by James Sibert.

Rudely awakened by an alarm clock at 03:30. Climbed in the truck with the other boys and went over to 04:00 breakfast. Ate and went over to 04:30 briefing. Was briefed for mission and I was the first to take off at 06:15. Circled the field and left the coast with 7 of the 9 — 2 didn't get off. Went through the Suez Corridor at 15:00 and then climbed to 23,250 before bombing Benghazi. Made bombing run with 6 ships, I had turned back. Ack-ack was deadly and accurate. I was hit in the left vertical fin and all ships but I were sprayed and hit. Unable to tell the effect of bombing. Came back and was shot at by German 88mm from the desert battle area. One went though Wilkinson's ship but didn't explode inside. Came back over field with my flight in an echelon at 200 ft and then broke away and landed.[19]

AAF TASK FORCE BR

Colonel Ted Timberlake lead a 24-airplane group on the 93rd's first combat mission against locomotive manufacturing facilities at Lille, France. The first mission was typical of things to come. German fighters attacked the formation as they were inbound to the target and the skies filled with flak as the Liberators began their bomb run. Several airplanes were hit by ground fire but, miraculously, only one B-24 failed to return from the mission.[20]

OPERATION VELVET

Churchill and Roosevelt reached an agreement, resulting in Churchill sending a message to Stalin. In particular, Churchill told Stalin that

1. approval had been given to form an Anglo-American air force that would be ready for combat early 1943
2. Britain would send additional Spitfires to Russia
3. Western Allied warships were being diverted to support Operation TORCH. Hence, they would not be available to provide convoy escort until 1943

Also, Churchill and Roosevelt asked if Russian air forces could do something about the Luftwaffe operations from Finland.[21]

Roosevelt was a bit less committed. "[W]e are going to move as rapidly as possible to place a force under your strategic command in the Caucasus."

The date was unstated. And the United States searched for additional planes to send.[22] Once again, Roosevelt was making promises that the United States could not keep.

1942 OCTOBER 10

OPERATION VELVET

Harry Hopkins sent a message to Marshall indicating that Roosevelt wanted to send more planes to Russia. Roosevelt wanted at least an additional 100 planes per month for the next three months. If Marshall could not find these planes from production, then the United States would need to divert planes from coastal patrol.[23]

Marshall did not even have to consult with his staff to respond, which he did.

1. If we divert planes from existing AAF units, it threatens Operation TORCH,
2. Diverting planes from coastal patrol exposed the United States to another attack from enemy carriers. Regardless, these planes were not equipped for combat operations.

Therefore, Marshall agreed to the commitment of the 376th HBG to the Caucasus.[24] Once again, Marshall was forced into a strategy that he opposed.

EASTERN FRONT

By the first week in October, Hitler realized that the capture of the Caucasus oilfields would not be accomplished in 1942. He therefore ordered the Luftwaffe to destroy the oil fields to deny their usage by the Russians. Luftflotte 4's Fliegerkorps IV was ordered to send every available bomber against the oilfields at Grozny. A second raid would be executed on October 12. (see Figure 57)

Unfortunately for the Germans, Grozny supplied only 10% of the Soviet oil. The rest of the oilfields were beyond the range of German fighters. Attacking these fields would require the bombers to fly unescorted. By now, the Russians had re-supplied their air defense units. Any attempt by the Germans to attack would be suicidal; and reminiscent of the Battle of Britain.

1942 OCTOBER 11

1ST PROVISIONAL GROUP

Nine B-24s attacked another supply convoy on the group's 55th mission, while seven B-17s continued the campaign against Rommel's supplies at Tobruk.[25] (see Figure 50)

1942 OCTOBER 12

OPERATION VELVET AND THE 376TH HBG
Marshall's agreement to Operation VELVET was enough for Roosevelt. He sent a message to Stalin:

> *Our heavy bombardment group has* been ordered mobilized immediately for the purpose of operating on your southern flank. ... [T[his movement will not be contingent on any other operation or commitment.[26]

Now that the United States was committed to Operation VELVET, Arnold sent the plans for the creation of the 376th HBG to the British and requested copies of the British plans.[27] Along with the creation of the 376th, the USAMEAF was also created.[28]

While the leaders made plans, the subordinates had to execute. Brereton was one of those so tasked.

> *The American ambassador to Russia,* Admiral Standley, arrived yesterday from Moscow and I had luncheon with him at the Mohamed Ali club and later a two-hour conference, at which were ACM Tedder; AM Drummond, and General Maxwell, on the subject of heavy bomber attacks to aid the Russians. Tedder and I emphasized that nothing must be done to jeopardize the strength of the heavy bombers in the battle of the Mediterranean, and Admiral Standley agreed. It was felt by all that, inasmuch as the Middle East is the only place where the United Nations can possibly conduct a victorious campaign this year, everything possible must be done to inflict defeat on the Axis in this theater. At best, all we can give Russia now is a token raid or two, and it was felt that this isn't the time for such a demonstration. Admiral Standley does not feel that a token to Stalin would have any effect whatsoever on the outcome of operations in Russia or on the determination of the Russians to continue fighting. It is perfectly evident that, no matter what happens, the Russians cannot stop fighting. If the Russian armies are defeated, or if a peace were concluded with the Axis, the Russian government and the Communist party would cease to exist, and Stalin knows this as well as the Communist party and the Army commanders. Admiral Standley, with a thorough grasp of the situation, assured us that he would present our view to the President in the strongest possible manner.[29]

EASTERN FRONT
The outcome of the conflict in North Africa was still unknown, and so was the Ger-

man campaign in Russia. *Time* correspondent James Aldridge filed a report from the Russian-Iranian border. After describing the conditions, he wrote:

> *If Stalingrad falls, most of* the German weight will be directed against the Caucasus. There are five ways through the Caucasus: one route on the Caspian Sea, one on the Black Sea, three passes across terrible mountains. They all come out into the southern Caucasus valley, stretching from Batum to Baku. It is most probable that, if Stalingrad falls, the German drive will be directed on Baku.[30]

Baku was a reference to the Russian oil fields. The issue was by no means decided.

1942 OCTOBER 13

OPERATION VELVET

Apparently, Stalin was not too impressed with Churchill's lengthy October 9 message. He merely said "Thank you."[31]

1942 OCTOBER 14

1ST PROVISIONAL GROUP

Six B-17s continued the campaign against Rommel's supplies at Tobruk.[32] (see Figure 50)

1942 OCTOBER 15

OPERATION VELVET

Stalin was only a bit more verbose in his reply to Roosevelt: "I am grateful for the information."[33]

1942 OCTOBER 16

1ST PROVISIONAL GROUP

Nine B-24s attacked Benghazi, while six B-17s attacked Rommel's supplies at Tobruk.[34] (see Figure 50)

1942 OCTOBER 18

OPERATION WATCHTOWER

Admiral William F. Halsey replaced Admiral Ghormley.[35]

1942 OCTOBER 19

OPERATION VELVET

The British sent their ideas for Operation VELVET a week after Arnold's request. They would provide fighters, medium bombers, and light bombers. They would not supply any heavy bombers. Since they would be providing the majority of the planes, they proposed that a British Air Marshall be put in overall command of the Anglo-American force. He, in turn, would report to the Russian high command. The mission objectives were:

1. Supporting Soviet ground forces
2. Attacking land and sea communications in the Black Sea area
3. Maintaining control of the Caspian Sea, and
4. Protecting such vital areas as Batum and Baku.[36]

1942 OCTOBER 20

NORTH AFRICA

Allied air activity focused on achieving air superiority. Montgomery wanted such superiority prior to the El Alamein offensive.[37]

1ST PROVISIONAL GROUP

The group continued flying separate bomber missions, but this time to the same target. Nine B-24s followed by six B-17s attacked Tobruk.[38] (see Figure 50)

1942 OCTOBER 22

NORTH AFRICA

United States Air Forces were organized into the Desert Air Task Force with Brereton in command. During the night of October 22/23, British Eighth Army moved into their assigned assault position.[39]

1942 OCTOBER 23

NORTH AFRICA

The Second Battle of El Alamein started with an artillery barrage by the British against Rommel's forces.[40] (see Figure 50)

1ST PROVISIONAL GROUP

Six B-17s attacked Canea, Crete.[41] The past few weeks of aerial bombing of Rommel's supply lines was having positive effects, even for the flight crews. Bill Mayhew wrote the following:

> *After a lot of hard* fighting by ground and air forces, we stopped Rommel at El Alamein, Egypt. The First Provisional Group was given credit for being instrumental in helping stop Rommel's drive by cutting off his flow of supplies with missions against the harbors of Tobruk and Benghazi, Libya, shipping in the Mediterranean, and other targets. Military authorities stated that the combination of RAF, B-17s, B-24s and B-25s of the U.S. Army Air Force had sunk two out of every three ships that tried to supply Rommel in North Africa. .... Two of the B-17s were cannibalized to maintain the rest of the B-17s.[42]

1942 OCTOBER 24

NORTH AFRICA

Following the artillery barrage of the previous day, British units began their assault on Axis positions.[43]

1942 OCTOBER 26

1ST PROVISIONAL GROUP

Nine B-24s followed by six B-17s attacked a German supply convoy.[44]

EASTERN FRONT

German Army Group A captured Nalchik, west of Grozny.[45] (see Figure 57)

1942 OCTOBER 27

NORTH AFRICA

Axis forces attempted a counter-attack. The attack was stopped after the British inflicted heavy losses.[46]

1942 OCTOBER 28

OPERATION VELVET

On October 6, Stalin had told General Bradley that he supported Operation VELVET and had authorized Bradley to examine Soviet airfields. On October 28, Bradley met with Molotov, who reiterated Stalin's support. Bradley responded that the airfield survey still needed to occur before any air units could be deployed. Molotov responsed with expressed consternation over the delays, but did little to expedite the surveys.[47]

1942 OCTOBER 30

1ST PROVISIONAL GROUP

The group flew to airfield targets on Crete. Six B-17s attacked the Tymbaki airfield, while nine B-24s attacked the Maleme airfield.[48]

HEAVY BOMBER PRODUCTION

During October, 288 heavy bombers were manufactured. One hundred thirty were B-24s.[49]

ENDNOTES

1. Clendenin, *Mission History of the 376th Bombardment Group*, p. 29.
2. ----, http://www.reddog1944.com/Missions_Targets_Combat_Action.htm , accessed November 23, 2015.
3. Williams, *Chronology 1941-1945 - 1942*, p. 57.
4. Craven, *The Army Air Forces in World War II, Vol. II*, p. 13.
5. Lukas, *The Velvet Project*, p. 152, Msg., Stalin to Churchill, October 3, 1942.
6. Churchill, *The Hinge of Fate*, pp. 576.
7. Churchill, *The Hinge of Fate*, pp. 576.
8. Williams, *Chronology 1941-1945 - 1942*, p. 58.
9. Clendenin, *Mission History of the 376th Bombardment Group*, p. 29.
10. Brown, *War Stories*.
11. Williams, *Chronology 1941-1945 - 1942*, p. 58.
12. Bevans, *Treaties*, p. 724-7.
13. Lukas, *The Velvet Project*, p. 155, Report of Bradley Mission to Russia, July 26-December 3, 1942. Bradley was in Moscow trying to negotiate an air route from the U.S. to Russia via Alaska and Siberia.
14. Lukas, *The Velvet Project*, p. 154, Memo, Brig. Gen. Anderson for Arnold, October 7, 1942.
15. Mayhew, *transcript of interview*.
16. Standley was the American naval member on the 1941 Beaverbrook-Harriman Special War Supply Mission to the USSR. In February 1942, he became the American Ambassador to the USSR.

17. Churchill, *The Hinge of Fate*, pp. 578.
18. Clendenin, *Mission History of the 376th Bombardment Group*, p. 30.
19. Sibert, personal diary.
20. ----, 93rd Bomb Group, http://www.93rdbombardmentgroup.com/ , accessed January 6, 2014.
21. Churchill, *The Hinge of Fate*, pp. 579.
22. Lukas, *The Velvet Project*, p. 153.
23. Lukas, *The Velvet Project*, p. 153, Ltr., Hopkins to Marshall, October 10, 1942.
24. Lukas, *The Velvet Project*, p. 153, Memo, Marshall for Roosevelt, October 10, 1942.
25. Clendenin, *Mission History of the 376th Bombardment Group*, p. 30.
26. Lukas, *The Velvet Project*, footnote 52, Msg, Roosevelt to Stalin, October 12, 1942, FRUS, 1942.
27. Lukas, *The Velvet Project*, p. 153-4.
28. Assistant Chief of Air Staff, *The AAF in the Middle East*, p. 53.
29. Brereton, *The Brereton Diaries*, p. 158.
30. Aldridge, James, *Time, A Song From The Caucasus, October* 12, 1942, http://www.time.com/time/magazine/article/0,9171,851497,00.html, accessed August 1, 2013.
31. Churchill, *The Hinge of Fate*, pp. 580.
32. Clendenin, *Mission History of the 376th Bombardment Group*, p. 31.
33. Lukas, *The Velvet Project*, p. 154.
34. Clendenin, *Mission History of the 376th Bombardment Group*, p. 31.
35. Williams, *Chronology 1941-1945 - 1942*, p. 59.
36. Lukas, *The Velvet Project*, p. 154.
37. Williams, *Chronology 1941-1945 - 1942*, p. 60.
38. Clendenin, *Mission History of the 376th Bombardment Group*, p. 32.
39. Williams, *Chronology 1941-1945 - 1942*, p. 60.
40. Williams, *Chronology 1941-1945 - 1942*, p. 60.
41. Clendenin, *Mission History of the 376th Bombardment Group*, p. 32.
42. Mayhew, *transcript of interview*.
43. Williams, *Chronology 1941-1945 - 1942*, p. 60.
44. Clendenin, *Mission History of the 376th Bombardment Group*, p. 33.
45. Williams, *Chronology 1941-1945 - 1942*, p. 61.
46. Williams, *Chronology 1941-1945 - 1942*, p. 62.
47. Lukas, *The Velvet Project*, p. 155.
48. Clendenin, *Mission History of the 376th Bombardment Group*, pp. 33 and 34.
49. ----, Army Air Force Statistical Digest, http://www.ibiblio.org/hyperwar/AAF/StatDigest/aafsd-3.html , accessed January 22, 2014, p. 112; B-24 production from Author's personal papers.

NOVEMBER 1942

November began with the Germans occupying most of Stalingrad. By the end of the month, they were trapped within the city limits.

The British Eighth Army began its campaign against Rommel. By the end of the month, they had recaptured Tobruk.

The western Allies executed their North African landings.

The commitment of United States air units to Russia resulted in the 1st Provisional Group being designated the 376th Heavy Bombardment Group. They would retain this designation for the duration of the war.

1942 BEGINNING OF NOVEMBER

EASTERN FRONT

The German 6th Army now occupied most of Stalingrad. (see Figure 54) However, despite major efforts, the Germans were unable to oust the Russians. In the Caucasus, the Russians had succeeded in stopping the Germans from reaching Grozny.[1] (see Figure 57)

1942 NOVEMBER 1

376TH HBG

The 1st Provisional Bombardment Group was reformed as the 376th Heavy Bombardment Group, the designation it retained for the duration of World War II.
Nine B-24s return to bomb the Maleme airfield on Crete, the group's 67th mission.[2]

1942 NOVEMBER 2

OPERATION SUPERCHARGE

Montgomery launched Operation SUPERCHARGE, the attack against the Afrika Korps with the intent of destroying the German armor.[3]

376TH HBG

Six B-17s bombed Rommel's supply port at Tobruk, the group's 68th mission.[4] (see Figure 50)

OPERATION VELVET

The Americans accepted the British proposals, except for one minor difference. Arnold thought that by the time VELVET was executed, American planes would outnumber British planes. Therefore, Arnold thought an American should be in command, not a British officer. But the point was moot, as Stalin had not given official approval of the project.[5]

1942 NOVEMBER 4

376TH HBG

Nine B-24s rendezvoused with the 98th HBG to bomb Benghazi, the group's 69th mission.[6] (see Figure 50)

OPERATION SUPERCHARGE

Operation SUPERCHARGE was so successful that the Afrika Korps was in full retreat.[7]

USAFIME

Lieutenant General Frank M. Andrews replaced Maxwell as commander of USAFIME.[8]

1942 NOVEMBER 6

376TH HBG

The 376th HBG resumed flying two separate daily missions. Six B-17s attacked Tobruk, while eight B-24s bomb Benghazi. (see Figure 50) These were the group's 70th and 71st missions.[9]

1942 NOVEMBER 7

OPERATION SUPERCHARGE

Heavy rains began on November 6. By the next day, British units were stopped due to the weather. The weather provided cover for the Axis to withdraw.[10]

1942 NOVEMBER 8

OPERATION SUPERCHARGE

British units cleared Marsa Matruh of Axis forces.[11] (see Figure 44)

OPERATION TORCH

United States forces landed in North West Africa, beginning Operation TORCH.[12] (see Figure 62)

Colonel Demas T. [Nick] Craw, who had been involved with the planning and execution of HALPRO's June 12 raid on Ploesti, was killed during the landing. He was posthumously awarded the Medal of Honor.

376TH HBG

With the retreat of Rommel in the face of British Eight Army advances, Egyptian airfields seem secure enough for Army Air Corps units to move from their Palestinian bases. The 376th HBG moved to Abu Suweir, just outside Cairo. (see Figure 50)

EASTERN FRONT

Units of the Luftflotte 4 were transferred to North Africa to support German resistance to the TORCH landings. (see Figure 62) While Luftflotte 4 had achieved air superiority over Stalingrad, continuous operations had reduced the combat effectiveness of the group. Removal of these units only exacerbated the situation. (see Figure 59)

OPERATION VELVET

On November 5, Churchill offered to send twenty Anglo-American air squadrons to the Eastern Front. But first, coordination meetings needed to take place between the squadron commanders and their Russian counterparts. On November 8, Stalin approved the visit of the Anglo-American team.[13]

1942 NOVEMBER 9

NORTH AFRICA

Apparently, the Free French did not put up enough resistance to the Allied landings in North West Africa to satisfy the Germans. So, German troops landed at the El Aouïna airport in Tunisia, the closest port to Sicily.[14]

OPERATION SUPERCHARGE

With an improvement in the weather, British forces resumed their pursuit of the retreating Germans.[15]

OPERATION TORCH

Once safely ashore, command of the Western Task Force was passed from the United States Navy to the United States Army under Maj. Gen. George S. Patton, Jr.[16]

1942 NOVEMBER 10

OPERATION TORCH

Admiral Jean Francois Darlan ordered his Free French forces to cease resistance to the Allied landings effective the next day.[17]

376TH HBG

Six B-17s attacked Canea, Crete, the group's 72nd mission.[18]

1942 NOVEMBER 11

OPERATION SUPERCHARGE

British forces crossed the Libyan border and occupied Bardia. The Germans had already abandoned Bardia.[19] (see Figure 5)

376TH HBG

Nine B-24s executed a mission near Benghazi, looking for a convoy, the group's 73rd mission.[20] Therman Brown described the mission:

> *I remember somewhere northeast of* Benghazi, on November 11, 1942, Armistice Day, nine of our B-24s bombed a troop ship carrying about 4,000 of Rommel's reinforcements. Each of our B-24s had six 1000 pound bombs. We approached the target, with its escort of destroyers, from about 19,000 feet. We were to bomb from 4,000 feet. At 4,000 feet, fighters were circling over the transport. Our flight leader chose to bomb from our present altitude of 19,000 feet. I don't think a bomb missed the troop ship. Our bombardier, Lt. Earl G. Matheny, reported the results over the intercom. At first he seemed elated and then sickened. He said, "Oh God, it is awful". A short way from the target we discovered that the bombs on our ship had not released. Something had malfunctioned. I think some of us felt relieved that our bombs had not contributed to the carnage three miles below.
>
> I don't think the fighters ever knew we were up there until the ship was hit. They never came up to attack us. We salvoed our bombs somewhere in the Med on the way home. We were not supposed to land with them on board.[21]

General Frank Andrews visited the 376th HBG for an awards ceremony. George Richardson recorded his impressions:

> *Lieutenant General Andrews, who had* newly assumed command of the theater, came up to make some awards. He was a splendid figure of a man and a thorough soldier. He flew his own plane. Everyone in the theater was glad to see General Frank Maxwell Andrews, the father of heavy bombardment. He was well over six feet in height, a general air of youthfulness, pleasant eyes and a searching glance that seemed to go right through to the heart of things. Surmounting all was a shock of graying hair. He was a "bomber's man". The assumption of command by General Andrews solved one thing in a hurry, once and for all. No longer would General Brereton have his Air Force authority chopped by General Maxwell who was being superseded as Theater Commander. Now, as Air Force Commander, General B[rereton] could go ahead, submitting his ideas to one who knew the whole business without trying to explain the mechanics of air power to one who did not understand.[22]

EUROPE

In response to Admiral Darlan's stand down order, German troops entered Vichy France.[23]

1942 NOVEMBER 12

9TH AIR FORCE

The United States Army Middle East Air Force was officially re-designated the 9th Air Force (AF). General Brereton was placed in command. The 98th and 376th HBG's were transferred to the 9th AF, IX Bomber Command.[24]

1942 NOVEMBER 13

OPERATION SUPERCHARGE

The British retook Tobruk.[25] (see Figure 50)

1942 NOVEMBER 14

376TH HBG

Six B-17s attacked Benghazi, the group's 74th mission.[26] (see Figure 50)

1942 NOVEMBER 15

376TH HBG

With the availability of Gambut for air operations, the United States heavy bombers expanded their reach. Ten 376th HBG B-24s joined with 98th HBG bombers to begin the air assault on Tripoli.[27] (see Figure 50) The mission was aborted due to weather, but that did not mean something exciting would not happen. This was Therman Brown's account:

> *Sometimes there is excitement even* when the mission has to abort because of weather. Lt. John Wilcox had a little on a mission to Tunis, November 15, 1942. The mission was aborted west of Benghazi when a heavy weather front was encountered. The leader of the flight and all the other B-24s, except John's plane, jettisoned their bombs after turning back.
>
> On the flight out, Wilcox and his crew saw a large German marshalling yard on the Benghazi - Tripoli highway, Rommel's escape route. The return route was over the same path. When they were near the marshalling yard, John eased his B-24 out of the formation, gained a little altitude, and when in range of the yard, made a successful bombing run scoring some direct hits.
>
> Almost immediately after the bomb drop, John's plane [Draggin Lady] was attacked by five ME 109's. They were flying at less than 10,000 feet altitude. They shot out the tail turret, immobilized the top turret, shot out the #3 and #4 engines and hit the controls of the #2 engine freezing the control at cruising r.p.m. The #1 engine was not hit and could reach full power. Some of the ME 109's bullets shattered the windshield of the plane, skimming the top of the head of the co-pilot, Lt. Foster.
>
> John regained his place in the formation. The altitude he had gained before the bomb drop made it possible. Somehow John managed to hold formation until the ME 109's broke off the attack for lack of fuel. It was also getting late.
>
> John's ability to hold formation was due to the excessive demand for power he made on the #1 engine. The engine temperature was dangerously high and now had to be throttled back. With loss of power and night coming on, a decision was made to belly land the plane in the desert while they could still see what they were doing. This was somewhere south of Tobruk. Although the desert air was quite cold, they managed to stay warm by burning engine oil.
>
> The following morning an RAF Dakota (DC-3), on its way to supply a hit and run Spitfire Squadron, spotted John's B-24. They landed nearby, collected the crew, completed its own mission, and deposited John and crew at their Suez Canal base that night.[28]

1942 NOVEMBER 16

AAF TASK FORCE BR

The Flying Fortresses of the 97th HBG had been ordered to North Africa to support Operation TORCH. They flew their first mission on November 16.[29]

OPERATION VELVET

The Anglo-American advance team left Cairo. British Air Marshall Sir R. M. Drummond headed the team. Brig. Gen. Elmer Adler represented the United States. The team became known as the Drummond-Adler mission.[30]

1942 NOVEMBER 18

376TH HBG

Six B-17s attacked Benghazi. This was the group's 76th mission.[31] (see Figure 50)

1942 NOVEMBER 19

EASTERN FRONT

The Russians launched a counter offensive against the German flanks at Stalingrad, Operation URANUS. On November 19, three Russian armies attacked on the northern flank.[32] (see Figure 61)

The Luftwaffe forces were tasked with destroying the oilfields in the Causasus region. They were now withdrawn to provide air support for Stalingrad.

1942 NOVEMBER 20

EASTERN FRONT

On November 20, two Russian armies started the second phase of Operation URANUS, attacking the German southern flank.[33] (see Figure 61)

OPERATION SUPERCHARGE

British retook Benghazi.[34] (see Figure 50)

1942 NOVEMBER 21

376TH HBG

Nine 376th HBG B-24s flew the group's 77th mission to Tripoli.[35] (see Figure 50) During the return George Whitlock, flying *Battleaxe*, ran into a bit of trouble.

James Barineau was the radio operator.

> *My most memorable mission was* over Tripoli, operating from a base near the Suez Canal. We received flak damage over the target, and while returning at night we were off course just enough to tangle with a barrage balloon over Port Said. A bomb exploded behind no. 1 engine, damaging one hydraulic system, which necessitated lowering the landing gear manually. At base they could only lower one main gear and the nose wheel, then could not raise them again for a belly landing. The Commanding Officer radioed Whitlock to head out to sea, and all crew bailout. Whitlock and Britt asked permission to bring it in, which was granted. The rest of the crew could bailout or ride it down. Two of us chose to jump. I was lucky enough to land a short distance from the runway, and saw them bring the plane in with skill and daring. They accomplished an almost impossible feat of holding the right wing up as long as possible, then letting it skid off the runway into the sand. The damage was repairable, and due to their bravery, this plane flew combat again. As we were short on planes, this was a great help.[36]

George Richardson described the same events:

> *On the return, after the* long flight of some thirteen to fifteen hours, Dag Whitlock ran off course due to a bad compass and got into the barrage balloons at Port Said, which extended 4500 feet in the air at the time. A bomb on the cable of one balloon hit his wing and almost blew it off but Dag came in without any trouble but memories of how big those balloons looked before he could pull away. His report, describing the incident as "similar to running into turbulent air", was a source of annoyment to the canal defenders who inspected the airplane and could not see how any plane that large had gotten through and away.[37]

EASTERN FRONT
The two pincers of the Russian Operation URANUS linked up near the town of Kalach. The German 6th Army was now trapped within Stalingrad.[38] (see Figure 61)

OPERATION VELVET
The Drummond-Adler mission arrived in Moscow.

1942 NOVEMBER 23

OPERATION SUPERCHARGE
Axis forces withdrew from Agedabia for El Agheila.[39] (see Figure 15)

1942 NOVEMBER 25

376TH HBG

After their return to the 1st Provisional Group, now the 376th, Bill Mayhew's crew continued to fly missions. That changed on November 25:

General Dwight Eisenhower landed the "Torch" troops in North Africa 8 November 1942. The Germans in North Africa now were caught between the Yanks in Algeria and the British in Libya. Major Max Fennell's crew was selected to take nearly all the high-ranking Air Force officers from Cairo, Egypt to Algiers, Algeria to confer with General Eisenhower. (We had been flying bombing missions in support of the British 8th Army from Palestine and Egypt since July 1942.) Needless to say, we felt rather honored.

We left Cairo (Landing Field 224) about 1:00 A.M. (0100) on 25 November 1942 in Number 9, our B-17E ("Fennell vs. Rommel"), headed for the island of Malta. Besides our crew, those aboard consisted of Air Chief Marshal Sir Arthur Tedder, Commander of all Air Forces in the Middle East (later General Eisenhower's Deputy Commander for the Normandy invasion), Major General Lewis Brereton, Commander of the U.S. Ninth Air Force, Brigadier General Patrick Timberlake, Commander of the Ninth Bomber Command, Brigadier General Airey (British 8th Army), Colonel Uzal Ent (later Brigadier General Ent of Ploesti "Tidal Wave" fame), Lieutenant Colonel Louis Hobbs (General Brereton's aide), and three R.A.F. officers (rank unknown). There was so much rank aboard; Colonel Ent sat in the back of the plane with Pat and me. We reached Malta about six hours later. These officers spent the day conferring with Lord Gort, Governor General of Malta.

The following morning at about 8:00 A.M. (0800), 26 November 1942 (Thanksgiving Day), we left Malta for Maison Blanche Airport in Algiers, Algeria. At that time Malta was completely surrounded by German territory, so we had to be rather careful in our flight. We were supposed to have a fighter escort for this flight, but it never showed up (we never had a fighter escort during our entire combat tour). We were directed (by radio) to cross the African coast at Sfax, Tunisia, then proceed to Algiers. The radio message advised us that the British 8th Army had just captured Sfax, so we should cross the coast there in our usual manner for returning to Allied lines. Therefore, we crossed the coast at 2,500 feet altitude (we always crossed below 3,000 feet upon return). I was in the ball turret of our plane, so I had a ring-side seat to the proceedings. I practiced tracking one of six gunboats in the harbor that were sailing east, presumably to Tripoli, Libya. We thought they were British, so we did not pay much attention to them. When we crossed the coast, Major Fennell was supposed to fire flares, indicating the colors of the day (which changed every 24 hours). However, we were so low, and our shiny B-17 was so well marked, he decided he did not need to do so (which probably saved our lives). From my position, I could see people plowing fields, riding bicycles down roads, etc. At that altitude, I could almost see the expressions on their faces as they watched us fly over. We flew unmolested over Tunisia on our way to Algiers. Once three fighters passed us about a mile to our right, but continued on their way. At the time, we assumed they were P-51's from General Eisenhower's forces, but we learned later they were Bf-109's (P-51's were not operating in that theater yet). Apparently the German pilots assumed the plane was a captured B-17 being flown by a German crew at that altitude. We continued to Algiers at that altitude, landing six hours and fifteen minutes after our take-off at Malta. When our navigator, Captain Lewellyn Daigle, checked in at Maison Blanche, which was 8 miles from Algiers, he learned that Sfax was still held by the Germans, and we had just flown over more than 300 miles of German territory below 3,000 feet (plus a couple of hundred miles of enemy Mediterranean Sea). Thus, we had been over German territory most of our trip!! Maybe you think we did not do some shaking then!!! We never have been able to figure out why we were not fired upon. That really was a Thanksgiving Day for us.[40]

EASTERN FRONT

Planning for two additional Russian offensives near Stalingrad began. Operation SATURN would be designed to destroy the Italian 8th Army and separate the German forces west of the Don River from the German Army Group A. Operation RING intended to capture Stalingrad. (see Figure 54)

Meanwhile, the Russians began Operation MARS, which was against German Army Group Center near Moscow. This operation had a dual purpose:

- add breathing room around Moscow, and
- prevent the diversion of potential German reinforcements to the Caucasus

1942 NOVEMBER 26

376TH HBG

Nine 376th HBG B-24s flew the group's 78th mission to Tripoli.[41] (see Figure 50)

1942 NOVEMBER 27

376TH HBG

Six 376th HBG B-17s flew the group's 79th mission to Leros Island, Greece.[42] This would be the last mission flown by these B-17s. They would be withdrawn for servicing. All future replacement crews and aircraft destined for the 376th HBG would be B-24s.

OPERATION VELVET

One of the few Americans to visit the Eastern Front was Maj. Gen. Patrick J. Hurley, who was in Moscow on a special mission for Roosevelt. He visited Stalingrad from November 27 until December 7. Later, he would visit the Caucasus front. Ironically, the Drummond-Adler mission was never allowed to visit the Caucasus.[43]

While Hurley was in Stalingrad, the Drummond-Adler mission was meeting with their Russian counterparts. Adler was so impressed with what the Russians were saying that he sent a proposal to Marshall that the United States consider asking the Russians for permission to use the Caucasus bases for operations against targets in southeastern Europe.[44]

EUROPE

Admiral Jean de Laborde ordered the Free French Fleet at Toulon to be scuttled, thus preventing it from falling into German control.[45]

1942 NOVEMBER 29

376TH HBG

Nine 376th HBG B-24s flew the group's 80th mission to Tripoli.[46] (see Figure 50)

HEAVY BOMBER PRODUCTION

During November, 304 heavy bombers were manufactured, 133 were B-24s.[47]

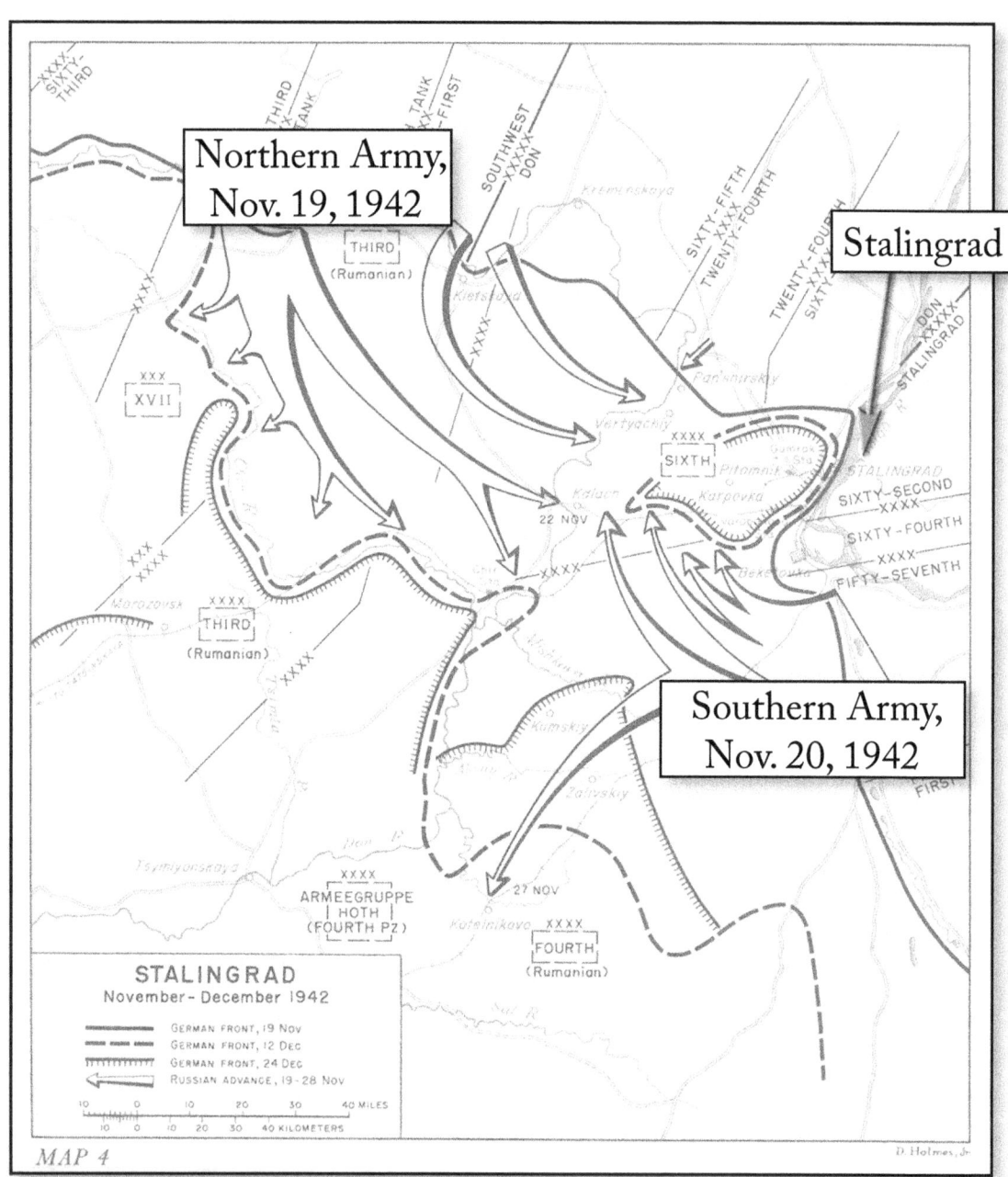

Figure 61 – Stalingrad Operation URANUS – November 1942

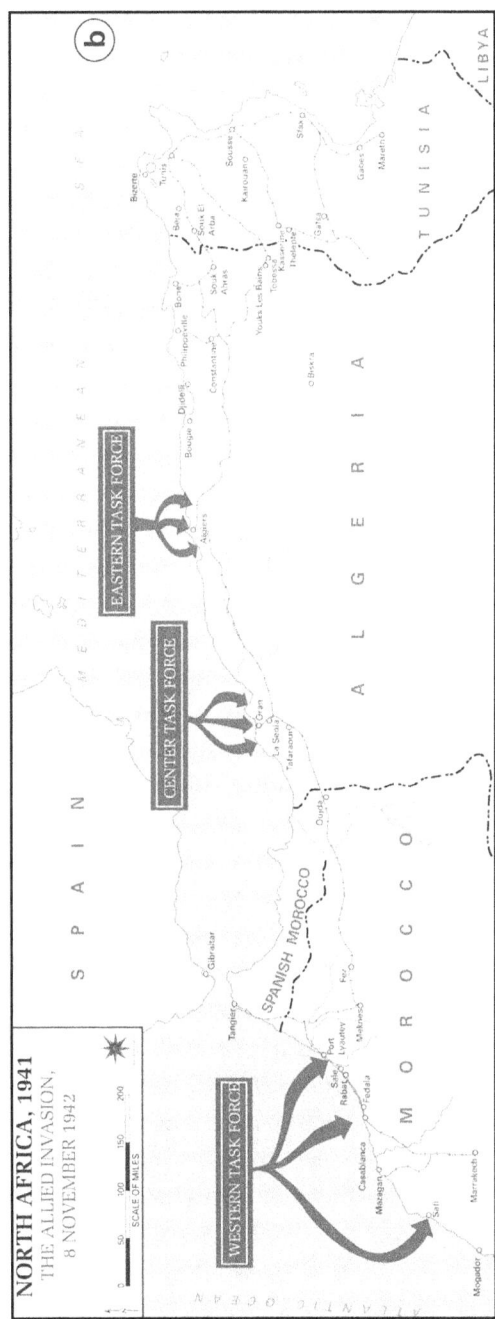

Figure 62 – Operation TORCH – November 1942

ENDNOTES

1. Williams, *Chronology 1941-1945 - 1942*, p. 62.
2. Clendenin, *Mission History of the 376th Bombardment Group*, p. 34.
3. Williams, *Chronology 1941-1945 - 1942*, p. 63.
4. Clendenin, *Mission History of the 376th Bombardment Group*, p. 35.
5. Lukas, *The Velvet Project*, p. 154, Ltr., Arnold to Evill, November 2, 1942.
6. Clendenin, *Mission History of the 376th Bombardment Group*, p. 35.
7. Williams, *Chronology 1941-1945 - 1942*, p. 64.
8. Williams, *Chronology 1941-1945 - 1942*, p. 64.
9. Clendenin, *Mission History of the 376th Bombardment Group*, pp. 35 and 36.
10. Williams, *Chronology 1941-1945 - 1942*, p. 63.
11. Williams, *Chronology 1941-1945 - 1942*, p. 64.
12. Williams, *Chronology 1941-1945 - 1942*, p. 64.
13. Lukas, *The Velvet Project*, p. 155, Msg., Stalin to Churchill, November 8, 1942.
14. Williams, *Chronology 1941-1945 - 1942*, p. 65.
15. Williams, *Chronology 1941-1945 - 1942*, p. 65.
16. Williams, *Chronology 1941-1945 - 1942*, p. 65.
17. Williams, *Chronology 1941-1945 - 1942*, p. 65.
18. Clendenin, *Mission History of the 376th Bombardment Group*, p. 36.
19. Williams, *Chronology 1941-1945 - 1942*, p. 66.
20. Clendenin, *Mission History of the 376th Bombardment Group*, p. 37.
21. Brown, *War Stories*.
22. Richardson, *The Forgotten Force*, p. 161-2.
23. Williams, *Chronology 1941-1945 - 1942*, p. 65.
24. Williams, *Chronology 1941-1945 - 1942*, p. 65.
25. Williams, *Chronology 1941-1945 - 1942*, p. 67.
26. Clendenin, *Mission History of the 376th Bombardment Group*, p. 37.
27. Clendenin, *Mission History of the 376th Bombardment Group*, p. 38.
28. Brown, *War Stories*.
29. ----, http://www.reddog1944.com/Missions_Targets_Combat_Action.htm , accessed November 23, 2015.
30. Lukas, *The Velvet Project*, p. 155.
31. Clendenin, *Mission History of the 376th Bombardment Group*, p. 38.
32. Williams, *Chronology 1941-1945 - 1942*, p. 68.
33. Williams, *Chronology 1941-1945 - 1942*, p. 68.
34. Williams, *Chronology 1941-1945 - 1942*, p. 68.
35. Clendenin, *Mission History of the 376th Bombardment Group*, p. 39.

36. Barineau, Letter, January 7, 1974.
37. Richardson, *The Forgotten Force*, p. 160.
38. Williams, *Chronology 1941-1945 - 1942*, p. 69.
39. Williams, *Chronology 1941-1945 - 1942*, p. 69.
40. Mayhew, *transcript of interview*.
41. Clendenin, *Mission History of the 376th Bombardment Group*, p. 40.
42. Clendenin, *Mission History of the 376th Bombardment Group*, p. 40.
43. Lukas, *The Velvet Project*, p. 161.
44. Lukas, *The Velvet Project*, p. 157, Msg., Adler to Marshall, November 27, 1942.
45. Williams, *Chronology 1941-1945 - 1942*, p. 70.
46. Clendenin, *Mission History of the 376th Bombardment Group*, p. 41.
47. ----, Army Air Force Statistical Digest, http://www.ibiblio.org/hyperwar/AAF/StatDigest/aafsd-3.html , accessed January 22, 2014, p. 112; B-24 production from Author's personal papers.

DECEMBER 1942

With the apparent success of the Russian forces at Stalingrad, Stalin no longer needed United States air units. He still wanted the planes, but not the men.

Rommel continued his strategic withdrawal. He was no longer being supplied due to the demands of German forces on the Eastern Front.

The 376th and 98th Bomb Groups began operations from forward air bases near Tobruk, enabling them to attack German supply bases in Tunisia. Units of the 93rd Bomb Group, which had been operating out of England, joined them.

1942 DECEMBER 2

OPERATION VELVET

After three meetings, it was obvious to the Drummond-Adler mission that the Russians were not going to accept Anglo-American air units into their territory as an independent force. They would accept the planes, but not the crews. When it was pointed out that Stalin had already agreed, the Russians offered to accept the air units, but only if they were integrated into the Soviet Air Force. The delegation reported this change to London and Washington. They recommended that they hold firm to Stalin's original agreement. They thought the Russians would eventually fold.[1]

1942 DECEMBER 3

OPERATION VELVET

By now, it was apparent to Anglo-American leadership that Operation VELVET was no longer needed. Allied success in North Africa and on the Russian front had stopped the Germans. Churchill told Roosevelt that this was his opinion.[2]

1942 DECEMBER 4

376TH HBG

After nearly six months of missions in support of the Allied effort to stop Rommel's advance, the 376th HBG directed its attention to the European mainland. Nine B-24s attacked the harbor at Naples, Italy. Gambut, near Tobruk, would be used as a re-fueling airfield.[3] This was the first assault by Allied aircraft on the Italian mainland. Therman Brown described the experience of one of the other crews.

Lt. Dick Miller, pilot, and Lt. Mayfield, co-pilot, flew such a mission to Naples on December 4, 1942. They had some engine trouble, but managed to reach the target for a successful raid.

They lost #3 engine on the homeward trip. They transferred gas to start it only to have engine #4 go out. Near Gambut (LG-139) in the Western Desert, they fired the colors of the day and flicked their landing lights but no runway lights came on. They went on to Cairo with the same results. With all hope gone, Dick ordered the crew to bail out. They all did.

As Dick Miller was preparing to jump, he thought he saw runway lights. He was correct. He put his wheels down. The #2 engine caught on fire and then sputtered out as he landed with only one engine. It was a perfect landing. It turned out to be Fayid near the Suez Canal. The next morning all the crew members were located. There were no injuries. He said he felt lonesome, there all by himself and everything was so quiet - it was an eerie feeling.[4]

George Richardson recounted the same story, but with a bit of humor.

On December 11 our raid against Naples was repeated. It was getting colder on the desert than could be imagined, and Gambut was not exactly a popular place. Crews would take any risk to by pass it on return by husbanding their fuel each way, saving three or four hours for the Gambut-Canal Area trip. In this particular raid our bombardier Little [Donald C. Little] figured. D. C. had parachuted with Wilhite [Irving J. Wilhite] on Mission I, missed his trip with John Kidd, and now had recovered enough from his illness to go out with Dick Miller and G. E. "Maypole" Mayfield on this mission. As the crew was returning from Naples after dark, Little asked Dick if he had enough gas to go on from Gambut. He told Dick that he had made a parachute jump once and was not too interested in repeating. Dick and Mayfield both reassured Little that there was gas aplenty as they passed on over Gambut. It was advantageous to get out of Gambut because German Ju 88's were based on Crete, 180 miles North.

As the ship neared the Suez Area the engines started to cough and kick out.

Lo and behold, the gas had ran low. Dick gave orders to bail out and Little said, "I told you so", telling the rest of the crew that the first jump was the hardest. "Maypole" put on his chute and opened the bomb bay doors, stood on the catwalk, and, holding his nose, plummeted into the darkness. Little sadly shook his head and stepped off into the void, followed by the remainder of the crew except Dick Miller.

Dick took one look around, was buckling his harness when he saw the flashing beacon of Kabrit, one of the 98th's bases below. He stayed with his ship and brought it in on their runway as the motors quit cold. Dick had made practically a "dead stick" landing at night and in a B-24.

The parachutists had landed on the desert below, wrapped up in their parachutes and slept till morning. Donald C. Little became "Ripcord" Little from then on. "Maypole" said he hoped the next jumps were never like the first. One of the boys had landed on a tent filled with British soldiers in the Gneifa Sub-Area below. All in all, it was the end of an amazing incident that could have been tragic but was not. Col. McGuire, John and the rest gave Dick a good, old-fashioned "eating out."[5]

1942 DECEMBER 6

AAF TASK FORCE BR

Three Squadrons of the 93rd HBG, currently stationed in England, were ordered to report to the 9th Air Force in North Africa. They would initially operate from Tafarouri Aerodrome, a former French airfield outside Oran in Algeria. After flying two missions from Tafarouri, the group re-deployed to Gambut, Libya.[6]

1942 DECEMBER 7

376TH HBG

The 376th HBG resumed its attack on Rommel's supply line. Nine B-24s attacked Tripoli.[7] (see Figure 50)

1942 DECEMBER 11

OPERATION SUPERCHARGE

After several days of major inaction, Montgomery ordered the assault on the German position at El Agheila. The campaign was to begin on December 14.[8] (see Figure 15)

376TH HBG

Nine B-24s repeated the December 4 attack on the Naples harbor.[9] Gambut would again be used as a re-fueling airfield.

1942 DECEMBER 12

EASTERN FRONT

The Germans launched Operation WINTERSTORM, an attempt to relieve the 6th Army trapped in Stalingrad. Initially, the Germans were able to thwart Russian counter-attacks.[10] (see Figure 54)

1942 DECEMBER 13

OPERATION SUPERCHARGE

Apparently anticipating the British attack, Axis forces began withdrawing from El Agheila.[11] (see Figure 15)

OPERATION VELVET

Molotov requested that Drummond visit the Kremlin. At this meeting, Molotov formally rejected the Anglo-American offer. He gave two reasons: too much time had elapsed, and supplying this force would interfere with other Lend-Lease supplies.[12]

1942 DECEMBER 16

OPERATION SUPERCHARGE

Despite attempts by the British to encircle the Axis troops, Axis forces evaded the British efforts. However, the Germans lost 20 badly needed tanks.[13]

376TH HBG

Nine B-24s attacked Sfax, Tunisia.[14] (see Figure 63) The 376th HBG was now supporting Operation TORCH.

At the end of the June 12 Ploesti mission, four HALPRO B-24s and crews had landed and were then interned in Turkey. A group of those servicemen managed to "steal" one of the B-24s, *Brooklyn Rambler*, and fly it to an RAF base on Cyprus. Al Story, who was one of those still interned in Turkey, filled in some of the blanks about the "theft."

> *According to the Geneva Convention,* Turkey was allowed to claim the B-24s that we had arrived in. Only one of the planes was in flying condition. The Turks asked if our people would teach them to fly it. Four pilots and two flight engineers agreed to help them. These people were accordingly moved to the City of Eskeshehir where the Turks had an airbase. We didn't hear much of these six for several weeks. One day we were elated to learn that the six Americans had flown out with the B-24. They had carefully monitored the fuel onboard and concluded

that they could make it to Cyprus so one morning while the Turks were being briefed in the hangar, the Americans all climbed aboard and took off. They were successful in reaching Cyprus …[15]

Therman Brown added similar facts.

The American crews took on the task of maintaining the B-24s for the Turks. Just about every day the crews would work on the B-24s, run up the engines and check them out completely. This took a lot of fuel.

Lt. Nesbitt's crew made it a point of always using less gas on each run up than they received from the Turks. Eventually, they saved enough to fly out of the country. One clear day they were running up all four engines at the same time. It just happened that all the crew were on board at the same moment. Nesbitt released his brakes and took off from the hangar without regard for runways or anything else. It worked. Before the Turks could get organized, Nesbitt and crew had landed in Cyprus.

The Turks were upset. The United States promised to return the B-24 and did. They didn't return the crew.[16]

EASTERN FRONT

The Russians launched Operation LITTLE SATURN, an attempt to drive through the Axis Armies to recapture Rostov. The Axis armies were primarily Italian. Consequently, the Germans gave up on any attempt to relieve the 6th Army trapped in Stalingrad.[17] (see Figure 54)

OPERATION VELVET

But Roosevelt was not willing to let the matter drop. He reiterated his offer to send air units to Russia and emphasized that they would be under Russian tactical command.[18]

1942 DECEMBER 18

OPERATION VELVET

With the victory at Stalingrad looming, Stalin sent a message to Roosevelt canceling plans to send an Anglo-American air unit to the Caucasus.[19] But he still wanted the planes:

I should be most grateful if you would expedite the dispatch of aircraft; especially fighters, but without crews, whom you now need badly for use in the areas mentioned. A feature of the Soviet Air Force is that we have more than enough pilots, but suffer from a shortage of machines.[20]

1942 DECEMBER 20

9TH AIR FORCE

B-24s bombed Sousse, Tunisia for the first time. On this raid, the 93rd HBG B-24s joined the 98th and the 376th.[21] (see Figure 63)

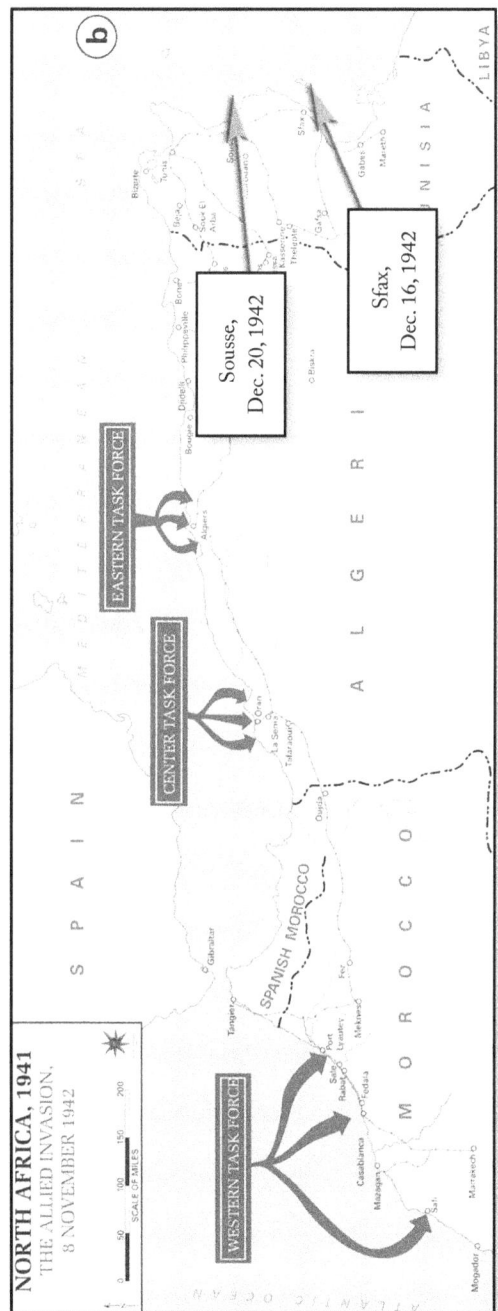

Figure 63 – Operaotion TORCH – December 1942

ENDNOTES

1. Lukas, *The Velvet Project*, p. 158, Adler to Marshall, December 2, 1942.
2. Lukas, *The Velvet Project*, p. 158-9, Msg., Churchill to Roosevelt, December 3, 1942.
3. Clendenin, *Mission History of the 376th Bombardment Group*, p. 42.
4. Brown, *War Stories*.
5. Richardson, *The Forgotten Force*, p. 177-8.
6. ----, 93rd Bomb Group, http://www.93rdbombardmentgroup.com/ , accessed January 6, 2014.
7. Clendenin, *Mission History of the 376th Bombardment Group*, p. 43.
8. Williams, *Chronology 1941-1945 - 1942*, p. 69.
9. Clendenin, *Mission History of the 376th Bombardment Group*, p. 43.
10. Williams, *Chronology 1941-1945 - 1942*, p. 74.
11. Williams, *Chronology 1941-1945 - 1942*, p. 74.
12. Lukas, *The Velvet Project*, p. 159, Msg., Drummond to Air Ministry, December 13, 1942.
13. Williams, *Chronology 1941-1945 - 1942*, p. 75.
14. Clendenin, *Mission History of the 376th Bombardment Group*, p. 44.
15. Story, *Private Story*.
16. Brown, *War Stories*.
17. Williams, *Chronology 1941-1945 - 1942*, p. 75.
18. Lukas, *The Velvet Project*, p. 160, Msg., Roosevelt to Stalin, December 16, 1942.
19. Kimball, *Stalingrad*, p. 111.
20. Lukas, *The Velvet Project*, p. 160.
21. Clendenin, *Mission History of the 376th Bombardment Group*, p. 45.

A SUMMARY OF 1942

The December 20 mission to Sousse and Stalin's rejection of Operation VELVET are fitting places to end the story of HALPRO. Three military events occurred which changed the destiny of the men and their planes:

1. The Doolittle raid had fulfilled Roosevelt's demand to retaliate for Pearl Harbor
2. Japanese aggression:

   - Moved prospective air bases in China beyond operational range
   - Removed the Burma Road as the vital supply line to any Chinese air field

3. The Battle of Midway stopped all Japanese aggression into India and eliminated any threat of a link-up between Japanese and German forces

In 1942, HALPRO, the 1st Provisional Group, and the 376th HBG carried out 87 missions. In addition to the June raids on Ploesti and the Italian Fleet, there were:

- 32 missions flown against Tobruk
- 20 flown against Benghazi
- 11 missions flown against supply convoys
- 9 against Crete, Canea (3), Maleme (2), Suda Bay (3), Tymbaki (1)
- 5 against Tripoli
- 3 against Tunisia - Sfax (1), Sousse (2)
- 3 against Naples
- 1 against Greece

HALPRO ceased to exist. It was now the 376th Heavy Bombardment Group. Along with the 98th HBG, they would form the nucleus of the 15th Air Force. Following the invasion of Italy in the autumn of 1943, both groups would move to bases in southern Italy, where they would remain for the duration of the war.

COMMITMENT TO CHINA

One author commented on the Allied commitment to support China:

> *Americans officially and unofficially promised* much to the Chinese government in 1942 but delivered very little to increase the size of American air units in China or to strengthen the Chinese air force. The dollar value of Lend-Lease aircraft, parts and accessories delivered to the British, Russian and Chinese governments in 1941 and 1942 reflected a great difference. Britain received $285.5 million in comparison to $2.7 million for China, a mere trickle of aircraft delivered to the Chinese. By the end of 1942 Washington had exported more than 2500 planes to Russia, far more than the Russian aircraft deployed at Stalingrad for their counter-offensive. In the meantime, China had received only 19 A-29 bombers and 27 P-40, P-41 P-43 and 82 P-66 pursuit planes.[1]

HISTORICAL PERSPECTIVE

In the 1920's, the United States was focused, militarily, on Japan. In the mid 1930's the focus shifted to Germany. The "Germany First" strategy was defined in 1941. However, in 1942, the military setbacks of the Allies forced planners to shift their focus away from the European Theater of Operations (ETO) to first, the Pacific and then, the Mediterranean Theater of Operations (MTO).

Numerous United States documents repeatedly emphasized that the MTO was a British area of interest. Commitment of American resources would be a diversion from the real objective – defeating Germany and Japan. Yet, three Air Corps Heavy Bombardment Groups (Brereton, HALPRO, and the 98th) were diverted from their original destinations to support the Allied efforts to stop Rommel's Afrika Korps.

The United States was now committed to the MTO. Additional military units were needed to support the invasion of Sicily, followed by the invasion of Italy in the fall of 1943. Three additional B-24 bomber groups were diverted from England to support the largest air operation of 1943 – Operation TIDALWAVE, the low-level raid against the Ploesti oilfields. Following Operation TIDALWAVE, the three diverted B-24 bomber groups were reassigned back to England. However, eighteen other heavy bomber groups ultimately joined the 98th and the 376th Heavy Bombardment Groups. These 20 groups comprised the 15th Air Force and operated from Italian bases.

This focus on the MTO should not have come as a surprise. As Lieutenant Colonel Wesley Frank Craven and Major James Lea Cate wrote in their seven-volume History of the Army Air Corps:

> *For all its awesome history* as a battleground between civilizations, the Middle East did not strike American strategists as an area in which the European war

could be expeditiously won. On the other hand, they recognized it as an area in which the global war could be very speedily lost. So, although large-scale U.S. offensives, air or ground, did not figure in the plans for the Middle East (the offensive function against the European Axis being largely reserved for the more convenient United Kingdom base), aid for its British defenders was never stinted. In fact, it was the large degree of logistical support afforded the Royal Air Force in the Middle East that finally, in the spring of 1942, brought the decision to commit an American air force there. The difficulties which shortly thereafter beset the British Eighth Army only advanced the date for that air force's appearance.[2]

ENDNOTES

1. Xu, Guangqiu, *The Issue of US Air Support*, p. 476. Information from: 64 (64 Lend-Lease Statistics and Report Branch of the Overseas Section, Lend-Lease Administration, 'Supplying the Air Forces of the United Nations', 1 January 1943, 201.38, AFA.); 65 (65 Richard C. Lukas, Eagles East: The Army Air Force and the Soviet Union, 1941-194s (Tallahassee, FL 1970), 232.) ; 66 (Hsu Long-hsuen and Chang Ming-Kai, History of the Sino-Japanese War, op. cit., 510.)
2. Craven, *The Army Air Forces in World War II, Vol. II*, p. 3. In 1945, the University of Chicago was asked to write the history of the Army Air Corps during World War II. The University was to take an academic approach. Craven and Cate were chosen as the editors. Between 1948 and 1958, they published seven volumes. Volume 2 of their work covered the period from August 1942 through December 1943. The first chapter of this volume was entitled: *Crisis in the Middle East*.

CONCLUSION

The title of this book, *The Other Doolittle Raid*, may have implied that Doolittle led another historic World War II battle. It was not my intent to mislead the reader or to minimize the importance of the actual Doolittle Raid and the heroic actions of the airmen who participated in the mission, especially those who gave their lives.

Rather, it is my hope that the sacrifices of the men of HALPRO will be made known and remembered. They deserve the same level of respect, honor and fame given to the Doolittle Raiders. Without the bravery of the HALPRO men, Rommel's Afrika Korps may have succeeded in its campaign. If Rommel had been able to reach the Middle Eastern oilfields, the history of World War II may have been quite different.

I am deeply honored and grateful to tell their story.

APPENDICES

APPENDIX A — INTERVIEWS

BRIGADIER GENERAL MERIAN COOPER

On May 22, 1970, retired Brigadier General Merian Cooper was interviewed about the missions to bomb China.

Question: Some detail on the Caleb V. Haynes mission. What was the original purpose of Operation Aquilla? How did it work out?

Answer: I have every detail on the Caleb V. Haynes mission. When the Doolittle raid was planned and the Haynes raid was planned, I was Executive Intelligence Officer to General Arnold. I asked to be relieved to go on the Haynes mission. So far as I know, I have obtained no clearance to detail the Haynes mission. I think the story should be told as it does great honor to General Arnold; but General Haynes who died 3 or 4 years ago told me he never obtained a clearance to talk about it, and neither have I. The one thing I can say about it is that Arnold was a driving fighter and once having given the order to hit Japan, to raise the morale of the United States, he called Caleb V. Haynes to do it with B-17s from China, and for Doolittle to do it with the Navy. If one of these missions failed, the other was to go on and do it. I can say that if General Haynes and I had not been halted in India, we would have been at least a week before General Doolittle in hitting Japan. But self-evidently, this might have endangered Halsey's carriers. I cannot say at this time whether this was the reason we were stopped or not. Some day I hope to write this story in full. It is fascinating and exciting; but I can say what I have said in order to show that General Arnold was a determined commander and he was not going to risk all on one mission. He had a right and left striking weapon in Doolittle and Haynes. The rest of the Air Force Aquilla under Colonel Halverson of course did not get to Egypt until three months after Haynes passed through there. I know all the details of that, too, but I have no clearance to write about it. Halverson was simply to be a follow up in more strength to Haynes' quick strike.

Question: Do you know why he chose you for this very important and sensitive mission?

Answer: Yes, I know why he chose me for the mission I was sent on in early 1942. The original idea to attack Japan as soon as possible after Pearl Harbor was President Roosevelt's. I have forgotten the day that this meeting in Roosevelt's office took place, but it was very shortly after Pearl Harbor. Most of the books I have read say that at the original conference there were only the President, Admiral King, General Marshall, and General Arnold. My remembrance is that the very first meeting was attended by the Secretary of the Army Atkinson and either Secretary of the Navy Knox or his deputy Forrestal, and Admiral King, Admiral Stark, General Marshall and General Arnold. President Roosevelt told both the Navy and the Army to come up with plans. My classmate at the Naval Academy, Captain Francis S. Low, an operations officer on King's staff, came up with the idea of flying Army planes off the new carrier Hornet. Details of this plan were made by Captain Donald B. "Wu" Duncan, who was King's Air Operations Officer. I believe a B-25 had taken off from a carrier before. I am not sure of that; but if so, Lt. Henry L. Miller (who retired as an Admiral USN) flew it. At any rate, he assisted Doolittle in training personnel at Eglin Field, Florida and later in California, before boarding the Hornet. He was also aboard the Hornet aiding in preparation for the raid. He is one of only two Honorary Tokyo Raiders. The other one was a Chinese civilian Tuno-Sheng Liu who risked his life, to my certain knowledge, to rescue American crews in imminent danger of being captured by the Japanese.[1]

According to various biographies, Merian Cooper was active in the Far East and China. One of those bios described his military service:

Though old enough to be free of service in World War II, [he had served in World War I] he enlisted anyway, commissioned as a colonel in the U.S. Army Air Forces, and accompanied Col. Robert L. Scott to India while serving as a logistics liaison for the Doolittle Raid. He and Scott traveled to Dinjan Airfield, Assam, where they assisted Col. Caleb V. Haynes, a bomber pilot, in setting up the Assam-Burma-China Ferrying Command, which was the origin of The Hump Airlift. He went on to serve in China as chief of staff for General Claire Chennault of the China Air Task Force — precursor of the Fourteenth Air Force — then from 1943 to 1945 in the Southwest Pacific as chief of staff for the Fifth Air Force's Bomber Command.

Leading many missions and carefully planning them to minimize loss of life, he was known for his hard work and relentless planning. At the end of the war, he was promoted to brigadier general. For his contributions, he was also aboard the USS Missouri to witness Japan's surrender.[2]

It seems Cooper would certainly have been knowledgeable of plans to bomb Japan.

BRIGADIER GENERAL EUGENE BEEBE

After talking with Cooper, the interviewer visited retired Brigadier General Eugene Beebe and asked him, in August of that year, about the bombing of Japan.

Question: Right after Pearl Harbor, President Roosevelt was putting the heat on Arnold to mount some kind of air attack against Japan. Were you familiar with that activity?

Answer: Well, only the preparations for Jimmy Doolittle.

Question: But it didn't start out as Jimmy Doolittle. It started out in several other directions.

Answer: Oh, they had a dozen different plans, of course.

Question: The Halverson HALPRO.

Answer: Yes.

Question: There was project AQUILLA. Do you remember a Merion Cooper?

Answer: Yes, he was in China before I got over there. He was a movie producer.

Question: I went to see him. Was he given some secret mission to try to line up airbases either in China or Siberia from which we could launch an attack?

Answer: I don't know about that, but I would say that if he were over in China, and I believe he was at the time Doolittle made his deal over there, that perhaps they had sent over inquiries as to where they could land in China, after they had come over from Japan. So he probably was involved in something, but I don't know about it.

Question: Well, you know in January, Arnold had several conferences with the President, and the President told him to let his imagination roam, to figure out some way they could get this attack doing. Of course, Roosevelt . . .

Answer: Well, we had to do something to save face, obviously.

Question: Roosevelt was thinking of the political benefits. Arnold was not enthusiastic about the Doolittle raid, because any time you lose most of your airplanes, this is not an efficient mission.

Answer: Well, of course, we didn't know, at that time, that we were going to, how it was going to turn out. That was an airplane that was not designed for the purpose.

Question: This may be the start of his so-called Advisory Council, or at least one of the reasons he dragged in Norstad and Cabell. Remember in February? According to Cabell, they were to do his thinking for him, on sort of these exotic type missions. Did he ever talk to you as to why he set up the Advisory Council?

Answer: He probably talked more about that problem than any other. We needed some vision. We needed to enlarge our field of thought and become hep with what was going on in the world, more than we had been, I suppose. I think, you know, we started to get a lot of scientists in along about that time, and I think the thought around there amongst some of the senior officers then was, if we could just get a few brains corralled in from somewhere, maybe we could come up with something to utilize our forces better than we have. This was our small attempt I think to get a few brains together to try to figure something out. Remember, they weren't harassed by all the daily work of running an office, or something, where they could go off in a room and try to study a matter out, and think it out.[3]

ENDNOTES

1. ----, AFHRA, Hap Arnold file, interview with BG Merian C. Cooper 22 May 70.
2. ----, *Together We Served*, Cooper, Merian Caldwell, Brig Gen, http://airforce.togetherweserved.com/usaf/servlet/tws.webapp.WebApp?cmd=ShadowBoxProfile&type=Person&ID=120615 , accessed July 13, 2012. See also ----, *Claire Lee Chennault Foundation of the Flying Tigers, Inc.*, Merian C. Cooper, http://www.chennaultfoundationflyingtigersinc.org/Merian%20Cooper.pdf , accessed July 13, 2013.
3. ----, AFHRA, Hap Arnold file, interview with Brig Gen Eugene Beebe - 12 August 70 - Long Beach. Calif

APPENDIX B — INDIVIDUAL STORIES

JUNE 12, 1942 MISSION TO PLOESTI

A few of the men kept memories about the mission:

Walter Shea, the navigator-bombardier on the crew of 1st Lt. John W. Wilkinson, Babe the Blue Ox

> *It was a night take-off,* June 11, 1942. He didn't remember the time but it was about right to arrive at dawn over the Black Sea. His ship was the last off — or damn near.
>
> The briefing: The officer in charge of the briefing was from the R.A.F. The HALPRO crews were instructed to fly from the R.A.F station at Fayid almost due north to Rosetta Point — at the mouth of the Nile River. From there they were to take a course to a lighthouse on the south coast of Turkey and then west, then north, and then east around Turkey into the Black Sea. Once in the Black Sea, they were to fly north to Constanta, one of Rumania's seaports. From there, they were to fly up the Danube to an island where the river split. Shea writes, "That from that point we were to assume a heading for Ploesti ... straight on in. We were told that our alternates in the event of an emergency would be Tiflis (Tbilisi), in Russia, and a Turkish base in south Turkey. I'm quite certain we had a telephone number to call in Turkey, and a code expression, all designed to allow us to get in — get refueled or whatever — and get out."
>
> Second briefing: Subsequent to the above, Lt. Bernie Rang assembled the navigators in one of the base quarters. There, Col. Halverson joined them. A map of the Eastern Mediterranean was posted. He very simply repeated the RAF briefing. He showed them how they could fly from Rosetta Point to a light house on the south coast of Turkey. With his finger on the lighthouse, he said, "Notice that you'll be flying this line of longitude ... can I help it if the line goes right across Turkey?" That is all he had to say. In leaving, he warned them to maintain plenty of altitude. "Those mountains really get up there."

Lt. Bernie Rang then took over in his straightforward way. He told them to ignore that bit about landing in Turkey. He warned that they would be interned if they landed there. He also discouraged landing in Russia. Bernie said, "If you must stop short some place, get your pilot heading for a base in southern Turkey. Having gotten him that far, talk him into Aleppo, in French Mandated Syria, just across the upper part of the Med."

This was a relief to us, for we damn well knew the B-24D couldn't fly the route laid out by the RAF. Don't know where they got their high opinion of our range.

Walter Shea states that the ultimate destination on return from the raid was Ramadi, Iraq, an RAF station way down on the Euphrates River. Shea believes there were actually two bases available.

The bomb load was six 500 pound bombs. The emphasis at the briefing was on the need to hit the Astra Roman refinery – the crux of the works. They were told that the war could end years early, if they could put the refineries out of commission permanently. Impossible, of course. Perhaps the raid had as much political significance as strategic or tactical value. Stalingrad was under siege at the time.

The Mission: The take-off and the flight across the Mediterranean were routine. Shea believes that part of the mission was flown at about 10,000 feet. They were saving their oxygen for the high altitude run into the target.

We hit the lighthouse on the south Turkish coast – right on the money – and I gave Wilkie a heading for the north coast of Turkey – to the lighthouse that I'd noticed on the map on that far side of Turkey. My own idea. Don't recall the northern lighthouse having been mentioned by Bernie or Col. Hal. Just spotted it on the map, as I recall. We were probably at 14,000 feet by then – for I believe that pilots were read-in on the 'route alteration' by then.

It was fine weather, and we just flew on north. We were using 'metro wind' at the time, as briefed to us by the weatherman, and amended by our flight to the south coast of Turkey. The metro wind was right on, as I recall. I don't believe I made any changes. After all, we did hit the southern lighthouse on the nose, so the wind had to be right.

Sometime later, as my ETA for the north coast of Turkey approached, I saw a light in the distance, to the right. Obviously, to me, the wind had shifted, and we had drifted left, toward Istanbul. But not knowing exactly where I was — in order to compute the current wind — I merely told Wilkie to head for the light (which I assumed was the northern lighthouse), at which point, with some calculation I could figure out what the wind effect had affected us, and then, from a new positive point, could redirect him to Constanta.

We flew on toward the light. It was the damnedest light! It never got closer. Just sat out there, bright as hell, no flickering. I became confused. My ETA was shot to hell. I had no idea of our ground speed. And then the revelation: The light was a star or planet! Venus, I believe. Damn thing kept rising as we flew towards it. At which point, dawn came to us and we discovered ourselves over the Black Sea! No time for recriminations. We all thought it was a light. Same problem we'd had off the coast of Fr. New Guinea — damn star/planet!

With a sort of left-handed navigational 'sense' of time/distance/heading, I gave Wilkie a 'heading' for Constanta. Which worked, for there it came, in the clear, right ahead. Time to check fuel, time in the air, new heading, etc. etc. — a new start.

But as we approached Constanta it was immediately apparent that heavy clouds, 10/10's, obscured the land area. Constanta was noted to have numerous ships in the harbor as we approached at 30,000 feet or nearly so, on oxygen now. I felt such temptation to make a long bomb run on these vessels. Tried to talk Wilkie into it, as bad as the intercom was at that altitude. Nothing doing. We had been told to bomb nothing but Ploesti, despite all. Wilkie wouldn't budge, so we flew on.

I recall seeing a light from the ground, and from all my earlier 'Flying Aces' magazines, assumed it was 'Archie' fire — ack-ack — anti aircraft. Strange feeling. Had no time for intricate computations, but just assumed it would take perhaps ten seconds for the shell to reach our altitude. Recall looking at my watch's sweep second hand and at ten seconds holding my hands over my ears! Maybe I hated noise?

We drifted over Constanta, now looking for the division of the Danube and that ISLAND. No go. The Danube was apparently in a flood stage, just a real Mississippi Delta effect. And that was our last sight of the ground as we sailed over the solid overcast.

Assuming a position, I gave Wilkie a heading for Ploesti.

Time passed. We just flew on. I had taken off my leather jacket and wrapped it around the bombsight, which I had running, to keep it from freezing-up. But I was damned cold! And always remembering the briefing by Major John Payne, our flight leader — that at 20,000 feet, you couldn't last long without oxygen. And we were at 30,000 feet! Felt that if I removed my mask for an instant, I would have had it! So I just shivered.

Suddenly, a cacophony on the intercom. We couldn't understand the transmission. Wilkie, in his usual calm way, tried to get the gunner to talk slowly and clearly. We finally learned that the tail gunner was trying to tell us that we had a fighter coming up on our tail. Wilkie merely told him to "keep your eyes on him." I peered thru the astrodome to see a B-24 coming up on us. We had been briefed to look for each other, and to pair up if we came together.

The B-24 came up on our right wing. I guess Wilkie throttled back. We paired off and flew side-by-side for perhaps fifteen minutes. I looked across and recognized Ted Bennett in the nose, Mark Mooty's navigator and my best friend.

We exchanged 'arms out – palms up, where the hell are we?' signals. But as briefed, maintained radio silence.

And then I had my greatest moment. I looked out ahead – about twenty degrees right, and saw bomb flashes! Having been the last, or near last to take off, I knew that the flashes came from the earlier ships bombing. Never having dropped anything more than a one hundred pound practice bomb, I just knew those flashes had to be from 500 pound bombs.

Not knowing just where the hell I was, atop the overcast, and not having a chance to check the wind drift and get a wind effect factor, it seemed that the quickest and slickest way out of the situation was to head for the bomb flashes. And I instructed Wilkie accordingly.

We didn't exactly peel off, but did swing out and head to the right. Mark Mooty and Ted just flew on and we lost them shortly.

After about fifteen minutes, I got the message. The flashes were lightning! A storm. Can't describe my emotions at that point. But checking back on when we changed course, for I was following the compass, and getting back onto the old wind factor I had in crossing the coast, I redirected Wilkie to Ploesti. Hopefully.

The ETA for Ploesti came up fast. Clouds were still 10/10's beneath us, perhaps at 10,000 feet. The temptation was to get down and SEE. But the briefing, again, was to under NO circumstances LET DOWN. The alternative was to bomb on our ETA. Which at this point was MOST problematical! Nevertheless, I had an ETA, for what it was worth. Sooooo opening the bomb bay doors, setting the bombs on train (forgot what interval) I just watched the ETA come up on my sweep hand.

At the last moment, I had an impulse. I set the damn bombs on SALVO. All together. A real dump!

We dropped, and peeled right, back toward the Med.

I understand someone got a mineral water factory. Always hoped it was me! I don't believe in mineral water.!

As we turned back to the southeast, I checked my maps, and as I really didn't have a very good map of the Eastern Med, I gave Wilkie a heading for the east portion of Turkey — toward Tiflis, toward Aleppo, etc. — hoping to skim the edge of the Med.

Found our altitude — for we let down — but the damn mountains were really up there. Took more than 14,000 feet to cross over them.

Taking my compass bearings from this little dumb map I had (SURELY I had been provided with better — must have lost it or some silly thing), I awaited Wilkie's asking me for a heading for a friendly Turkish airfield, which I was supposed to divert him from, talking him into Aleppo, etc., as briefed by Bernie. But Wilkie never questioned me. GREAT PILOT.

But he did question me when we ran out of mountains and found ourselves over desert! We had missed the whole goddam Med! I REALLY skimmed it close! We were breezing down the darn Euphrates, as briefed, but we were REALLY low on gas. Wilkie wanted an alternate. Of course, Aleppo. But where in the hell was it?

No stars. Nor could I EVER shoot stars. But I could shoot the damn sun. Which I did. And got a line of position that showed me south of Aleppo. No way to get an EXACT course back to Aleppo, so told Wilkie to fly west until we hit the Med and then north to Aleppo. And that's what he did. Came in and landed. Can't recall the amount of gas remaining.

DO recall getting out of the nose and finding a few French soldiers accosting me at bayonet point. Mumbled something about other American airplanes, and had one guard point over my head. Turning around, I found myself standing right UNDER the tail of another B-24! Never saw the damn thing.

Somehow, we ended up in town, at a hotel — relatively speaking, quite modern.

After that, I recall Wilkie and Wilcox, I assume, etc., on some roof garden restaurant, facing a big bowl of cherries on ice and falling asleep on the table, head in hand. Having to roust out a chambermaid at the hotel — had collapsed (from overwork?) on one of our beds (she was maybe seventy years old?).

Took off — next day? Not sure — and back to Fayid.

Took notice of the biblical features of the geography as we flew on. Mt. Ararat? Etc.

Wilkie — for the first time — told me that I had the damnedest habit of holding down the mike button after I'd transmitted, which killed his transmission! He also, when I told a 'joke' on the intercom, responded by just pressing his mike button. I complained at his cold reception of a good joke, and he asked me, "What do you want me to do? Say, 'Ha Ha'!" I felt that he had a point.[1]

JAMES SIBERT, PILOT OF QUEEN B

JUNE 11, 1942

Ploesti #1 Up at 07:30. "Had a spot of tea" and ate breakfast. Went out to ship and made a last minute check on bomb rack etc. Ate dinner and was notified that briefing would be at 14:00. Went to first briefing and Col. Halverson told us about our objective. Was given maps and data on the Ploesti Rumanian oil fields from which 1/3 of oil was being supplied for the Russian front. Plan was to go around Turkey through the Dardanelles. Was given Ramadi, Iraq as field to return to with several alternates in case of trouble or gas shortage. Ate supper and had a pilot meeting in Col. Halverson's quarters, before going to ship for takeoff. Took off at 22:00 with 6 - 500 lb bombs and 3100 gal of gas (2 bomb bay tanks). Flew over Turkey at 14,000' for three hours without oxygen in order to save it for 25,000' operation.

6-12-42

Hit the coast of Rumania just before daylight and headed inland at 25,000' indicated temp -20 C. Encountered ack-ack from the coast in until we hit a solid overcast 8Qmiles from our target. Decided to let down through it so nosed her down and airspeed hit 250 mph through 11,000' of cloud and I busted out in the clear at 12,000' right over Bucharest. Was nearly shot out of the sky by ack-ack and one burst just under the ship and bounced us up 50 ft higher.

Two men passed out due to ice in their oxygen masks and the tail turret froze and wouldn't operate. Pulled up into overcast and headed north on instruments. Let down between Bucharest and Ploesti and sighted a pumping station for the refinery with 15-20 storage tanks. Dropped on it and saw it go up in smoke. Climbed back to 22,000' and returned toward Turkey after firing on 2 ME109's that left us. My co-pilot gave the two men oxygen and kept them alive until I could let down over the Black Sea. Checked our gas and figured we could make the alternate field of Aleppo, Syria. Landed after a search for field and was practically out of gas. Had flown 14 1/2 hours and was dead tired. Greeted by RAF Officers who fed us and took us into Aleppo to a hotel to sleep that night. Slept from 17:00 to 07:00 without any supper and my clothes on. Woke up with eyes bloodshot. Went back to field and cleared for Fayid.

6-13-42

Arrived at home base at 11:45. Greeted by Col. Feldman and was interrogated by S-2. Ate dinner and went to bed. Arose, ate supper and went to bed again. Woke up at 08:00 and as the others had returned home from Ramadi I learned four ships had landed in neutral Turkey. Two had cracked up due to lack of fuel near Ramadi. No one injured. Nesbitt was in Turkey. Report of result of bombing was favorable. Spent most of day just relaxing and taking it easy. Ate dinner and went out to ship to see crew and check on the ship.[2]

RICHARD MILLER, CO-PILOT QUEEN B

Well, we went. Had as our target the most vital oil refineries and fuel storage that the Germans have — Ploesti. This town supplies one-third of the fuel oil for the German Army, and the majority of it for the Russian front. Since it is so important, and rather inaccessible, the place had elaborate camouflage and not much pursuit. We left here (Fayid) at 10:30 Thursday night (15 of us) and flew over part of Mediterranean over Turkey, over the Black Sea, then the coast and over the Danube and then we hit trouble. By the way each plane went as an individual trouble shooter — just 50 miles from the target, we had an overcast under us. We flew over this, hoping to find a break in it, but had no luck; decided to go down through the overcast, so let down gradually.

    Before we could get down the engineer staggered up and said the waist gunner was "out," and since his own hands were frozen he couldn't go back to help him. I clambered back to find the airman at his gun but a dark purple color, worked over him and brought him to on my oxygen, fixed another mask for him and went back to the cabin to find the engineer out. Had to give artificial resuscitation for ten minutes or so, while I worked with just gulps of oxygen. When finally got engineer to, he was too sick to be of further use. We went down to 12,000 feet to find two M.E. 109s waiting. Our waist gunner, just visibly alive, fired a good burst of tracer and incendiaries and scared them off. We looked about and found Bucharest Rumania, which was south of our original target. Ack-ack opened up on us, and one shell burst so close we gained some 50 feet on its force. I sincerely believe this was one of the very few times I have ever been scared. We zoomed for cloud protection and headed for our target. Then our tail gunner called and said his turret was frozen and wouldn't operate. Since we now had no rear gun, and a sick engineer, we wondered what to attempt, but continued for our target.

When our top turret went out for lack of oxygen, we decided we had best drop our bombs and scram out. Found a cluster of 10 to 15 fuel storage tanks and blew them to hell. Big billows and rolls of black smoke rolled out, an intriguing sight. We hit for altitude when we ran into more ack-ack and had one rear gunner go out because of lack of oxygen.

All this time, I had to work with only gasps of oxygen for I had to work on each of the men and give him my mask. I really believe I saved the lives of our engineer and armorer, for they could not have lasted but a few minutes longer, and we couldn't go down. All of our trouble was not helped by the fact that our wings and props iced up. The pilot had a tremendous jolt flying by himself and the navigator had all he could handle in locating us, and changing his bombsight so often. This time I can modestly say the officers truly came through. All the icing of oxygen came because the men did not take care of their equipment properly. I warned them no less than 20 times to check for ice, but the excitement was too much for them. On leaving enemy territory, I snoozed a bit, nearly passed out from the extreme effort of working at high altitudes with very little oxygen, and awoke to find we were low on gas and had only one field to get to. Found a beautiful airport, started to land and discovered that it was closed with oil barrels stacked all over the runways. Flew blindly on and at the last second, found a runway at Aleppo, Syria — 'twas a grand sight for we had been debating what type of field to try and land in. We landed to find a British garrison of R.A.F. ground officers, probably was the first time I've been happy with the sight of the R.A.F. The Group Captain knew one of our colonels, therefore treated us royally. They had gas but we had to strain it through a chamois cloth and pump it in, gallon by gallon with a crude hand pump. We stayed in Aleppo overnight and had quite a time. The Armenians there had queer customs — many wore the traditional fez and the majority wore black bloomers, very baggy affairs. The bagginess is for some belief of reincarnation. The women were veiled in black. The city was quite clean except for the bazaars. The buildings were many stories and built of an attractive, local stone. The bazaar presented the usual filth of Mexican markets and one sickening sight was a small boy with a shaved head. He had sores (open) all over his face and hands, and one leg was half-eaten away with an open running sore. The flies nearly filled it and in all was a most gruesome sight. In searching the bazaar, bought a camel's hair-skin or piece of cloth. Had a good meal near the hotel — had our first good fresh fruit since the States. Left Aleppo (Alep) and flew back to Fayid today. On the way back, we flew over the seat of our Christian religion.

Flew over the Garden of Eden (wanted to bailout, but the Garden didn't appear to be bearing) over the Sea of Galilee, City of Damascus, City of Nazareth (where Christ was raised) and many famous Biblical spots. Landed here at Fayid at noon; got to land myself — great sport. Heard that we might get another mission tomorrow morning. Hope so!!

June 14. In writing up the action of our first raid, I failed to put in an incident that was very effective in calming us down so that we came through with our minds clicking instead of blowing up. It was in the heat of the battle when our tail turret went out. Cpl. Fillipi, a young lad of 18 fresh from the Bronx, called most frantically in a voice that denoted complete despair, "Lt. Miller, Lt. Miller, my tail turret won't work!" Guess it was the tone of his voice and his accent that struck the humorous chord, but whatever it was, it relieved a lot of tension and let us settle down to the grim business of bringing men to dropping our bombs. After talking to most of the men who went on the mission, I am of firm conviction that we did more damage than any other ship and even more so considering our tribulations. Was a bit disappointed to learn that not one of our Senior Officers (over Captain) took the risk of going down under the overcast to try and find the target. Instead they bombed on an E.T.A. (Estimated Time of Arrival) at 30,000 feet. Seems to me that in the business of war a certain amount of guts and daring is essential or else we'll soon hold the short straw. One of the ships got one M.E.109 and this same ship was saved at the crucial moment by German ack-ack — an M.E. 109 dived in and was firing a burst at what appeared to be an impossible miss when he went up in one of the black puffs. Ack-ack fire is an odd sight. The first I saw was at the coast of Rumania at dawn; at first I thought it was a signal light, then it began to burst in orange flames off to one side. Reminds one of a Roman candle or sky rocket — a pretty sight at a distance. During the day it looks like a dark puff, like a clay pigeon that is "powdered" by a good shoot in skeet; sometimes it is a similar puff, only white.[3]

ALBERT STORY, THE TAIL GUNNER ON THE ED CAVE CREW, BLUE GOOSE

When we arrived at our base at Fayid, Rommel's Afrika Corps was only about seventy miles from Alexander. His spies must have reported the presence of the large bombers that had just arrived at Fayid. One night just before we left there was an interesting display over the Suez Canal area. Evidently it was a high-flying German observation plane. The sky was lit up with hundreds of searchlights. We never heard what happened to the German plane. Pursuit planes had taken off and the plane flew out of range of the searchlights and disappeared. The Germans never bombed the field while we were there, so I guess we got out before they could get a raid organized.

On the night of June 10, 1942 we all were called in and were briefed on the details of the mission. We were to fly straight across Turkey, a neutral country, at 25000 to 30000 ft. altitude, head northwest for the point where the Danube River meets the Black Sea and from there go directly to Ploesti drop our bombs and try to get back as best we could. Our navigators quickly calculated the round-trip distance to be around 2600 miles. This was the very maximum limit for a B-24s range. Two days before the scheduled trip to Ploesti we took Hellzapoppin out over the desert on a practice bomb run. After dropping our bombs, we returned to the base. Upon landing our front wheel collapsed and old no.23 skidded along on her nose to the end of the runway. Since the plane could not be repaired in time for the Ploesti trip, we were assigned another plane named Blue Goose. The pilot of this plane was ill and returned back to the States so our crew replaced the crew on Blue Goose.

The night of June 11 was totally dark. There was no moon at all and everything on the ground was blacked out when we began to take off at eleven p.m. We could not fly in formation so everyone was on his own. We climbed to our designated altitude of somewhere near 30000 ft. At this altitude the temperature was minus forty degrees Fahrenheit. Sometime during the night I noticed a searchlight beam piercing the darkness. This I reported to the pilot. We kept on flying as other beams came on. We then noticed flashes of light from the ground. Somebody on board shouted out," Hey, they're shooting at us." I myself was trying to shrink myself up into the smallest ball that I could. The searchlights never found us so we kept flying until finally everything was total blackness again. Some time during the night as we headed west I noticed a dull glow in the east. It appeared that the sun was coming up, but it had a strange reddish hue. We later learned that was the glow from the Russian city of Sevastopol (see Figure 8) under siege by the Germans.

We approached the Rumanian coast just as the sun was rising. We were ordered to re-check our guns and that is when I discovered that after several hours at minus forty degrees, the hydraulic fluid had become so viscous that the turret would barely creep around. The sun was now up and shining squarely into my eyes. I could see nothing. By looking down I did get a glimpse of a river below us. I assumed it was the Danube. Shortly after this we noticed that there was a low cloud covering everything below. The pilot and bombardier decided we should try to get below the clouds in order to see the target so we made a rapid descent through the clouds. As we did so, our cold airplane entered the moist clouds and everything iced up. The inside of my turret was coated with ice. As we approached what the navigator assumed was the target he called out, "Bombs away!" I heard some explosions as if there was some sporadic shooting from below but I still could see nothing until looking out my side window I could see smoke pouring out from one of our engines. I reported this to the pilot and he ordered the engine shut down. By this time we had turned and were heading east.

The flight to Ankara was surprisingly uneventful. We had got back above the clouds and no German planes followed us. After leaving Rumania, we let down to around eight thousand feet altitude and headed for Ankara. Our number two engine smoked all the way. When we reached the Ankara airport and landed we were greeted by Turkish soldiers carrying guns with very long bayonets attached. The bayonets were pointed at us. The main thing I noticed was that they were not smiling. They seemed very serious. Finally someone who spoke English approached and our pilot began talking with him. Actually there was not much explaining to do because two B-24s had landed before us. They, like we, were also out of gas. After a while, the U.S. ambassador showed up and things became more amicable. We were all gathered for a photo shoot and then we were escorted into the terminal restaurant where we ate our first Turkish meal. Later we learned that our government was presented with a bill for this show of hospitality. The end of this flight was the beginning of our stay in Turkey, which for me lasted eleven months.[4]

HOWARD E. WALKER, PILOT LITTLE EVA

Concerning the Ploesti mission: The basic briefing information was essentially as you find from the statements of the four crews provided to me before I left Turkey in November of 1942. My memory follows that of Fred Nesbitt. We were briefed (after the navigators had had a private briefing by Bernie Rang and which Davis and Humphreys described). All crew members were included in this one. The background of the mission was provided. The British didn't have Libs or Wellingtons with sufficient fuel capacity to reach up into the Ploesti area which was the fuel supply of Hitler's armys. We did have the extra 800 gallon fuel capacity in our planes that had been installed to assist in the long hops required to get us across the Atlantic and they had not been removed yet. Therefore, we had what the British didn't have. The mission was to be a high level attack and if we could accomplish it to any degree we would create considerable havoc with the Nazi battle plan.

The British briefed us on the planned route of approach from Fayid; north but east of Turkey, over the Dodecanese Islands and then cut into a route up over the Black Sea until approximately east of Bucharest and then onto the target area (which was supposed to be heavily camouflaged and possibly to have a complete twin visible but with no actual or real refinerys or storage areas). Ploesti was the primary target with Constanza as the secondary target, but we weren't supposed to have to deviate from Ploesti. Our actual route was straight up the 30th Parallel which would take us over Turkey which wasn't kosher as far as the British were concerned. After the British had given us their best information, Colonel Nick Craw was introduced as a man who was well qualified to provide us first hand information of the Middle East and Rumanian oil fields since he had worked in and around them before the war forced him out. He described the area around Ploesti and adding that information he had concerning the twin sister city that had been built as a decoy. He also had considerable information concerning the political climate in Turkey and their relationship with British. He gave us a list of four airports in various parts of Turkey that could handle (if you call a 2600 foot runway adequate) our B-24s in case of a forced landing in Turkey. The physical dimensions of the airfields were in reverse order to the political advisability of landing. In other words, the best facility was at Ankara but the political situation was the least viable to the accomplishment of obtaining release of the crews and their aircraft. We were advised that it would only take about three or four days for the British to get us out of Turkey and if we didn't get to take our planes with us, they would have them out shortly thereafter. We were also given telephone numbers at each of these locations to call. (At one time I had my little notebook with this data in, but am not sure that I can locate it now. Will keep looking and if found will forward it you.) The rendezvous point given us was not, to my recollection, Habbaniya but I don't remember what it was. It was around Damascus, I believe. I will assume that McGuire's notebook is correct as to my position for takeoff was 4th in line. I have on my itinerary a time of 10:32 which must mean 2232 Fayid time on the night of June 11,1942.

The flight after takeoff from Fayid was uneventful until we found ourselves just east of Istanbul before heading out over the Black Sea. We climbed to a high enough altitude to clear the 14000 foot peaks, then on up to 30000 feet which was to have been our bombing altitude. We reached our offshore position before dawn and circled to await some light. As everybody else did at that time, we saw a low deck of overcast, moving in from the northwest. Land was not completely covered but the target area was not visible. Davis set us up on a dead reckoning course to the target area. Just shortly before this we were advised by Albertson, the radio operator and waist gunner that O'Connor had passed out, apparently from lack of oxygen. Sgt. Lippencott, using the long oxygen hose he had connected, took a spare mask and oxygen bottle back to O'Connor and succeeded in bringing him around. He found a small pinhole in the bladder of O'Connor's mask and gave him the spare mask we had. Shortly after this mishap, we were again on our run toward Ploesti. All four engines quit and it got awfully quiet. Sgt. Lippencott had taken off his headset when he had gone back to help O'Connor and had not put it on when he returned to the upper turret position. He was just beyond the reach of either West or myself from the cockpit. Wilbur got out of his seat and did reach him but by that time Lippencott was heading for the gas gauges. They all read empty. Then he realized with the rest of us that we had not shifted over to our bomb bay tanks which still had approximately half capacity. We succeeded in getting all engines turning over again and Davis made a quick estimate of our situation from the standpoint of how far we could go from that point to any point of landing we might select or hope to reach. His calculations said we might just possibly make it back across Turkey if we were lucky. So, we headed down from about 28000 feet on a reverse course to get back over water and hit the coast line at approximately 22000 feet. We had not unloaded our bombs and started to look for a likely target where we could inflict some damage with out hurting the civilian population. Below us we saw some docks at what we determined later to be Constanza and Davis salvoed the bombs just as we crossed the edge of the Black Sea. As Davis noted we found out later through a Vichy French newscast that we had inflicted damage on more than the fishing industry in Constanza. Precisely what we hit I don't know. At some time near when we lost our engines, I saw a blue biplane flying off our right wing. He did nothing but sat out there for a while and then disappeared. (I don't know whether anyone else saw it or not). Later while we were in Turkey I saw pictures of Rumanian Air Force planes and one of them was exactly like the one I saw.

One thing I didn't mention that was covered in the briefing for the mission was that we were advised that the American government and the Russian government had reached an agreement whereby any planes that were in trouble in the Black Sea area could come in to Batum which was down near the Turkish border on the west side of the Black Sea, and expect to be welcomed. The only problem was that we did not have very definite approach signal information and figured that it was better to take a chance on Turkey if we had to make a forced landing. We considered the Russian deal but like Nesbitt's decision not to accept, headed south back to Turkey. When we arrived in the vicinity of Istanbul, we were advised that they didn't consider us welcome because some ack-ack was sent up. We avoided this and started to follow the railroad tracks into the middle of Turkey. We were watching our gas consumption very closely and decided that we had insufficient to attempt to clear the mountains between Ankara and the Syrian border town of Aleppo where some of the others landed. We apparently passed over the airfield at Adipazari where Nesbitt set his plane down. By the time we reached Ankara, we calculated that we had about 75 gallons of fuel left and would be stupid to try to continue.

We made one pass over the runway and found men working on it. Made another pass coming in pretty hot and with West putting all his weight and more too on the brakes, got stopped before reaching the end of the 2600 foot runway. Taxied back up to the terminal and cut the engines. Shortly after we landed, Nate Brown and his crew landed followed by Ed Cave and his ship. They each had jugs blown and were low on gas. We (crew #4) had no enemy damage to the aircraft, we just didn't have enough gasoline to continue and this was primarily attributable to our circling over the Black Sea waiting for dawn but also aided by my lack of tight fuel control enroute to the target area.

Our stay in Turkey is probably pretty well chronicled by the information you have already received from the other fellows who were there so I won't belabour that point unless you ask for more info after you have been able to wade through this. My information on the escape from Turkey is all second hand; I was in the 38th General Hospital at Helmopolis, Kilo 13 or El Dabb Hagg as the place was called. I left Ankara on the 27th of November 1942 after having been advised that I had been traded for an ailing German professor and would be repatriated back to the US. This took place about the middle of November but there was no space available on the train until the date noted. I left in the company of Bob Parker, head of the US Information Service office in Turkey, and several other members of his staff. I was the first to leave Turkey of the group interned. Everyone spent some time in the Turkish military hospital with malaria and other illnesses of one sort or another. I contracted malaria which after the third trip to the hospital in less than two months turned in to infectious hepatitis (yellow jaundice) which proceeded to take my weight way down and got me in a condition that was figured to make me no longer fit for military service. Though I left Turkey near the end of November 1942 I did not arrive back in the States until the end of January, 1943. When I got to Egypt, I went by Abu Suehr where the 376th Bomb Group was stationed and was there over night then went into Cairo with (I believe it was Gardner). By the time I reached there, after a long and slow ride in a command car because the Aussies were returning from El Alamein, I was pretty sick. I was checked into Col. Kendrick's office and from there shipped immediately to the 38th General Hospital where I remained for about seven weeks until they concluded I was fit to move on.[5]

EDWARD CAVE, PILOT BLUE GOOSE

We had a briefing in Fayid and were told that we were going to bomb the oil refineries at Ploesti, Rumania and would be the first US aircraft to bomb Europe. About two days before the mission the navigators were briefed and were told where we were going and that was over Turkey. The next day we had a general briefing. The British were there and we were given escape equipment and so on. The navigators were not to tell the pilots or any one else about their briefing. At chow about 6 p.m. the pilots were called to Col. Allison's private room. He closed the door and on the inside was a map. He said you were told at the briefing you were going to fly around Turkey — he said you are going to fly over it — and the navigators have already been instructed. He also said that if Col. Craw — to say that if we had a problem to land in Turkey that we would be out in about 24 hours.

About 9 o'clock on June 11 1942 the 13 operational airplanes were taking off to go to Ploesti. I think I got off the ground around 9:15 to 9:30 — had a little problem with #2 engine — had a little fire coming out of it but it soon extinguished and we continued on since we did not fly in formation it was sort of a dog eat dog situation and each individual could fly at his own altitude that he wanted to go into the target with, but stay as high as he could. We went in at about 25,000' the next morning, we couldn't see a thing due to a heavy overcast. So we dropped down and bombed at about 8500' — and come to think about it we were lucky we didn't get hit with another airplane from above us. We had no radio contact with any of the other planes and had no idea where they were since we were not in formation.

I believe we went over Turkey that night about 10,000' since I know we called in on oxygen at that time. Early morning on the 12th of June we were over the Black Sea and began to climb to high altitude and look for a horseshoe bend in the Danube River, which would take us into the target. Ploesti is about 80 air miles north and slightly west of Bucharest. It was quite cold at 25,000' and the type oxygen masks we had in those days had a tendency to freeze up so you have to pull the bottom plug and squeeze out the ice.

When we dropped down to make the run on the target, I got the plane all squared away and called Wicklund on the intercom and told him to get the bomb sight set and take over for the run. The rear bomb bay doors were open and I kept waiting for the plane to lift as the bombs dropped out. About that time #2 engine was smoking and running real bad, so I had to use the fire extinguisher and feather the engine and of course I didn't know what was wrong or what caused the problem but later found the #9 cylinder head on the front engine bank had been blown off. I called Wick again and asked why he had not dropped the bombs. He replied that he saw a little bit better target. I said man we just lost an engine and if you don't drop the bombs I'm going to salvo them.

For a few seconds I felt the plane lift and knew he was making his drop. We were heading in the direction of the Black Sea and I called all stations to see if anyone had gotten hit, and thank goodness there were no casualties. In about 30 minutes my engineer Perrone who had to stay with the airplane and had to take the Blue Goose came forward and said we did not have enough gas left to even think about trying to make Syria. The decision was then made to try to make somewhere in Turkey - possibly Ankara since the American Embassy was located there. We threw out several cases of .50 cal machine ammo to lighten our load along with some other things that we didn't think we needed.

We kept constant watch on our remaining fuel and transferred what gas was left in the auxiliary tanks to the wing tanks and pumped out the gas in #2 engine to #1, 3 and 4 wing tanks. Wicklund was busy making calculations to distance that we had to go to get into Ankara. About 11 o'clock or so the morning of June 12, we had Ankara in our sights. We were at about 3,000' and picked out a runway since we had no contact. I saw two B-24s on the parking ramp and got all squared away for landing. The wheels had been dropped and checked a little farther out than usual because I knew that we had no chance to make another approach.

As we were getting in to touch down there were some Turks working there on the end of the runway. I knew I had to cut the thing as close as I could cause the runway was only 3,000'. When we did hit down I told Eugene Ziesel (my copilot) that I had it under control — which I didn't — my legs were jumping on the brakes and shaking and another engine stuck and pulled us off the runway about two feet. We sank down about to the hub and didn't do any other damage, but we were stuck. And I had a gut feeling that there was something wrong because none of the fellows from the other B-24s came out there to give us some jazz about not being able to run an airplane properly etc. About that time a fellow walked out and got to talking to one of the crew and saying where is the pilot of this plane. They called me and this fellow was Jewish looking — dark complexioned — he was Lars Steinhardt and he was the Ambassador at that time.

So he said I want you to squat down and I want to ask you some questions. Well I thought that was rather strange because I always had heard that your Ambassador was your mouthpiece and you could just tell him whatever you wanted to tell him.

Anyway I cut him real short and he didn't like it and stormed off and went back to the Terminal. At that time the Norden bombsight was still up in the nose because in the briefing we were told we were going to get out and there wasn't any point in trying to hide it and besides my money was locked up in the vault. After we got the airplane out after being stuck we went inside and still didn't see any of the other gang. They took us upstairs interrogating us what we were doing and where we were etc. Anyway they wanted to know where my money was and I told them I didn't have any.

The next day some Turkish Officers came to the barracks where we were staying at the time and with a representative from the Embassy. Then they made me go back out to the airport and unlock the safe where the bombsight was kept and give them the money, which I didn't get a receipt for and I didn't get a chance to count it. And that followed me around forever while I was in the service because they said I was $117 short with funds. They took us into town and gave us some lunch and we lay around awhile and I think the reason was that they were trying to get enough beds put in one room in the barracks where they had an infantry and also a cavalry outfit. So we got there all in the same room and it was a lousy setup. I think we stayed there about three or four days and then they took us out to a government farm, out of Turkey in an old schoolhouse and we stayed there until winter when it got cold because we had no heat or hot water and they moved us into town.

We moved into a hotel about two blocks off the main drag. This hotel looked pretty decent from the outside but inside it was a mess — infested with bedbugs, which we had to fight constantly. We were put up on the second floor of a three-story building. I believe it was about the first or middle part of October. They decided they wanted two of the airplanes taken over to a place called Eskisehir, which was the so-called Randolph Field of Turkey. None of the Turks could fly the B-24 and all they wanted was a pilot, co-pilot and navigator. So we knew that we ought to try to get more people out and get squared away so we could make an escape if it became at all possible. So we told them we wouldn't fly the airplane until they agreed to let at least five people on each airplane go. They finally agreed to - they had two officers on each airplane we sat on so we didn't try to do anything - we were told by the Embassy too that if we did try anything or escape at that time that we would be brought back was about half way between Ankara and Istanbul. It was an old town looked like it still had marks from WWI.

We would walk from the place where we were staying out to the airport and we decided at that time that we could make an escape since the Turks wanted us to work on the airplanes and get them in shape and try to give them some instruction on it. About a week or so after I was there I flew one of the airplanes. I don't know why I was picked - but I flew Little Eva to _____. At lunchtime they would and everybody would drop what they were doing so they could go to chow without being fined. I think Enlisted men were making about 7cents American a week.

I don't know how long it was that I was there in Eska but I got a terrible case of almost death threatening jaundice and they put me in a hospital which was an old run down place and nobody in the hospital could speak English and I could speak very little Turkish so one day I got hold of a fellow who came in from the air field and told him to get hold of Lt. Brown — he flew the other airplane over there — and tell him that I needed some help. So he came to see me and the second night after that two guards took me on a train at midnight and took me to Istanbul to the American Hospital, which was run, by Dr. Shepherd and his wife who had also had a stint in WWI. They had a thorough hospital ranking. Their daughter and son were over in the States. One was going to Harvard. They had Armenian nurses and people that had been purged out of Russia way back in WWI.

I stayed there for about a month and a half in the hospital, which was real nice. They decided that they were going to try to get me released by the Turks on medical basis. Well I went before a board with Dr. Shepherd and they interrogated me about my condition and at that time I weighed about 118 lbs. I had lost a lot of weight. They turned me down. They said I would be taken back over to Ankara and be with the rest of the fellows. There was a fella named Robert Brown who lives in California now — he was rather young — but he was supposed to meet me at the train when they brought me back from Istanbul. He didn't show up, so I was at the train station for about two or three hours and finally somebody came from the Embassy and picked me up and said that I was going to live with Bob Brown and a couple of other fellows, one named Coffman who was there on lend-lease business and he's now a federal judge someplace.

I stayed with them about two weeks and one day Bob Brown came in and said they were going to try to get two or three people out and did I want to stay or did I want to get out. Of course I said I want to get the hell out of this place. So we made the arrangements. They took me back to the hotel where the rest of the guys were and arrangements were made from there with Bob Brown. There was a fella who was a colonel in civilian clothes who came from the United States over to Ankara and he was going to teach the tricks of military tactics. But he needed somebody that could use a typewriter and one of our fellas who was the navigator/bombardier — Charlie Davis (who died about a year ago) could type. He was a reporter for the Pittsburgh Gazette. So he had the background for taking down information that the Col. needed.

I think at this point I'm going to clarify a couple of things. When I left the States my airplane was called Hellzapoppin and after we got over to Fayid, Egypt we went out about three days before the mission for a run in the desert and when we came back to land my nose wheel collapsed on landing. I had no indication that anything was wrong. Anyway it took the drum up front and just turned it up and made a skid out of it and busted the front of the glass out of the navigator's observation window. So we put the wing jacks under it and started the engines back up and screwed the nose wheel down and taxied it over to the ramp and had to take another airplane on the mission, which was called the Blue Goose. That's when I had to take the Engineer that was on the Blue Goose instead of taking the one I had on the original airplane. That would clear that up.

Another thing when we were interned in Turkey when we first got there we didn't take parole (??) and wouldn't sign anything that we wouldn't try to get out and that's the reason we were under guard all the time. We bought civilian clothes and actually later on we were reimbursed and the information we got was that the enlisted men would be reimbursed for their clothes but that the officers had to furnish their own uniforms. Figure that one out. Another thing is while we lived out on the farm there was a small restaurant and we ate most of our meals outside on bench type tables. The food was terrible. We couldn't get them to boil the potatoes. They had to use grease. They didn't eat anything in the way of meat except lamb. They had a big five gallon can by the stove and they would dip in there and get some lamb grease and cook something and what was left over they would put it right in the can with the grease and after a few days it would get rancid. Best thing we had while we were on the farm — we were close to a brewery owned by the government — we would have the guard go over to the brewery and have sent over some kilos of beer.

Going back to our escape — I can't remember the night we got out — but we had gotten word from Bob Brown that we were to go out the window in the toilet room. This window was high and it folded back inside and made it very difficult to get out of there. But my tail gunner had tied a rope to a radiator and threw it out the window. I had told Ziesel and Anderson and Wicklund to go ahead whenever they were ready and we counted the steps that the guards made in the hall to where we knew when they got down to the end and we could slip over into the toilet room and go out. None of the three of us had any overcoats and it was cold. I guess it was about the 15th of January. There was about a foot and a half of snow on the ground. When I went out the window I figured I was going to be the last one and started running and ran down town to meet Bob Brown. We got in the car with him and incidentally we had to wait because we didn't all take the same route downtown. Anyway, Bob Brown had our tickets to ride on the express, which left at midnight. We were to ride second-class and he had overcoats for us and suitcases.

We drove over to the train station, sat there in the car for about 10 or 15 minutes, but before we got out of the car Bob told us that a person was going to meet us about half way between Ankara and Adana, which is down on the Mediterranean. We bid him goodbye and got in our compartment and we didn't have anything in the suitcases except some brown bread that the Turks make and some cheese. We proceeded and stayed in the car. I don't think I went to the toilet but once during that whole time because when we missed our contact we were to go on to Adana and someone would meet us there. We finally got to Adana. It was from midnight one night till about 10 o'clock or so the next night. So we took off walking and we saw a fella that had a horse and a carriage. It was about 25 miles from the camp that we were supposed to go to. He said he couldn't take the horses that far because it would be against the law and told us to go over to the fire station close by and see if they could tell us where we could get a cab. Wicklund by the way could speak pretty good Turkish and was quite a help to us in getting out. Anyway they told us there was a fella that had a car, and we went there and there was a restaurant.

There was a Ford car sitting out under a tree. It had side curtains on it ,might have been about a 1929 model. Anyway Wick went in and contacted the owner and told him that we wanted to go to this road camp. We told him we had some money and would pay him. Anderson and I were sitting in the car and the fella came out and had his son with him cause he wanted somebody to ride back with him. He had to get some water and put in the radiator because it didn't have any antifreeze so every time they went to town and stop they had to drain the water out. Off we go. We stopped at a hotel. It had a red light out front. Wick went in to use the telephone there. Meantime a policeman had come by and shined his light in there at us and the driver told him we were going to a road camp. Wick came out and said he couldn't use the telephone in there.

They told him the telephone was about a block down the road. So we paid the guy off and walked down the street. Wick went upstairs, the door open and a stairway. Virgil and I kind of stayed in the stairwell. Wick made the phone call to the British Consulate and they said they would send a man down to escort us. He came down and got in behind the steps with us. Pretty soon here came a fella and he looked like a policeman. You know you can think of everything in the world that's going to happen to you with the prison camp right there. We didn't walk out and tell him who we were or anything because we were afraid of him. But he walked on upstairs. Pretty soon he came back down and left.

Well we didn't know what to do then so Wick went back up an called this other number and the fella was getting a little irritated and told him not to call again that he was sending a man back down there. So we waited for him and when this same guy came back, he had the same hat on and we figured he was not a policeman so we stepped out from under the stairway. And he told us to follow him about a half block back. So we followed him to the American Consulate (I mean to the British Consulate).

When we got in there, man I tell you, talk about a relief, they sent some kind of food — I've forgotten what it was — and they had some vodka and some cognac. After a while they took us out the back door, put us in a car and took us out to this road camp run by the British. We were introduced to the fellas out there and that's where we were going to stay until we could get on a ship, which was called the SS Inviken, a Norwegian ship. I'm enclosing a copy of my identification card, which I kept.

The Inviken was out in the Mediterranean and you could see it from land. We decided to use some used gas cans and make sort of a raft to get out over the breakers so that somebody could come in from the ship and pick us up. Then they decided that was going to be too risky so I don't know how they made contact with Captain Olson, who was the Norwegian captain of the ship. And incidentally the ship had a Norwegian engineer and the rest of the crew was Chinese. Even the cabin boy the captain had was Chinese. So they got word to the captain to send somebody in for us. They took us from the camp into town late in the afternoon and we walked out on the dock and there was a customs place in the water, and we walked further on out and here came a fella with a rowboat from the ship. We just walked on through the customs place. We didn't see anybody.

When we got in the boat with the Norwegian engineer and just started to push off and here came the customs man running like crazy down the board walk and said no, no only one person came over and four people couldn't go back — so we had to get off. Well, we walked back through the customs building there and the fella that brought us there — I walked by him and told him we didn't make it — they wouldn't let us go. So he said go back to the car, which was stationed about a block or so away, and we will have to come up with something else. Wick and Addison followed me and we went back to the car and they took us back to the road camp.

About half hour or a little longer after we returned to the road camp a knock came on the door and someone said "you have three American airmen in here" and it scared the daylights out of us. Then he started laughing — said he was going to come ashore and go to the and said that he had three fellas that had jumped ship and come into town and got drunk and he wanted them back. So they poured a little cognac on us and got some on our breath and we acted like we were about half shot. They took us back into town and there was Capt. Olson. We didn't know him and he didn't know us. We were standing under a streetlight. So we walked up there and he started giving us the devil about why we didn't get back on the ship and all that for whoever's benefit that was listening and we walked on in and got in the customs place and stayed in there about ten minutes, I don't know why, and finally got clearance and went on out and got in the row boat and went on out to the ship.

We stayed in the hole. They were loading [the ship] to bring [material] back to the United States. We were about two days out there that we had to stay hidden until we could pull anchor. I think it was the third day that we pulled anchor and Capt. Olsen went topside and we took over his quarters, which was pretty nice, and the food was good, and it was great. When we got underway we were to join up with a convoy in a place called Iskenduren, which is a little bit east of Adana and we did join up with some other ships which were going down to Port Said in Egypt.

Everything was going along fine and the weather wasn't too bad. We were a little short on armament. We had one gun .50 cal. and two .30 cal machine guns and that's all the armament we had. I can't remember how long we were on the water there coming across the Mediterranean but one night we had engine trouble. We lost the convoy because we were only doing about probably 3 knots at best and we knew we couldn't hold them back so we were just out there by ourselves. We were going to try to get into Haifa harbor two days later to see about getting repairs maybe.

We didn't get down there in time. After about 4 o'clock in the afternoon they closed the sub nets and you have to have a captain come from Haifa harbor to get you through. We had to stay out there all night long. Of course we had black out. The three of us were watching each night to see who was going to sleep in the captain's bed. On this particular night I was the odd one.

Along about 4 o'clock in the morning I heard a bullhorn. They wanted the colors of the day and the code of the day. The first mate was on duty. He was Chinese and I don't know whether he could understand too well what they were talking about. I looked out the window and told Ziesel and Anson don't say a word, there's a submarine out there and they got a dead aim on us. I went topside and finally found the colors of the day and the code of the day. Because the captain, as soon as we took anchor up at Adana he started drinking and he was drunk and couldn't be of any help. But I finally found the stuff and had a megaphone and hollered back over what the color of the day and the code of the day was. Then they identified themselves that they were English battleships and that they were going to stay alongside of us until morning so the captain could come out and take us into the harbor. That's how we got down to Haifa.

We decided that while we were in there we would walk up to Mt. Carmel. There was a restaurant up there and we just went up there to see what was going on. We started walking up there and then we heard some whistles — I'm sure they weren't whistling at us — but we took off and went back down to the ship and we didn't leave that thing again til we got down to Heliopolis which is outside Cairo. We landed there and waited awhile and some MP's came aboard and I assure that they paid the captain for our passage. They took charge of us and took us into MF headquarters and stayed there awhile. They made a phone call or two and then took us on into Cairo.

We were running around in civilian clothes and ran into some of the other fellas that were on R&R. I saw Col. Kendricks. He was a medic (head of our medics) when we went overseas. He said, "Cave you don't look too good," and I said, "Well I don't feel too good." So I ran around there for about 4 or 5 days in civilian clothes and the Col. got hold of me. We didn't have any mess halls to go to and hell, I was broke. So I had to have some money. So I went to the finance office and told them I wanted some back pay. So they said OK. They said we got a thing on you here that you owe $117 for money that you were short when you were in Turkey. I said "my god, well go ahead and take it out of what I've got coming and that will take care of that". And so they did.

When I ran into Col. Kendricks again I told him "Damned if I can figure this out they charged me $117 for what I was supposed to be short in that money they took off me in Turkey." He said, "Well I don't think they can do that. Go back in and get your money back." So I went back to finance and said I want my money back. He said well all right and he gave it to me. And that situation bothered me every place I went after that and finally when I got out at Scott Field Illinois after the war was over they made me pay that $117 before I could be discharged.

Hotel there in Cairo and that's when Col. Kendricks decided I better get back to the states. I have enclosed another piece of paper here, passage and transportation record that they gave me in Cairo, April 15, 1943 to be flown from Air Transport Command from Cairo to Miami.

I don't know whether I'm repeating myself or not but on the 4th of July [1942] they sent a bus for us to take us to the American Embassy. To take some pictures for Life magazine. They are in the issue of Sept. 14, 1942, if you would be interested. They are on page 97. We had a party at the embassy and we had a drink or two and had a little something to eat and after they got their pictures they put us back on the bus and took us back out to the farm. That one picture of me skipping rope there was in this magazine and the fella on the left is Nathan Brown who was a pilot and the other fella is Conrad Pierce who was my armament man. We took as much exercise as we could and didn't have anything else to do and it was pretty boring.

There is one picture there that I might have mentioned of me sitting at the table with Bob Brown, who would come to the farm about every other week and give us a little money so that when the guards would take six of us at a time into town we could go to a place and eat and maybe go to a show. When they would take us to a movie, after the movie got going, there were only two guards and six of us so we could, when they got interested in it we would just jump up and take off and they could only follow two people and we'd split up and we would go to the restaurant and go down in the basement. Old man knew who we were and he would send us down some caviar and all sorts of goodies.

Sometimes when the guard would take six of us in town and I happened to be one of them, we would go to the Ankara Palace Hotel and the bartender there knew us and we would go down in the basement and he would send us down some gin fizzes. I happened to be down there one day when Wendell Wilkie came over to Turkey and they were having a big reception upstairs so the bartender told us that he thought we'd like to come up and see the table they set for him. So we went upstairs and man they had food spread out everywhere. So we took a little bit of everything and went back downstairs and enjoyed it. But we never did get to see Mr. Wilkie. It's been so long ago I don't remember whether Wilkie was Vice Pres. at the time or whether he was running for president or just what but he was quite a figure back then and you're probably not old enough to remember him.

I left the states with #23 Hellzapoppin then after a nose wheel collapse in Fayid, Egypt just 3 days before Ploesti I had to take the Blue Goose #19 on the mission and landed it in Turkey. Then when the Turks wanted to take 2 of the good planes to Eskisehir I flew Little Eva #6 with Wilbur West as co-pilot. I might have told you that I gave Ankara one of the best low level fly bys it ever had. I knew the Turkish officers aboard couldn't fly the thing so there was nothing they could do about it (really I think they enjoyed it). The Blue Goose had 2 engines out when I landed in Turkey so I don't know what happened to it. I was in the hospital in Istanbul when Wilbur "HIJACKED" " Brooklyn Rambler" but guess you know that story.[6]

THE *"HIJACKING" OF* BROOKLYN RAMBLER

As previously noted by Ed Cave, several Turkish internees managed to fly *Brooklyn Rambler* from its base in Turkey to the RAF airfield on Cyprus. This was Wilbur West's recollection of what happened:

In November, the nine of us who flew out of Turkey moved to Eski Sehir, a military base southwest of Ankara. Being a graduate engineer, I was engineering officer in charge of getting the four B-24s ready to fly. They had been given to Turkey through Lend-Lease agreement, (we flew the B-24s from Ankara to Eski Sehir accompanied by Turkish pilots.) and were to teach the Turks to fly them. Close examination of the airplanes found that #2 (Brown's airplane) in best condition to get ready first. Sgt's Taylor and Briscoe and myself were allowed liberal access to the B-24s, Turkish tools and mechanics. The three of us worked on the airplanes daily, making minor repairs and transferring gasoline from the other ships to #2. I had a silk escape map of the area (Turkey and Cyprus) and talked over the possibility of escape with the others. Most of them thought I was kidding, but Briscoe, Taylor and myself continued to prepare the airplane for the 400 mile trip. When we finally got about 400 gallons of fuel in the ship, we thought we were ready to go. The weather was beautiful, but only one mag was working on #1 engine. We had no Mae Wests, no parachutes, no guns, no radio, no alcohol in the compass, remote compass out, so we decided to repair the mag on #1 before leaving.

Because of the poor food available, we always skipped lunch and, as was our custom, worked through the lunch hour. We had run the engines several times on the ship without attracting any unusual attention so with repair of the left mag on #1, which I did myself, we decided we were ready to go. I had talked this over with an RAF pilot who had been loaned to the Turks and was stationed at Eski Sehir, and he advised me of the proper approach corridor to the island of Cyprus. I then told the others of our plan for the following day, December 15, 1942. Everyone took all his personal belongings he could put in his pockets to the filed the following morning. At noon Taylor and I began to start the engines, and one by one the others came out to the ship and boarded it. Just as we were about to close the bomb bay door, a Turkish mechanic stuck his head in the bomb bay. Quick thinking Sgt. Taylor sent him after a wrench, closed the doors, and we gave the engine full throttle from the ramp at the end of the hangar. I had stepped the distance across the sod field to the runway and figured we would clear the ground just as we crossed the runway. Being December, the ground was frozen hard and it was no problem getting airborne as calculated. The one guard, with rifle in hand, standing in rear of #4 engine, disappeared in a cloud of dust from the ramp. Lt. Brown and I flew the airplane. We stayed just above ground level until a few minutes out of the area and began to climb to get over the 11,000 ft. mountains between us and the Mediterranean. Lt. R. P. Humphreys of Marfa, Texas used the silk escape map I had been able to hang on to and a thumbnail-sized compass I had also kept to do a perfect job of navigating us to Cyprus.

When we leveled out at 11,000 ft., reduced power and leaned the engines as far as we could without over heating, our gas gauges all read "0". When we had been flying about 1 1/2 hours, and before we could see land, #4 engine fuel pressure dropped and it stopped. Just as we feathered it, Spitfires from Cyprus appeared on the scene and made some close passes at us as if they were about to close in for the kill. Boys in the waist opened the waist windows to motion we were friends. (We had Turkish Insignia on the airplanes) With no radio contact we decided to put the airplane in the water before we would allow them to shoot us down. As we started our descent, Humphreys sighted land so we lowered our landing gear indicating our intentions to land. On three low powered engines our airspeed with gear down was only about 140. We knew we would have to ignore the proper corridor to enter Cyprus and we crossed the North coast just west of Nicosia. In a very few minutes we sighted the airfield there. Ignoring traffic, we landed straight ahead on the runway nearest our heading. The nose wheel shimmy dampner was dry, and on landing, the wheel turned and blew out the nose wheel tire. We abandoned the ship in the middle of the runway there. One of the fellows took a picture of it while a British lorry was coming out to pick us up.[7]

JUNE 15, 1942 MISSION AGAINST THE ITALIAN FLEET

WALTER SHEA

At the time the RAF came to Col. Halverson for aid and succor in deterring the Italian Fleet from intercepting convoys (from both Alexandria and Gibraltar to

Malta, which was almost destitute of aviation fuel, etc.), we were back from the Ploesti mess and reloading bomb bays for the onward trip to China. Whether we were fully loaded or not, I can't recall. If we were, a few of us were unloaded to accept bombs.

The briefing was, again, an RAF affair, post punctuated by remarks from Col. Halverson. You must realize we didn't know from shinola in those days. It was an RAF all the way. We were just there, trying to learn.

A mid-day briefing. The figures that come to mind were;
- 2 battleships
- 5 cruisers
- Assorted destroyers in perimeter
- 300 fighters for cover of the fleet

Don't recall much other than Col. Hal saying, 'Send five of the boys.' The sixth to be an RAF B-24 lacking our high altitude carburetor fittings (turbo superchargers), but having only a high speed blower good for 14,000 feet or so.

Wilkie (me) was to fly left wing after (Navigator) Bryant? (Olen C. Bryant – N/B John H. Payne crew) in the lead – with the RAF kite on the right wing. We were the second Vic of three, following Kalberer's lead Vic. We were to formulate over a spit of land west of Alexandria labeled as Ras El Kenayas (Kenias?) – at dawn – following a night takeoff.

The Pilot – Co-Pilot – Navigator/Bombardier of the two Vics were as follows:

HALPRO #1

Kalberer – Rhodes – Rang

HALPRO #9	*HALPRO #18*	*HALPRO #3*
Sibert – R. Miller – Ebert	Payne – Patterson – Bryant	P. Davis – Crouchley – Joyner

HALPRO #21	RAF B-24
Wilkinson – Wilcox – Shea	

I insist that I recall waking in my mosquito-netted bunk and recalling that we should be up and about by then. But all was silence. Maybe others were awake, but not in my hut. I called Wilkie (after examining my soul to see if silence wouldn't prove to be golden, for five bombers against a fleet didn't make much sense to me. We did scramble to get ready.

Guess someone else must have been on the stick, for we did get into the trucks and down to the flight line. The bomb loading was six 500 pound semi-armor piercing bombs from the RAF, which had been stored a long time in the desert.

Takeoff was normal, as was the flight to Ras El Kenayas, but to a young navigator, simply a flight into sheer darkness, only mitigated by a compass that if properly compensated for winds, deviation, and elevation, would bring us there.

At stroke of dawn or so, we were there. And there were other aircraft circling around. Somehow, we 'formated.'

I, at least, and I am quite certain all the other Nav-Bomb's had British Naval Observers accompanying us in the nose. Mine had a big, black beard. A nice guy. His job: to point out the battleships, that is, the key targets.

On we flew, WNW, with Bernie Rang (and Kalberer) in the lead ship of the lead Vic. We other navigators had the task of 'following the pilot'; that is, watching the compass to see where they were leading us, and tracking ourselves in a negative sense of the word. I do recall wishing I could run a fast 45 to 45 degree off-shoot to get a drift reading for wind, but simply calculated a 'no wind' effect. The weather was superb.

We got to 14,000 feet as I recall (that number sticks in my mind) and before too long, spotted funnel smoke on the horizon. WOW! Bomb bay doors open! Clutch-in the bomb sight. Too far as yet. Unclutch. Set the bomb train! 20? 50? foot interval.

Here they come – a 'group' of ships. Long range look thru the bombsight — long before 'clutch-in' feasible — WOW — a cruiser, or something. NOW clutch-in, freeze the over-under horizontal hair and double clutch to hold it there along the mid of the cruiser. Look up and out — FLAK — they were shooting at us! Kalberer weaving his Vic to avoid the flak. We, the 2nd Vic, trying to do something appropriate. What the hell!

Clutch-in now, the horizontal cross-hair would follow the cruiser. Double-clutch to set the rate of continual depression HOLDING! Right on. Left to right in the deviation — that was the problem of the lead navigator in our Vic. If he was a little left or right, that is what we wingmen were for! To compensate. We dropped on the over-short only! Get it?

Then Kalberer started to fire off Verey signals. The 'colors of the day.' Look at us, we're friendly. 'Bull,' said the Royal Navy. And the flak came up heavier than ever.

Kal peeled his Vic off to the left and we in the second Vic followed.

It had been a close call, for I had my bombsight hairs on that damn cruiser, or whatever, for quite a while. As I looked up for seconds from time to time, watching the indices creep closer together (there were two zeros as they crossed, that meant bombs away), and my RAF observer told me, 'Hey, that's the British convoy!' I barely had time to 'unclutch' the whole damn sequence! Maybe a half-inch? Five seconds? Hell, the flak was there, etc. They had to be baddies.

Slightly 'miffed' we flew on. Recall questioning my plot of Bernie's navigation, thinking we were too damn far south, until there they were!

First view was unbelievable.

Two large vessels, one astern of the other, straight (or nearly so) at us, with curving wakes all around them from maneuvering cruisers and destroyers. Obviously, busy as hell with something! (As we later learned, being attacked by, I believe the number is correct, 15 Beauforts, twin engined torpedo bombers out of Malta).

I put the horizontal cross-hairs on the foremost battleship (didn't need the Naval observer to point that out) and held it at the long range setting until close enough to clutch-in. Kal had, by then, peeled his Vic off to the left in an obvious attack on the rear battleship. We just bored in at 14,000 feet.

There was flak, but not at us. That I noticed. It was dream run-in. Straight and level forever! Finally, we were close enough for me to clutch-in the bombsight, and I had the problem of deciding where to start my train of bombs. No one had briefed us.

Start at the front and run it all the way down the deck. Nope, suppose you're 'long' or short. Play it safe, aim for the middle. But wait. Ammo lockers are usually in the rear. How about a salvo at the rear? Forget it – play it safe. Aim for the middle.

Which I did. And as my hairs held and the indices crossed, down they went. The lead Nav-Bomb, Bryant, had done a superb job. His bombs fell tight-in alongside the right side of the battleship. The RAF bomber, on his right, who had bombed beautifully on an over-short basis, obviously fell to the far right. (My copy is so poor at this point that I can't make it all out, but it is pretty obvious that Shea's bombs went down the middle of the battleship and three of his bombs went off the end and didn't explode. His bomb train started where he aimed — at the middle of the ship.)

The battleship stopped right there. The second battleship, the target of Kal's Vic, had been hit by a salvo from Bernie Rang. It too stopped dead in the water. At this point, it all became quite confusing, for Wilkie peeled off to the right and dove for the water surface – as best you can dive a B-24!

I merely recall coming 'out of it' someplace at a rather low altitude, with our ship following five others in a rather messy formation, more horizontal than anything, each trying to find a hole to fill in as we raced back east.

Somehow we formed some sort of formation and headed home. I was elated! When telling Wilkie of my absolutely phenomenal skill in bombing, he merely asked how I could be sure that I had hit the ship! Killjoy. The formation bombing pattern proved it was mine that went down the ship. (All thanks to Bryant!)

I stood behind Wilkie and Wilcox on the flight deck as we (heroically) flew home and sort of half-assed realized we were cruising south of the island of Crete, which was occupied by the Germans. I think I mumbled something about the need to look out for fighters from Crete when I suddenly noticed black puffs off (I think it was the left) wing. Wilkie dispatched me to my (ridiculous) nose gun and I scrambled past the nose wheel into the greenhouse. My Navy friend was huddled behind the nose wheel, and I soon joined him.

I recall looking back thru my astrodome to see a fighter (F-109 or Italian Macchi) on our tail, actually the tail of the whole horizontal flying formation. The damn fool flew in close to see what type of albatross we were and got hit by a mess of tail gunners! He went down.

With that victory (to boot) behind us, we flew gloriously on toward the Delta.

Gloriously, there were our friends! The English convoy from Alexandria who had shot at us earlier. Boy, would they be proud of us. So we headed for them in some sort of victory whatever and got our ass shot off! We were low level, firing the colors of the day, etc., but no soap. Here came the flak. Right down our throat. (The Royal Navy takes no chances.)

Someone must have tempered judgment or whatever, for we swung aside unscathed and made it into Fayid. Can't recall much of the debriefing, except Kal's famous remark about 'shooting fish in a barrel.' Some questions about who hit what, etc. But no one cared. Just glad to be back. A British sub in the vicinity at the time got into the act on reporting who hit what, and I believe claimed the sinking of a cruiser. As I recall, all 15 Beauforts were shot down! They had most successfully, if not unintentionally, diverted the German/Italian fighters from our formation. I truly believe we caught them completely by surprise, not ever knowing we were in the . . . (copy too poor to read) . . . with a prisoner, Italian general, confirmed this. No one, he said, could hit anything from 14,000 feet! [8]

JAMES SIBERT

Up at 02:00 and then over for breakfast. Was briefed the night before and learned the Italian fleet was coming out of Taranto and we were to keep it from intercepting a British convoy headed for Malta. Our ships and British Liberators took off and rendezvoused over the coast headed out to sea. Passed over a British convoy at 14,000' that nearly brought one of our ships down until we identified ourselves. Intercepted the Italian fleet a couple of hours later. Cruisers, destroyers and battleships all looked beautiful with the wake left behind. We made a run for the battleships and dropped in pattern bombing. Encountered ack-ack but was not too effective. Hit two battleships and one destroyer. Left them smoking and altering course. Headed for home losing altitude and indicating 200 mph. Let down to just off the water. Had flown 30 minutes when we encountered six ME109's, one of which we shot down and another damaged. We suffered no damage and kept on the coal at 35"-2300. Returned to Fayid, landed, was interrogated by S-2, ate and went to bed. Got up, ate supper, and went to bed again.[9]

RICHARD MILLER

Went on another raid yesterday; went after the Italian fleet that had shipped out of Taranto harbor to intercept a large convoy that was headed for Malta. If this convoy hadn't gotten through, the British would have had to quit Malta, for every other convoy had been knocked out. In desperation, the British sent a convoy from the west and one from the east; I understand the one from the west was having a bloody hard time and was being butchered by someone, but the convoy from the east got through after our raid turned the Italian fleet back.

...

Through the remarkably accurate British intelligence, we knew that the Italian fleet was to leave to stop the British convoy. We knew the time they left, their direction, and their speed. On the way, we received a radio message in code, of course, telling us to alter our course 30 miles; this we did, and we found them without a bobble. We worked on our planes all day, got briefed, taxied into position til 2:00 a.m. and then took off at 3:00 a.m. Flew to the coast and rendezvoused; went in three flights of three, 7 of our ships and 2 Liberators (not much armour). We flew till 8:00 a.m. and sighted a fleet of ships; though we had British naval observers on each plane to facilitate identification of our target, we couldn't determine whose ships we were flying over, so we had to nearly make a run on them. They opened up with very accurate ack-ack fire and even put a splinter through the leading edge of our lead ship. Finally, we decided we were over the convoy and not the Italian fleet so we went on. We have radio equipment known as I.F.F. (Identification Friend or Foe) which sends out a wavelength only known to the British and on approaching a British base or ship, we turn it on so as to call off the fighters. But in our present case, the rule of the sea says no naval vessel will be flown over unless you intend to bomb it. Consequently any ship will fire at aircraft flying directly at it.

We flew till 9:00 a.m. and all of a sudden saw the Italian fleet. Excitement was way up and we were all quite eager to have a really successful mission. We completely surprised them. They were slow in getting their ack-ack into action, then they were very poor shots. They were 4,000 ft. low. By the way, we were flying at 14,000 feet. We had hoped to hit the target in a coordinated attack with two groups of Beaufort fighters (torpedo planes), one from Malta and one from the western desert. However, I saw only one low-flying plane, but most of the men said one group had just finished their attack. We veered off and made one run dropping our load of six 500 lb bombs in salvo.

Our plane got a direct hit on one of the two big battleships. This set it afire and from then on we were really in high spirits. The other ships in our formation managed a few hits because we fired both capital ships and caused them to turn and run for Italy. On heading home, we were flying low and fast, nearly 250 with a small tail wind. I was flying then and moved into a tight formation, much to the pilot's displeasure, but it was the only practice at defensive work (tight formations) I'd had. We had quite an argument over it, but I managed to stay in. About 10:30 a few M.E. 109's got on our tails and the formation moved in beautifully without a moment's loss. We went right down over the Mediterranean and scooted for home. By being low, we had naught to worry about but our rather vulnerable belly. The M.E.s didn't know how to take all this and tried varied attacks which resulted in only a few bullet holes in our ship and the loss of one of theirs. One of the tail gunners got credit for it. We proceeded home and had quite a time buzzing lone ships. They'd futilely fire at us, but a short burst of our guns over their nose usually silenced them. We came into the field and from the cocky appearance of the one-pass landings, everyone knew we were successful. We landed about 2:00 p.m. and got to bed about 9 p.m., nearly 40 hours at steady work. Today, we learned for sure that we fired both battleships and got one pursuit. A sub witnessed it all and gave us a favorable report. And so we saved Malta. I understand Malta has been bombed some 2,000 times, at least 4 times every day. It is hard to see what good it is, but it is vital. Just learned an interesting report: the Italian radio just announced that their fleet intercepted and completely destroyed a British convoy, and one lone Italian pursuit plane knocked down 49 Allied planes. Some story!

Seems hard to believe that anyone could possibly put out such puerile propaganda and hope any of it could be swallowed.[10]

JUNE 21-22, 1942 NIGHT MISSION TO BENGHAZI

THERMAN BROWN, PILOT

I remember my first mission. This was on the night of June 21-22, 1942. The target was the shipping in the harbor at Benghazi. We were to bomb in the middle of the night and return to our base at Fayid — a ten hour mission.

Spotting the target at night would have been difficult if only the enemy had doused the search lights and held their fire. In some cases, we would never have found the target at all. You could depend on them turning on the lights and start shooting at something.

There was someone ahead of us. Turning onto the target was no problem. All hell was breaking loose up ahead. I don't remember being afraid but I can remember my feet shaking a little on the rudders as we made the turn onto the inferno above the target. Nothing happened to us. The anti-aircraft fire continued to follow the poor souls ahead of us. We essentially got a free pass. Getting caught in the search lights isn't fun but the lights don't hurt you. The only help the anti-aircraft gunners improve their accuracy.[11]

Figure 64 – *Brooklyn Rambler* and the Hi-jacking Crew

ENDNOTES

1. Brown, *War Stories*.
2. Sibert, *personal diary*.
3. Miller, personal diary, pp. 22–29.
4. Story, *Private Story*.
5. Walker, letter to Brooks,
6. Cave, *transcript of interview*,
7. West, Wilbur, letter to Robert Storz, pp 2-3.
8. Brown, *War Stories*.
9. Sibert, *personal diary*.
10. Miller, personal diary, pp. 29-34.
11. Brown, *War Stories*.

APPENDIX C — HALPRO CREWS & PLANES

"A" FLIGHT

AIRCRAFT NO. 1 — *Ole Faithful*
SERIAL NO. 41-11595

Pilot	Capt. Alfred F. Kalberer
Co-Pilot	2/Lt. Richard L. Rhoades
NavBomb	1/Lt. Francis B. Rang
Engineer	S/Sgt. Lacey A. Whitley
Radio Op.	Sgt. Anderson T. Patrick
Armorer	Corp. Robert J. Coutre
Gunner	Cpl. James R. Peterson
Pass.	Col. Harry A. Halverson
	1/Lt. Arthur L. Cox
	T/Sgt. Carl W. Edwards

AIRCRAFT NO. 2 — *Brooklyn Rambler*
SERIAL NO. 41-11596

Pilot	1/Lt. Nathan K. Brown, Jr.
Co-Pilot	2/Lt. Walter C. Swarner
NavBomb	2/Lt. Robert P. Humphreys
Engineer	S/Sgt. Ula H. Taylor
Radio Op.	Sgt. George A. Charles
Armorer	Cpl. David D. Coward
Gunner	Cpl. Charles E. Collum
Pass.	S/Sgt. Karl A. Seidel
	Capt. Ulysses S. Nero
	Major Jack R. Naylor

AIRCRAFT NO. 3 — *Yank*
SERIAL NO. 41-11625

Pilot	Capt. Paul Davis
Co-Pilot	2/Lt. Edward A. Crouchley
NavBomb	2/Lt. William R. Joyner
Engineer	Sgt. Joe C. Saia
Radio Op.	Cpl. Raul R. Venegas
Armorer	S/Sgt. Ralph Alexander
Gunner	Sgt. Edward F. Weingart
Pass.	Lt. Col. Floyd N. Shumaker,
	Gen. Ping Han Whang,
	S/Sgt. Donald Ward

AIRCRAFT NO. 4 — *Ole Rock*
SERIAL NO. 41-11618

Pilot	2/Lt. George A. Uhrich
Co-Pilot	2/Lt. Ferdinand R. Schmidt
NavBomb	2/Lt. Allen V. Hopkins
Engineer	S/Sgt. Louie L. Walters
Radio Op.	S/Sgt. Douglas H. Williams
Armorer	Sgt. Harold C. Vanness
Gunner	S/Sgt. David A. Tunno
Pass.	Cpl. John M. Thompson
	Capt. Edgar W. Gardner
	Sgt. Leo F. Fanning

AIRCRAFT NO. 5 — *Town Hall*
SERIAL NO. 41-11622

Pilot	1/Lt. Frederick W. Nesbitt, Jr.
Co-Pilot	2/Lt. Virgil D. Anderson
NavBomb	1 Lt. Lyman S. Smith
Engineer	S/Sgt. Jimmie Briscoe
Radio Op.	Sgt. Harold J. Jackson
Armorer	S/Sgt. Enoch G. Kusilavge
Gunner	Cpl. Frank B. Pearson
Pass.	Lt. Col. George F. McGuire
	Lt. Col. Edward J. Kendricks
	M/Sgt. Gordon H. Hadlow

AIRCRAFT NO. 6 — *Little Ev*
SERIAL NO. 41-11609

Pilot	1/Lt. Howard E. Walker
Co-Pilot	2/Lt. Wilbur C. West
NavBomb	2/Lt. Charles T. Davis, Jr.
Engineer	S/Sgt. Garland A. Lippencott
Radio Op.	Cpl. Robert L. Albertson
Armorer	Cpl. John E. O'Connor
Gunner	Cpl. Thomas G. Owens
Pass.	M/Sgt. George J. DeCisneros
	S/Sgt. Wilbur C. McNeal

AIRCRAFT NO. 7 — *Eager Beaver*
SERIAL NO. 41-11600

Pilot	1/Lt. Charles O. Brown
Co-Pilot	2/Lt. John R. Taylor
NavBomb	2/Lt. Malcolm R. Anderson
Engineer	S/Sgt. Morris A. Cannon
Radio Op.	S/Sgt. Irving Cutler
Armorer	Cpl. Harold L. Osgood
Gunner	Cpl. Robert E. Thompson
Pass.	1/Lt. Wilfred J. Smith
	S/Sgt. Charles L. Hazlett

"B" FLIGHT

AIRCRAFT NO. 1 — *Queen B*
SERIAL NO. 41-11591

Pilot	1/Lt. James W. Sibert
Co-Pilot	2/Lt. Richard G. Miller
NavBomb	2/Lt. Harry W. Ebert, Jr.
Engineer	S/Sgt. Claude F. Anglin
Radio Op.	S/Sgt. Noel W. Meek
Armorer	S/Sgt. James R. Milliren
Gunner	Cpl. Anthony Filippi
Pass.	Lt. Col. Carl L. Feldmann
	M/Sgt. Jessy C. McConnell
	S/Sgt. Francis J. Laney

AIRCRAFT NO. 2 — *Florine Ju Ju*
SERIAL NO. 41-11613

Pilot	1/Lt. Kenneth W. Butler
Co-Pilot	2/Lt. Herbert A. Kysar
NavBomb	2/Lt. Walker H. Hiatt
Engineer	S/Sgt. Dewey J. Williams
Radio Op.	S/Sgt. Matthew Cwikiel
Armorer	S/Sgt. Thomas A. Blair, Jr.
Gunner	Cpl. Peter Bedrosian
Pass.	S/Sgt. Nathan C. Drown

AIRCRAFT NO. 3 — *Ball of Fire*
SERIAL NO. 41-11624

Pilot	1/Lt. Walter Clark
Co-Pilot	2/Lt. Jackson B. Clayton
NavBomb	2/Lt. Robert W. Helms
Engineer	S/Sgt. Grover L. Knox
Radio Op.	S/Sgt. John A. Cook
Armorer	Sgt. William S. Buckwalter
Gunner	Cpl. John Nappi
Pass.	T/Sgt. Harry W. Dewald
	S/Sgt. Richard J. McKee

AIRCRAFT NO. 4 — *Old King Solomon*
SERIAL NO. 41-11617

Pilot	1/Lt. James S. Solomon
Co-Pilot	2/Lt. John C. Medford
NavBomb	2/Lt. Douglas S. Welfare
Engineer	S/Sgt. Thomas A. Smith
Radio Op.	S/Sgt. Jack B. Lavender
Armorer	Sgt. Hugh S. Powell
Gunner	Cpl. Perley W. Spaulding
Pass.	S/Sgt. Charles H. Watts
	M/Sgt. James C. Woodyard

AIRCRAFT NO. 5 — *Arkansas Traveler*
SERIAL NO. 41-11616

Pilot	1/Lt. Homer E. Adams
Co-Pilot	2/Lt. Lin Parker, Jr.
NavBomb	1/Lt. Robert B. Kirkaldy
Engineer	Sgt. Harry F. Orris
Radio Op.	Sgt. Dillon W. Waters
Armorer	Cpl. Jesse R. LaRue
Gunner	Cpl. Roy O. Woody
Pass.	Major Jessie C. Williams
	T/Sgt. Martin E. Metzler

AIRCRAFT NO. 6 — *Ripper the First*
SERIAL NO. 41-11614

Pilot	1/Lt. Robert I. Paullin
Co-Pilot	2/Lt. Charles O. Peek
NavBomb	2/Lt. Thomas A. Shumaker
Engineer	S/Sgt. Edwin R. Sparks
Radio Op.	S/Sgt. Robert T. Wysong
Armorer	S/Sgt. Talbert A. DeHaven
Gunner	Sgt. Baxter C. Luton
Pass.	T/Sgt. Vitus Hrubes
	S/Sgt. James W. Kiser

AIRCRAFT NO. 7 — *Edna Elizabeth*
SERIAL NO. 41-11620

Pilot	1/Lt. Sam R. Oglesby
Co-Pilot	2/Lt. John W. Kidd
NavBomb	2/Lt. Ernest M. Duckworth
Engineer	Sgt. Henry R. Ballantine
Radio Op.	Sgt. John R. Walker
Armorer	Sgt. Eldon B. Pickert
Gunner	Cpl. John E. Kaminska
Pass.	T/Sgt. Paul W. Fitzsimmons
	Sgt. William E. Fields

AIRCRAFT NO. 8 — *Draggin Lady*
SERIAL NO. 41-11592

Pilot	1/Lt. Therman D. Brown
Co-Pilot	2/Lt. William P. Dwyer
NavBomb	2/Lt. Norman Davis
Engineer	S/Sgt. Kenneth R. DeLong
Radio Op.	S/Sgt. Taylor E. Van Gilder
Armorer	Sgt. Robert F. Rendell
Gunner	Cpl. Alphonse Izzo
	Sgt. Howard C. Muse

"C" FLIGHT

AIRCRAFT NO. 1 — *Black Mariah II*
SERIAL NO. 41-11593

Pilot	Capt. John H. Payne
Co-Pilot	2/Lt. Cecil E. Patterson, Jr.
NavBomb	2/Lt. Olen C. Bryant
Engineer	T/Sgt. Gus D. Portl
Radio Op.	Sgt. Robert Kessler
Armorer	S/Sgt. James H. Leaman
Gunner	Cpl. John J. Beatty, Jr.
Pass.	Capt. George S. Richardson
	T/Sgt. George D. McNelly

AIRCRAFT NO. 2 — *Blue Goose*
SERIAL NO. 41-11597

Pilot	1/Lt. Andrew M. Moore
Co-Pilot	2/Lt. George B. Whitlock
NavBomb	2/Lt. William O. Mally
Engineer	S/Sgt. Frank Perrone
Radio Op.	S/Sgt. James L. Barineau
Armorer	S/Sgt. Frederic S. Moran
Gunner	Cpl. Robert R. Kramer
Pass.	M/Sgt. Harry L. Hines
	S/Sgt. Clair E. Harshbarger

AIRCRAFT NO. 3 — *Mona the Lame Duck*
SERIAL NO. 41-11615

Pilot	1/Lt. Francis E. Nestor
Co-Pilot	2/Lt. Charles A. Shaw
NavBomb	2/Lt. Marshall L. Phillips
Engineer	Sgt. Charles W. Hunter
Radio Op.	S/Sgt. Arnold M. Umstead
Armorer	Sgt. Stanley B. Rosanski
Gunner	Cpl. Harold W. Kramer
Pass.	T/Sgt. James C. Owen.
	Sgt. William G. Richards

AIRCRAFT NO. 4 — *Babe the Big Blue Ox*
SERIAL NO. 41-11602

Pilot	1/Lt. John W. Wilkinson
Co-Pilot	2/Lt. John R. Wilcox
NavBomb	2/Lt. Walter L. Shea
Engineer	S/Sgt. Albert S. Fisher
Radio Op.	S/Sgt. Roy R. Taylor
Armorer	Sgt. Charles E. Salmon, Jr.
Gunner	Cpl. Joseph Troyanowski
Pass.	S/Sgt. William J. Donovan
	S/Sgt. Alfred C. Colt

AIRCRAFT NO. 5 — *Malicious*
SERIAL NO. 41-11603

Pilot	Capt. Richard C. Sanders
Co-Pilot	2/Lt. Louis A. Prchal
NavBomb	1/Lt. Francis H. Smith
Engineer	S/Sgt. Joseph S. Domino
Radio Op	S/Sgt. Joseph J. Bolen
Armorer	S/Sgt. Harold R. Vasquez
Gunner	Sgt. Maryon R. Lunsford
Pass.	1/Lt. Ralph S. Royce,
	S/Sgt. William C. Plyler

AIRCRAFT NO. 6 — *Hellzapoppin*
SERIAL NO. 41-11601

Pilot	1/Lt. Edward A. Cave, Jr.
Co-Pilot	2/Lt. Eugene L. Ziesel
NavBomb	2/Lt. Harold A. Wicklund
Engineer	S/Sgt. Glen H. Pearce
Radio Op	S/Sgt. Carl J. Dupree
Armorer	Sgt. Conrad R. Pearce
Gunner	Cpl. Albert H. Story
Pass.	M/Sgt. Sidney J. Willis
	S/Sgt. Stephen T. Pundzak

AIRCRAFT NO. 7 — *Wash's Tub*
SERIAL NO. 41-11636

Pilot	1/Lt. Martin R. Walsh, Jr.
Co-Pilot	2/Lt. Meech Tahsequah
NavBomb	2/Lt. Alfred L. Schwanabeck
Engineer	S/Sgt. Coy B. Payne
Radio Op	S/Sgt. Elmer E. Withan
Armorer	Cpl. Charles C. Ruppert
Gunner	Cpl. Frank: W. Mahboub
Pass.	S/Sgt. Robert H. McComb
	Cpl. Richard C. Hebert

AIRCRAFT NO. 8 — *Jap Trap*
SERIAL NO. 41-11629

Pilot	2/Lt. Mark T. Mooty
Co-Pilot	2/Lt. James L. Yelvington
NavBomb	2/Lt. Theodore E. Bennett
Engineer	Sgt. Blane L. Eagon
Radio Op	Cpl. John W. Kinnane, Jr.
Armorer	Cpl. Albert F. Osterhaus
Gunner	Cpl. Earl C. Parr
Pass.	S/Sgt. John M. Winecoff
	T/Sgt. Luke L. Britton

APPENDIX D — ORAL HISTORIES

At the 2007 376 Veterans Association reunion, oral history interviews were conducted. Also, some interviews were done later by telephone. These oral history recordings are housed in the Digital Media Repository at Ball State University in Muncie, Indiana. The on-line link to access the collection is libx.bsu.edu/cdm/search/collection/376OrHis

The oral history collection includes interviews of men who were part of the 1942 operations of HALPRO, the 1st Provisional Group, and the 376th Heavy Bombardment Group are contained in this collection. They are:

- Ebert, Harry W. (http://libx.bsu.edu/cdm/singleitem/collection/376OrHis/id/62/rec/14)
- Izzo, Alphonse (http://libx.bsu.edu/cdm/singleitem/collection/376OrHis/id/10/rec/22)
- Sibert, James (http://libx.bsu.edu/cdm/singleitem/collection/376OrHis/id/64/rec/44)
- Story, Albert H. (http://libx.bsu.edu/cdm/singleitem/collection/376OrHis/id/31/rec/48)

REFERENCES

1. ----, *11th Bomb Group (H), The Grey Geese*, Paducah, Kentucky, Turner Publishing Company, 1996
2. ----, Air Force Historical Research Agency (AFHRA), Maxwell Air Force Base, Maxwell, Alabama, visited 2007.
3. ----, Daily Status Reports, copies in 376 Archives in possession of author.
4. ----, Experimental Engineering Section, Memorandum Report, March 21, 1942 in possession of author.
5. ----, Franklin D. Roosevelt Presidential Library and Museum (FDRL), visited September 2009.
6. ----, Interview of Ulysses S. Nero, May 21, 1974.
7. ----, *Necessary Modifications to B-24D on HALPRO as of March 28, 1942*, copy in 376 Archives in possession of author.
8. Adams, Homer, Contact Report.
9. Adams, Homer, Letter to Robert Brooks, August 3, 1974.
10. Arnold, H. H., *Global Mission*, Harper & Brothers, New York, 1949.
11. Baker, Nicholson, *Human Smoke*, Simon & Schuster, New York, 2002.
12. Barineau, James L., Letter to Richard Brooks, January 7, 1974.
13. Baroni, George, editor, *The Pyramiders, The Story of the 98th, 1942-1945*.
14. Bevans, Charles I.,LL.B., editor, *Treaties and Other International Agreements of the United States of America, 1776-1949, Vol. 3, 1931-1945*, published by United States State Department, November 1969.
15. Brand, H. W., *FDR: Traitor to His Class*, Doubleday, New York, 2008.
16. Brereton, Lewis H., *The Brereton Diaries, The War in the Air in the Pacific, Middle East, and Europe*, William Morrow and Company, New York, 1946.
17. Brown, Therman, *War Stories,* unpublished, copy in 376 Archives in possession of author.
18. Buchanan, Andrew, *A Friend Indeed? From Tobruk to El Alamein: The American Contribution to Victory in the Desert*, Diplomacy and Statecraft 15, pp. 279-301.
19. Cave, Edward, *transcript of interview*, copy in 376 Archives in possession of author.
20. Christensen, Harold, *transcript of interview*, copy in 376 Archives in possession of author.
21. Churchill, Winston S., *The Hinge of Fate,* Boston: Houghton Mifflin Co., 1950.

22. Clendenin, Edward, *Mission History of the 376th Bombardment Group*, self-published, 2002.
23. Conlee, Howard, *transcript of interview*, copy in 376 Archives in possession of author.
24. Craven, W. F. and Cate, J. L. *The Army Air Forces in World War II, Vol. 1, Plans and Early Operations January 1939 to August 1942*, The University of Chicago Press, Chicago, 1948.
25. Craven, W. F. and Cate, J. L. *The Army Air Forces in World War II, Vol. 2, Europe: Torch to Pointblank August 1942 to December 1943*, The University of Chicago Press, Chicago, 1948.
26. Davis, Richard G., *Carl A. Spaatz and the Air War in Europe*, Center for Air Force History, Washington D.C., 1993.
27. Dorr, Robert F. *7th Bombardment Group/Wing, 1918-1995*, Turner Publishing, Paducah, 1996.
28. Fleming, Thomas, *The New Dealer's War, F.D.R. and the War Within World War II*, Basic Books, New York, 2001.
29. Frank, Richard B., *Guadalcanal, The Definitive Account of the Landmark Battle*, New York, Penguin Books, 1990.
30. Glines, Carroll V., *The Doolittle Raid: America's Daring First Strike Against Japan*, New York, Orion Books, 1988.
31. Hannah, William, letters to the author, copy in 376 Archives in possession of author.
32. Herring, George C., *From Colony to Superpower, U.S. Foreign Relations since 1776*, Oxford University Press, Oxford, 2008.
33. Herman, Arthur, *To Rule The Waves, How the British Navy Shaped the Modern World*, Harper Perennial, New York, 2004.
34. Holloway, Harry, letters to the author, copy in 376 Archives in possession of author.
35. Hoyt, Edwin P., *Japan's War – The Great Pacific Conflict*, Cooper Square Press, New York, 2001
36. Ickes, Harold, *The Secret Diary of Harold L. Ickes, Vol. III The Lowering Clouds*, Simon and Schuster, New York, 1955.
37. Jackson, W. G. F., *The Battle For North Africa 1940-43*, Mason/Charter, New York, 1975.
38. Kimball, Warren F., *Stalingrad: A Chance for Choices*, The Journal of Military History, Vol. 60, No. 1 (Jan., 1996), pp. 89-114, Published by: Society for Military History.
39. King, Ernest J., *Fleet Admiral King, A Naval Record*, Da Capo Press, New York, 1976.
40. Lash, Joseph P., *Roosevelt and Churchill 1939-1941*, W. W. Norton & Company Inc., New York, 1976.
41. Lowenthal, Mark M., *Roosevelt and the Coming War: The Search for United States Policy 1937-42*, Source: Journal of Contemporary History, Vol. 16, No.3, The Second World War: Part 2, (Jul., 1981), pp. 413-440
42. Mayhew, Wilbur, *transcript of interview*, copy in 376 Archives in possession of author.
43. McClain, James L., *Japan, A Modern History*, W. W. Norton & Company, New York, 2002.
44. Miller, Edward S., *War Plan Orange – The U.S. Strategy to Defeat Japan, 1897-1945*, Naval Institute Press, Annapolis, 1991.
45. Mitcham, Jr., Samuel W., *Rommel's Desert War*, Stein and Day, New York, 1982.

46. Morison, Samuel Eliot, *The Two-Ocean War, A Short History of the United States Navy in the Second World War*, Little, Brown and Company, Boston, 1963.
47. Morton, Lewis, *The Fall of the Philippines*, Center of Military History, United States Army, Washington D.C., 1953.
48. Pearson, Drew, his personal papers, Lyndon Baines Johnson Library & Museum, 2313 Red River Street, Austin, Texas.
49. Perry, Mark, *The Most Dangerous Man in America*, Basic Books, New York, 2014.
50. Porch, Douglas, *The Path To Victory – The Mediterranean Theater in World War II*, Farrar, Straus and Giroux, New York, 2004.
51. Richardson, George S., *The Forgotten Force*, unpublished personal diary.
52. Royce, Ralph "Scotty," Transcript of an interview, in author's possession,
53. Rust, Kenn C., *The 9th Air Force in World War II*, Aero Publishers, Inc, Fallbrook, California, 1970.
54. Sibert, James, *personal diary*, copy in 376 Archives in possession of author.
55. Spector, Ronald H., *Eagle Against the Sun, The American Was With Japan*, Random House, New York, 1985.
56. Stimson, Henry L., *Stimson Diaries*, copy accessed at the Franklin D. Roosevelt Library, Hyde Park, New York, 1950.
57. Story, Albert, *Private Story and World War II*, unpublished, copy in 376 Archives in possession of author.
58. Thomas, Rowan, *Born in Battle*, The John C. Winston Company, Philadelphia, 1944.
59. Tuchman, Barbara W., *Stilwell and the American Experience in China, 1911-45*, The Macmillan Company, New York, 1970.
60. Wadle, Ryan David, *United States Navy Fleet Problems And The Development Of Carrier Aviation, 1929-1933*, A Thesis, Texas A&M University, August 2005.
61. Walker, Howard E., letter to Robert Brooks, 30 December 1973, in author's possession.
62. Wedemeyer, Albert C., General, *Wedemeyer Reports!*, Henry Holt & Company, New York, 1958.
63. West, Wilbur, letter to Robert Storz, May 29, 1962, copy in author's files.

ONLINE REFERENCES

1. ----, 93rd Bomb Group, http://www.93rdbombardmentgroup.com/, accessed December 6, 2013.
2. ----, Army Air Force Forum, http://forum.armyairforces.com/default.aspx, accessed July 12, 2012.
3. ----, Army Air Force Statistical Digest, http://www.ibiblio.org/hyperwar/AAF/StatDigest/aafsd-3.html, accessed January 22, 2014.
4. ----, *Claire Lee Chennault Foundation of the Flying Tigers, Inc.*, http://www.chennaultfoundationflyingtigersinc.org/, accessed July 13, 2013.
5. ----, *Events Leading Up To World War II*, United States Government Printing Office, Washington, 1944, http://www.ibiblio.org/pha/events/, accessed June 29, 2012.
6. ----, *Fireside Chats of Franklin D. Roosevelt*, http://www.mhric.org/fdr/, accessed July 26, 2012.

7. ----, *Free Republic*, http://www.freerepublic.com/home.htm, accessed 15 August 2015.
8. ----. Franklin D. Roosevelt Presidential Library and Museum (FDRL), President's Secretary File (PSF), http://docs.fdrlibrary.marist.edu/PSF/, accessed July 10, 2012.
9. ----. Franklin D. Roosevelt Presidential Library and Museum (FDRL), Day by Day, http://www.fdrlibrary.marist.edu/daybyday/, accessed July 10, 2012.
10. ----, "The "Kangaroo Squadron," http://www.ozatwar.com/usaaf/435th.htm, accessed August 11, 2015.
11. ----, *League of Nations Treaty Series*, http://www.worldlii.org/int/other/LNTSer/1941/37.html, accessed October 21, 2015.
12. ----, *Matinee Classics*, http://matineeclassics.com/, accessed July 13, 2012.
13. ----, Military Times Hall of Valor, http://valor.militarytimes.com/recipient.php?recipientid=676, accessed August 11, 2015.
14. ----, National Archives at College Park, Record Group 225.2: Records of the Joint Board (1903 - 1947), Joint Board File No. 325 (War Plans), Serial 19, http://strategytheory.org/military/us/joint_board/Symbols%20to%20Represent%20Foreign%20Countries%20(1904).pdf, accessed June 25, 2012.
15. ----, *The New York Times* On-line Archives, http://spiderbites.nytimes.com/, accessed June 27, 2012.
16. ----, *Operation 'Rutter' - The Planned Attack on Dieppe 7 July 1942*, http://www.historyofwar.org/articles/battles_rutter.html accessed December 5, 2013.
17. ----, *Pacific Wrecks*, http://www.pacificwrecks.com/aircraft/b-17/41-2446.html, accessed February 11, 2014.
18. ----, *Peace and War, United States Foreign Policy, 1931-1941*, United States Government Printing Office, 1943, http://www.ibiblio.org/hyperwar/Dip/PaW/, accessed June 25, 2012.
19. ----, *Pearl Harbor Attack, Hearings Before The Joint Committee On The Investigation Of The Pearl Harbor Attack*; http://www.ibiblio.org/pha/pha/invest.html, accessed July 1, 2012.
20. ----, *Royal Air Force 1939-1945, Volume II The Fight Avails*, http://www.ibiblio.org/hyperwar/UN/UK/UK-RAF-II/index.html, accessed October 21, 2015.
21. ----, *U.S. Marine Corps in World War II, Guadalcanal Campaign*, http://www.ibiblio.org/hyperwar/USMC/, accessed January 20, 2014.
22. ----, *State of the Union addresses by Franklin D Roosevelt*, http://www2.hn.psu.edu/faculty/jmanis/poldocs/uspressu/SUaddressFRoosevelt.pdf, accessed July 13, 2012.
23. ----, *Texas State Historical Association*, http://www.tshaonline.org/, accessed July 6, 2012.
24. ----, The Argus On-line Archives, http://trove.nla.gov.au/ndp/del/article/8157900, accessed June 27, 2012.
25. ----, The American Presidency Project, http://www.presidency.ucsb.edu/index.php, accessed July 12, 2012.
26. ----, *The Campaigns of the Pacific War*: United States Strategic Bombing Survey (Pacific): Naval

27. ---, *The China-Burma-India Theater*, http://www.ibiblio.org/hyperwar/USA/USA-CBI-Mission/, accessed January 8, 2014.
28. ----, The Public Papers of the Presidents of the United States, *The public papers and addresses of Franklin D. Roosevelt. 1941 volume*, http://quod.lib.umich.edu/p/ppotpus/4926590.1941.001?rgn=main;view=fulltext, accessed July 12, 2012.
29. ----, *Time* archive, http://www.time.com/time/magazine/, accessed July 11, 2012.
30. ----, *Together We Served*, http://airforce.togetherweserved.com/usaf/landing/index.jsp?mode=Verify, accessed July 13, 2012.
31. ----, *TROVE digitized newspapers and more*, http://trove.nla.gov.au/ndp/del/article/50155133, accessed January 20, 2014.
32. ----, *The U.S. Army Center of Military History*, http://www.history.army.mil/index.html, accessed March 11, 2014.
33. ----, World War II Database, http://ww2db.com/ship_spec.php?ship_id=10
34. Akin, Gary R., *Early Air Force Pioneer, Father of U.S. Precision Bombing – Msgt (later Major) Ulysses S. Nero*, Air Force Enlisted Heritage Research Institute, Maxwell Air Fore Base, Montgomery, Alabama, 1998, http://afehri.maxwell.af.mil/Documents/EnlistedHistory/nero.pdf, accessed July 15, 2012.
35. Assistant Chief of Air Staff, *Army Air Action in the Philippines and Netherlands East Indies, 1941-1942*, United States Air Force Historical Study No. 111, (AFHS-111), Air Force Historical Research Agency (AFHRA), Maxwell Air Force Base, Maxwell, AL, March 1945, http://www.afhra.af.mil/shared/media/document/AFD-090522-048.pdf, accessed June 28, 2012.
36. Assistant Chief of Air Staff, *The AAF in the Middle East - A Study Of The Origins Of The Ninth Air Force*, United States Air Force Historical Study No: 108, USAF Historical Division, Archives Branch, Bldg. 91 Maxwell Air Force Base, Alabama, 1945, http://www.afhra.af.mil/shared/media/document/AFD-090522-044.pdf, accessed July 16, 2012.
37. Assistant Chief of Air Staff, *The Tenth Air Force 1942*, United States Air Force Historical Study No. 12, Air Force Historical Research Agency (AFHRA), Maxwell Air Force Base, Maxwell, AL, August 1944, http://www.afhra.af.mil/shared/media/document/AFD-090522-049.pdf, accessed July 18, 2012.
38. Assistant Chief of Air Staff, *The Ploesti Mission of 1 August 1943*, United States Air Force Historical Study No. 103, Air Force Historical Research Agency (AFHRA), Maxwell Air Force Base, Maxwell, AL, June 1944, http://www.afhra.af.mil/shared/media/document/AFD-090522-039.pdf, accessed July 16, 2012.
39. Ben-Moshe, Tuvia, *Winston Churchill and the "Second Front": A Reappraisal*, The Journal of Modern History, Vol. 62, No. 3 (Sep. 1990), pp. 503-537, http://links.jstor.org/sici?sici=0022-2801%281999 009%2962%3A3%3C503%3AWCAT%22F%3E2.0.CO%3B2-Z, accessed August 30, 2007.

40. Cilio, John and Royce, Scott, *Two Plans-One Target*, Atlantic Flyer, January 2009, http://www.aflyer.com/0812_f_twoplanes.html, accessed July 31, 2012.
41. Clancey, Patrick, *Peace and War, United States Foreign Policy, 1931-1941*, HyperWar Foundation, http://www.ibiblio.org/hyperwar/Dip/PaW/197.html, accessed June 25, 2012.
42. Cline, Ray S., *United States Army in World War II, The War Department, Washington Command Post: The Operations Division*, HyperWar Foundation, http://www.ibiblio.org/hyperwar/USA/USA-WD-Ops/index.html, accessed July 29, 2012.
43. Clodfelter, Mark, *Pinpointing Devastation: American Air Campaign Planning Before Pearl Harbor*, The Journal of Military History, Vol. 58, No. 1, (Jan. 1994) pp. 75-101, http://www.jstor.org/stable/2944180, accessed February 2, 2009.
44. Colomb, P. H., *The United States Navy Under The New Conditions Of National Life*, The North American Review, Vol. 167, No. 503 (Oct 1898), pp. 434-444, http://www.jstor.org/stable/25119076, accessed July 8, 2012.
45. Davis, Richard, *Carl A. Spaatz and the Development of the Royal Air Force-U. S. Army Air Corps*, The Journal of Military History, Vol. 54, No. 4 (Oct 1990), pp. 453-472, http://www.jstor.org/stable/1986066, accessed July 7, 2008.
46. Eckerly, Paul W., *A Pilot's Story*, http://eckleyaviationart.com/project-x.html, accessed July 11, 2017.
47. Fleming, Thomas, *February 7,1932, A Date That Would Live In… Amnesia*, American Heritage, July/August 2011, Vol. 52, Issue 5, http://www.americanheritage.com/content/early-warning, accessed August 1, 2012.
48. Harrington, Daniel F., *A Careless Hope: American Air Power and Japan, 1941*, The Pacific Historical Review, Vol. 48, No. 2, (May, 1979), pp. 217-238, http://www.jstor.org/stable/3639273, accessed August 31, 2008.
49. Headrick, LCDR Alan C., *Bicycle Blitzkrieg: The Malayan Campaign and the Fall of Singapore*, Naval War College, http://www.ibiblio.org/hyperwar/PTO/RisingSun/BicycleBlitz/, accessed October 15, 2015.
50. Herzog, James H., *Influence of the United States Navy in the Embargo of Oil to Japan, 1940-1941*, The Pacific Historical Review, Vol. 35, No. 3, (Aug. 1966), pp. 317-328, http://links.jstor.org/sici?sici=0030-8684%28196608%2935%3A3%3C317%3AIOTUSN%3E2.0.CO%3B2-0, accessed November 10, 2007.
51. Hosoya, Chihiro, *Miscalculations in Deterrent Policy: Japanese-U.S. Relations, 1938-1941*, Journal of Peace Research, Vol. 5, No. 2, (1968) pp. 97-115, http://www.jstor.org/stable/423231, accessed September 7 2008.
52. Huston, John, *American Airpower Comes of Age, Vol. 1*, Air University, Maxwell Air Base, Alabama, 2002, http://aupress.au.af.mil/digital/pdf/book/b_0084_huston_american_airpower_diaries.pdf, accessed June 29, 2012.
53. Leiser, Gary, *The U.S. Military And Palestine In 1942*, Air Power History, Spring 2000, Vol. 47, Issue 1, pp. 12-23.

54. Lukas, Richard C., *The Velvet Project: Hope and Frustration*, Military Affairs, Vol. 28, No. 4, pp. 145-162, http://www.jstor.org/stable/1984385, accessed July 3, 2008.
55. Marshall, George, *Biennial Reports of the Chief of Staff of the United States Army. July 1, 1939 - June 30, 1945*, http://ibiblio.org/hyperwar/USA/COS-Biennial/, accessed June 27, 2012.
56. Matoloff, Maurice, and Snell, Edwin M., *Strategic Planning for Coalition Warfare, 1941-1942*, http://www.ibiblio.org/hyperwar/USA/USA-WD-Strategic1/index.html, accessed January 8, 2014.
57. Miller, John R., *The Chiang-Stilwell Conflict, 1942-1944*, Military Affairs, Vol. 43, No. 2 (Apr. 1979), pp. 59-62, http://links.jstor.org/sici?sici=0026-3931%28197904%2943%3A2%3C59%3ATCC1%3E2.0.CO%3B2-W, accessed August 30, 2007.
58. Morton, Louis, *War Plan Orange: Evolution of a Strategy*, World Politics, Vol. 11, No. 2, (Jan. 1959), pp. 221-250, http://links.jstor.org/sici?sici=0043-8871%28195901%2911%3A2%3C221%3AWPOEOA%3E2.0.CO%3B2-R, accessed November 10, 2007.
59. Morton, Louis, *Strategy And Command: The First Two Years*, United States Army in World War II, (1962), http://www.history.army.mil/html/books/005/5-1/index.html, accessed November 8, 2011.
60. Rickard, J, *Battle of Wake Island, 8-23 December 1941*, http://www.historyofwar.org/articles/battles_wake_island.html, accessed July 12, 2012.
61. Roosevelt, Franklin D, Address to Congress Requesting a Declaration of War, December 8, 1941, http://millercenter.org/president/fdroosevelt/speeches/speech-3324, accessed August 19, 2016.
62. Roskill, S.W., *War At Sea 1939-1945, Volume Ii, The Period Of Balance*, http://www.ibiblio.org/hyperwar/UN/UK/UK-RN-II/, accessed August 3, 2013.
63. Sagan, Scott D., *The Origins of the Pacific War*, Journal of Interdisciplinary History, Vol. 18, No. 4, The Origin and Prevention of Major Wars (Spring 1988), pp. 893-922, http://www.jstor.org/stable/204828, accessed August 31, 2008.
64. Sebrega, John J., *The Anticolonial Policies of Franklin D. Roosevelt- A Reappraisal*, Political Science Quarterly, Vol. 101, No. 1, 1986, pp. 65-84, http://links.jstor.org/sici?sici=0032-3195%281986%29101%3A1%3C65%3ATAPOFD%3E2.0.CO%3B2-B, accessed November 11, 2007.
65. Schaller, Michael, *American Air Strategy in China, 1939-1941: The Origins of Clandestine Air Warfare*, American Quarterly, Vol. 28, No. 1 (Spring, 1976) pp. 3-19, http://links.jstor.org/sici?sici=0003-0678%28197621%2928%3A1%3C3%3AAASIC1%3E2.0.CO%3B2-V, accessed August 30, 2007.
66. Steele, Richard W., *American Popular Opinion and the War against Germany: The Issue of Negotiated Peace, 1942*, The Journal of American History, Vol. 65, No. 3, (Dec. 1978), pp. 704-723, http://links.jstor.org/sici?sici=0021-8723%28197812%2965%3A3%3C704%3AAPOOATW%3E2.0.CO%3B2-O, accessed September 18, 2007.
67. Watson, Mark, *The War Department, Chief of Staff: Prewar Plans and Preparations*, Center of Military History, United States Army, Washington D.C., 1991, http://www.history.army.mil/html/books/001/1-1/index.html, accessed June 25, 2012.

68. Williams, Mary H., *United States Army In World War II, Special Studies, Chronology 1941-1945,* Center of Military History, United States Army, Washington D.C., 1989, http://www.ibiblio.org/hyperwar/USA/USA-SS-Chronology/index.html#index, accessed October 15, 2015.

69. Xu, Guangqiu, *The Issue of U.S. Air Support for China during the Second World War, 1942-45,* Journal of Contemporary History, Vol. 36, No. 3, (Jul. 2001) pp. 459-484, http://links.jstor.org/sici?sici=0022-0094%28200107%2936%3A3%3C459%3ATIOUAS%3E2.0.CO%3B2-Y, accessed August 30, 2007.

FIGURE REFERENCES

1. AAF-I-18 Doolittle Landing Sites (Army Air Forces in World War II; Vol. I: Plans & Early Operations, January 1939 to August 1942, p. 443.).
2. The ABDACOM Area – January- February 1942, (http://www.history.army.mil/brochures/eindies/p11%28map%29.jpg), Courtesy of the U.S. Army Center of Military History
3. Burma 1942 (http://www.history.army.mil/brochures/burma42/p07(map).jpg) Courtesy of the U.S. Army Center of Military History
4. British dispositions in Malaya and Singapore on 7 Dec 1941 (http://www.westpoint.edu/history/SiteAssets/SitePages/World%20War%20II%20Pacific/ww2%20asia%20map%2007.jpg) Courtesy of the USMA, Department of History.
5. Egypt-Libya 1942 (http://www.history.army.mil/brochures/egypt/p12(map).jpg.) Courtesy of the U.S. Army Center of Military History
6. Advance Japanese Landings 8–20 December 1941, from "The War in the Pacific – The Fall of the Philippines", by Louis Morton, CMH-Pub 5-2, Courtesy of the U.S. Army Center of Military History
7. Soviet 1941 Winter counteroffensive (http://www.westpoint.edu/history/SiteAssets/SitePages/World%20War%20II%20Europe/WWIIEurope21Combined.gif) Courtesy of the USMA, Department of History
8. German Advance To Stalingrad, 24 July-18 November 1942 (http://www.westpoint.edu/history/SiteAssets/SitePages/World%20War%20II%20Europe/WWIIEurope23.gif) Courtesy of the USMA, Department of History'
9. Photo credit – Albert Story
10. Pacific and the Far East (http://www.westpoint.edu/history/SiteAssets/SitePages/World%20War%20II%20Pacific/WWIIAsia01.gif) Courtesy of the USMA, Department of History.
11. Approaches to the Middle East (http://www.history.army.mil/books/wwii/persian/photos/map1.jpg) Courtesy of the U.S. Army Center of Military History
12. The ABDACOM Area – January- February 1942, (http://www.history.army.mil/brochures/eindies/p11%28map%29.jpg), Courtesy of the U.S. Army Center of Military History
13. Burma 1942 (http://www.history.army.mil/brochures/burma42/p07(map).jpg) Courtesy of the U.S. Army Center of Military History
14. British dispositions in Malaya and Singapore on 7 Dec 1941 (http://www.westpoint.edu/history/

SiteAssets/SitePages/World%20War%20II%20Pacific/ww2%20asia%20map%2007.jpg) Courtesy of the USMA, Department of History.

15. Egypt-Libya 1942 (http://www.history.army.mil/brochures/egypt/p12(map).jpg.) Courtesy of the U.S. Army Center of Military History
16. Advance Japanese Landings 8–20 December 1941, from "The War in the Pacific – The Fall of the Philippines", by Louis Morton, CMH-Pub 5-2, Courtesy of the U.S. Army Center of Military History
17. Soviet 1941 Winter counteroffensive (http://www.westpoint.edu/history/SiteAssets/SitePages/World%20War%20II%20Europe/WWIIEurope21Combined.gif) Courtesy of the USMA, Department of History
18. German Advance To Stalingrad, 24 July-18 November 1942 (http://www.westpoint.edu/history/SiteAssets/SitePages/World%20War%20II%20Europe/WWIIEurope23.gif) Courtesy of the USMA, Department of History
19. AAF-I-17 Island Chain of the South Pacific (Army Air Forces in World War II; Vol. I: Plans & Early Operations, January 1939 to August 1942).
20. The ABDACOM Area – January- February 1942, (http://www.history.army.mil/brochures/eindies/p11%28map%29.jpg), Courtesy of the U.S. Army Center of Military History
21. Approaches to the Middle East (http://www.history.army.mil/books/wwii/persian/photos/map1.jpg) Courtesy of the U.S. Army Center of Military History
22. British dispositions in Malaya and Singapore on 7 Dec 1941 (http://www.westpoint.edu/history/SiteAssets/SitePages/World%20War%20II%20Pacific/ww2%20asia%20map%2007.jpg) Courtesy of the USMA, Department of History.
23. Advance Japanese Landings 8–20 December 1941, from "The War in the Pacific – The Fall of the Philippines", by Louis Morton, CMH-Pub 5-2, Courtesy of the U.S. Army Center of Military History
24. Soviet 1941 Winter counteroffensive (http://www.westpoint.edu/history/SiteAssets/SitePages/World%20War%20II%20Europe/WWIIEurope21Combined.gif) Courtesy of the USMA, Department of History
25. AAF-I-17 Island Chain of the South Pacific (Army Air Forces in World War II; Vol. I: Plans & Early Operations, January 1939 to August 1942).
26. The Middle East 1942 (http://www.history.army.mil/brochures/egypt/p10(map).jpg) Courtesy of the U.S. Army Center of Military History
27. Approaches to the Middle East (http://www.history.army.mil/books/wwii/persian/photos/map1.jpg) Courtesy of the U.S. Army Center of Military History
28. The ABDACOM Area – January- February 1942, (http://www.history.army.mil/brochures/eindies/p11%28map%29.jpg), Courtesy of the U.S. Army Center of Military History
29. Burma 1942 (http://www.history.army.mil/brochures/burma42/p07(map).jpg) Courtesy of the U.S. Army Center of Military History

30. AAF-I-17 Island Chain of the South Pacific (Army Air Forces in World War II; Vol. I: Plans & Early Operations, January 1939 to August 1942).
31. Burma 1942 (http://www.history.army.mil/brochures/burma42/p07(map).jpg) Courtesy of the U.S. Army Center of Military History
32. Approaches to the Middle East (http://www.history.army.mil/books/wwii/persian/photos/map1.jpg) Courtesy of the U.S. Army Center of Military History
33. AAF-I-17 Island Chain of the South Pacific (Army Air Forces in World War II; Vol. I: Plans & Early Operations, January 1939 to August 1942).
34. AAF-I-18 Doolittle Landing Sites (Army Air Forces in World War II; Vol. I: Plans & Early Operations, January 1939 to August 1942, p. 443.).
35. Egypt-Libya 1942 (http://www.history.army.mil/brochures/egypt/p12(map).jpg.) Courtesy of the U.S. Army Center of Military History.
36. Photo credit – Albert Story
37. Burma 1942 (http://www.history.army.mil/brochures/burma42/p07(map).jpg) Courtesy of the U.S. Army Center of Military History
38. German Advance To Stalingrad, 24 July-18 November 1942 (http://www.westpoint.edu/history/SiteAssets/SitePages/World%20War%20II%20Europe/WWIIEurope23.gif) Courtesy of the USMA, Department of History
39. AAF-I-17 Island Chain of the South Pacific (Army Air Forces in World War II; Vol. I: Plans & Early Operations, January 1939 to August 1942).
40. Burma 1942 (http://www.history.army.mil/brochures/burma42/p07(map).jpg) Courtesy of the U.S. Army Center of Military History
41. Egypt-Libya 1942 (http://www.history.army.mil/brochures/egypt/p12(map).jpg.) Courtesy of the U.S. Army Center of Military History
42. German Advance To Stalingrad, 24 July-18 November 1942 (http://www.westpoint.edu/history/SiteAssets/SitePages/World%20War%20II%20Europe/WWIIEurope23.gif) Courtesy of the USMA, Department of History
43. Europe And The Middle East (http://www.westpoint.edu/history/SiteAssets/SitePages/World%20War%20II%20Europe/WWIIEurope43.gif) Courtesy of the USMA, Department of History
44. Egypt-Libya 1942 (http://www.history.army.mil/brochures/egypt/p12(map).jpg.) Courtesy of the U.S. Army Center of Military History
45. AAF-I-17 Island Chain of the South Pacific (Army Air Forces in World War II; Vol. I: Plans & Early Operations, January 1939 to August 1942).
46. The Middle East 1942 (http://www.history.army.mil/brochures/egypt/p10(map).jpg) Courtesy of the U.S. Army Center of Military History
47. Gazala and Vicinity, Libya Africa 1942 (http://www.westpoint.edu/history/SiteAssets/SitePages/World%20War%20II%20Europe/WWIIEurope37.pdf), Courtesy of the U.S. Army Center of Military History.

REFERENCES | 473

48. Photo credit – Albert Story.
49. Europe And The Middle East (http://www.westpoint.edu/history/SiteAssets/SitePages/World%20War%20II%20Europe/WWIIEurope43.gif) Courtesy of the USMA, Department of History
50. Egypt-Libya 1942 (http://www.history.army.mil/brochures/egypt/p12(map).jpg.) Courtesy of the U.S. Army Center of Military History
51. German Advance To Stalingrad, 24 July-18 November 1942 (http://www.westpoint.edu/history/SiteAssets/SitePages/World%20War%20II%20Europe/WWIIEurope23.gif) Courtesy of the USMA, Department of History
52. Europe And The Middle East (http://www.westpoint.edu/history/SiteAssets/SitePages/World%20War%20II%20Europe/WWIIEurope43.gif) Courtesy of the USMA, Department of History
53. The ABDACOM Area – January- February 1942, (http://www.history.army.mil/brochures/eindies/p11%28map%29.jpg), Courtesy of the U.S. Army Center of Military History
54. German Advance To Stalingrad, 24 July-18 November 1942 (http://www.westpoint.edu/history/SiteAssets/SitePages/World%20War%20II%20Europe/WWIIEurope23.gif) Courtesy of the USMA, Department of History
55. AAF-I-17 Island Chain of the South Pacific (Army Air Forces in World War II; Vol. I: Plans & Early Operations, January 1939 to August 1942).
56. Europe And The Middle East (http://www.westpoint.edu/history/SiteAssets/SitePages/World%20War%20II%20Europe/WWIIEurope43.gif) Courtesy of the USMA, Department of History
57. German Advance To Stalingrad, 24 July-18 November 1942 (http://www.westpoint.edu/history/SiteAssets/SitePages/World%20War%20II%20Europe/WWIIEurope23.gif) Courtesy of the USMA, Department of History
58. AAF-I-17 Island Chain of the South Pacific (Army Air Forces in World War II; Vol. I: Plans & Early Operations, January 1939 to August 1942).
59. German Advance To Stalingrad, 24 July-18 November 1942 (http://www.westpoint.edu/history/SiteAssets/SitePages/World%20War%20II%20Europe/WWIIEurope23.gif) Courtesy of the USMA, Department of History
60. AAF-I-17 Island Chain of the South Pacific (Army Air Forces in World War II; Vol. I: Plans & Early Operations, January 1939 to August 1942).
61. Stalingrad November-December 1942, (from "Stalingrad to Berlin" by Earl F. Ziemke, CMH-Pub 30-5-1), Courtesy of the USMA, Department of History
62. Operation Torch landings in North Africa, 8 Nov 1942 (http://www.westpoint.edu/history/SiteAssets/SitePages/World%20War%20II%20Europe/WWIIEurope38Combined.gif) Courtesy of the USMA, Department of History
63. Operation Torch landings in North Africa, 8 Nov 1942 (http://www.westpoint.edu/history/Site-

Assets/SitePages/World%20War%20II%20Europe/WWIIEurope38Combined.gif) Courtesy of the USMA, Department of History.
64. Photo credit – Albert Story.

INDEX

AAF

AIR FORCE

10th Air Force 130, 133, 138, 140, 142, 156, 160, 162, 165, 169, 173, 175, 185, 186, 187, 191, 193, 194, 195, 198, 199, 200, 202, 223, 233, 237, 238, 239, 240, 275, 291, 294, 295, 307, 309, 311, 312, 347
7th Air Force 234
9th Air Force 268, 296, 385, 389, 399, 402
Australia (USAFIA) 56
Far East Air Force (USFEAF) ... 15, 16, 35, 56, 155
Hawaiian Air Force 17
Middle East (USAFIME) 271, 382
Middle East Air Force (USAMEAF) 295, 358, 359, 375

GROUP

11th Bombardment 324, 331, 341
12th Bombardment (M) 359
17th Bombardment 124
17th Pursuit 155
19th Bombardment 12, 15, 42, 56, 155, 164, 167, 325
1st Provisional Group 305, 311, 318, 322, 323, 324, 326, 328, 329, 331, 339, 341, 347, 348, 350, 351, 357, 358, 359, 360, 362, 363, 364, 369, 370, 371, 373, 374, 376, 377, 378, 379, 381, 389, 405, 461
27th Bombardment 34
376th Heavy Bombardment 371, 374, 375, 381, 382, 383, 384, 385, 386, 387, 389, 390, 391, 397, 399, 400, 402, 405, 432, 461
42nd Bombardment 90
44th Bombardment 194
57th Fighter 317, 359
58th Pursuit 87
78th Pursuit 87
7th Bombardment 15, 42, 56, 135, 136, 137, 166, 198, 308
93rd Bombardment 100, 156, 359, 373, 397, 399, 402
97th Bombardment 222, 339, 344, 347, 387
98th Bombardment 132, 184, 186, 189, 257, 293, 305, 319, 320, 326, 327, 330, 339, 346, 347, 359, 382, 385, 386, 397, 399, 402, 405

SQUADRON

11th Bombardment 55
14th Bombardment 12
20th Pursuit 32
22nd Bombardment 55
30th Bombardment 15

38th Recon 26
3rd Pursuit 32
40th Bombardment 155, 164, 167, 198
431st Bombardment 259
435th Bombardment 198, 218, 274, 321
436th Bombardment....................................... 198
88th Recon 15, 26, 27, 55, 123, 125, 135, 136, 137, 165, 166, 198
93rd Bombardment... 15
9th Bombardment ... 55, 123, 125, 173, 221, 257, 298, 308, 319, 323

TASK FORCE

BR 83, 84, 85, 87, 99, 100, 222, 223, 306, 339, 344, 345, 347, 348, 349, 350, 359, 369, 373, 387, 399

CAIRO 84, 85, 87, 136

Desert Air Task Force 377

Doolittle Raid 1, 50, 81, 88, 92, 96, 100, 123, 124, 126, 157, 164, 172, 174, 175, 183, 185, 190, 192, 195, 196, 213, 232, 237, 258, 372, 405, 413, 414, 416

FIVE ISLANDS.. 85

Halverson Project 1, 2, 76, 80, 87, 89, 90, 92, 93, 94, 95, 98, 99, 100, 102, 104, 119, 122, 126, 129, 132, 133, 140, 155, 156, 160, 164, 166, 168, 170, 171, 174, 175, 183, 184, 186, 188, 189, 192, 194, 196, 197, 200, 201, 213, 214, 215, 217, 219, 220, 223, 231, 232, 233, 234, 236, 237, 238, 239, 240, 241, 243, 244, 245, 246, 247, 248, 257, 258, 259, 260, 261, 262, 263, 264, 265, 266, 267, 268, 269, 270, 271, 272, 273, 275, 276, 277, 292, 294, 295, 296, 297, 298, 305, 306, 307, 309, 312, 315, 318, 319, 323, 324, 327, 330, 339, 341, 344, 347, 383, 400, 405, 415, 419, 461

Project X........ 41, 42, 44, 50, 56, 76, 78, 85, 132, 176, 275

TASKFORCE

Halverson Project 122

ADAMS

Homer.................................. 293, 313, 342, 458

ADLER

Elmer E........................309, 387, 388, 391, 397

AGREEMENT

Arnold-Portal.............................. 73, 83, 84, 87
Arnold-Portal-Towers 358, 359
Moscow Protocol... 371

AIREY

Terence 389

ALBERTSON

Robert L.. 431, 456

ALDRIDGE

James 376

ALEXANDER

E. H.............56, 73, 79, 121, 122, 161, 191, 201
Harold 339, 343, 344
Ralph... 455

ALLEN

Robert S.. 269

ANDERSON

Charles 317
Malcolm R.. 456
Virgil D... 456

ANDREWS

Frank M.. 382, 385

INDEX

ANGLIN
    Claude F. 457

APPOLD
    Norman 315, 316, 317, 330

ARNOLD
    Henry 'Hap' 9, 12, 17, 36, 41, 50, 51, 53, 56, 73, 74, 75, 76, 79, 80, 82, 84, 88, 89, 98, 99, 100, 101, 102, 103, 104, 121, 122, 123, 125, 126, 127, 129, 133, 140, 156, 161, 163, 166, 168, 169, 172, 183, 185, 188, 192, 193, 195, 197, 199, 200, 201, 216, 217, 218, 219, 220, 222, 231, 233, 234, 235, 236, 237, 244, 245, 247, 257, 258, 260, 261, 264, 265, 266, 269, 271, 274, 293, 296, 307, 314, 349, 371, 375, 377, 382, 413, 414, 415, 416

AUCHINLECK
    Sir Claude ..37, 82, 89, 121, 261, 275, 291, 292, 296, 329, 339, 343

AVG
    .... *See* Flying Tigers

B-17
    Fennell vs. Rommel 258, 308, 389
    Swamp Ghost 44, 130, 136, 137
    Topper 74, 275

B-24
    Arkansas Traveler 242, 246, 458
    Babe the Big Blue Ox 459
    Ball of Fire 457
    Black Mariah II 248, 459
    Blue Goose 265, 428, 434, 437, 442, 459
    Brooklyn Rambler 400, 443, 455
    Draggin Lady 246, 326, 386, 458
    Eager Beaver 318, 456
    Edna Elizabeth 458
    Florine Jo Jo 457
    Hellzapoppin 238, 243, 248, 265, 339, 428, 437, 442, 460
    Jap Trap 243, 460
    Little Eva 442, 456
    Malicious 242, 460
    Mona The Lame Duck 244, 295, 324, 325, 459
    Old King Solomon 457
    Old Spareribs 387
    Ole Faithful 242, 293, 455
    Ole Rock 242, 313, 455
    Queen B 238, 240, 241, 243, 244, 246, 262, 340, 457
    Ripper the First 458
    Town Hall 242, 456
    Wash's Tub 238, 460
    Yank 455

BALDWIN
    Hanson W. 47, 48, 134

BALLANTINE
    Henry R. 458

BARINEAU
    James L. 388, 459

BEATTY
    John J. Jr. 459

BEDROSIAN
    Peter 457

BELYAEV
    Alexander I. 220, 260, 264

BENNETT
    Theodore E. 460

BISSELL
Clayton 122, 190, 192

BLINDAIR
Thomas A. Jr. 457

BOLEN
Joseph J. 460

BOWAN
Andrew 117

BRADLEY
Follett 346, 371, 378

BRERETON
Lewis H. 13, 15, 16, 17, 18, 32, 40, 43, 56, 78, 128, 132, 136, 140, 142, 156, 157, 160, 171, 172, 173, 186, 187, 191, 194, 195, 197, 199, 200, 221, 223, 239, 240, 274, 275, 291, 292, 294, 295, 296, 298, 305, 307, 308, 323, 327, 341, 343, 347, 348, 349, 357, 358, 360, 363, 375, 385, 389

BRETANDAS
Victor 126

BRETT
George H. 39, 41, 53, 55, 56, 79, 140

BRISCOE
Jimmie 443, 456

BRITT
James O. 388

BRITTON
Luke L. 460

BROCK
Ray 54

BROWN
Charles O. 318, 456
Nathan K. Jr. 432, 436, 442, 443, 444, 455
Robert 436, 437, 442
Therman 186, 246, 277, 298, 305, 326, 360, 370, 384, 398, 401, 450, 458

BRYANT
Olen C. 459

BUCKWALTER
William S. 457

BULLITT
William C. 49, 75, 166, 169

BUTLER
Kenneth W. 273, 313, 457

CABELL
Charles P. 216, 219, 234, 237, 416

CANNON
Morris A. 456

CARMICHAEL
John 129, 135

CARTER
John F. 101, 236

CATE
James L. 202

INDEX

CAVE
  Edward A. Jr. "Hawk"....38, 188, 194, 201, 241, 243, 265, 432, 433, 441, 443, 460

CHARLES
  George A.. 455

CHENNAULT
  Claire L. ... 5, 6, 7, 34, 39, 88, 98, 101, 122, 129, 185, 221, 344, 414

CHIANG
  Kai-shek..............................5, 6, 34, 37, 39, 50, 52, 54, 80, 81, 88, 93, 94, 97, 101, 122, 125, 127, 141, 172, 174, 185, 191, 192, 195, 197, 200, 201, 239, 269, 274, 276, 293, 295, 296, 312, 322, 344
  Madame Kai-shek................. 88, 191, 248, 293

CHRISTIANSEN
  Harold 315

CHUIKOV
  Vasiliy 360

CHURCHILL
  Winston S.................................... 31, 44, 46, 47, 50, 51, 52, 75, 84, 85, 87, 97, 101, 102, 125, 128, 133, 135, 138, 139, 141, 155, 157, 158, 159, 161, 188, 191, 193, 195, 257, 261, 271, 274, 275, 276, 278, 279, 305, 306, 311, 318, 321, 323, 328, 329, 330, 339, 342, 343, 345, 346, 348, 350, 352, 357, 358, 362, 363, 364, 372, 373, 376, 383, 397

CLARK
  Walter... 457

CLAYTON
  Jackson B... 457

COBB
  James O..74, 86

COLLUM
  Charles E... 455

COLT
  ALfred C.. 117, 459

CONFERENCES
  1941 December Arcadia..42, 44, 46, 47, 50, 84, 85, 87, 278
  1942 April London 189
  1942 August Moscow................................. 346
  1942 July London JCS 318, 321, 323, 324, 326, 327, 328
  1942 June Washington 276

CONLEE
  Howard 315, 336, 464

CONRAD
  James 122

CONVOY
  P.Q. 16 .... 237, 247
  P.Q. 17 ....295, 306, 314
  P.Q. 18 .... 330, 358
  P.Q. 19 .... 370, 372

COOK
  John A. 457

COUTRE
  Robert J.. 455

COWARD
  David D.. 455

COX
  Arthur L. 117, 455

CRAIG
  Howard 220, 237

CRAVEN
  Wesley F. 202

CRAW
  Demas 'Nick' 266, 267, 272, 383, 430, 433

CROUCHLEY
  Edward A. 313, 445, 455

CURRIE
  Lauchlin 7, 31, 34, 36, 78, 95, 98, 101, 122, 166, 191, 197, 240, 322, 344

CUTLER
  Irving 456

CWIKIEL
  Matthew 457

DAIGLE
  Lewellyn 257, 390

DARLAN
  Jean Francois 384, 385

DAVIS
  Charles T. Jr. 429, 431, 437, 456
  Harvey 236
  Norman 361, 458
  Paul 312, 313, 445, 455

DE LABORDE
  Jean 391

DECISNEROS
  George J. 456

DEHAVEN
  Talbert A. 458

DELONG
  Kenneth 186, 189, 298, 361, 458

DEWALD
  Harry W. 117, 457

DILL
  Sir John 47, 318, 319

DOMINO
  Joseph S. 460

DONOVAN
  William J. 459

DOOLITTLE
  James .. 88, 92, 97, 123, 124, 132, 157, 172, 175, 258, 372, 414, 415, 416

DOORMAN
  Karel 125, 155

DROWN
  Nathan C. 457

DRUMMOND
  Peter 375, 387, 388, 391, 397, 400, 404

DUCKWORTH
  Ernest M. 458

DUPREE
  Carl J. 460

DWYER
William P. 458

EAGON
Blaine L. 460

EATON
Frederick 130, 137

EBERT
Harry W. 240, 243, 246, 445, 457

ECKERLY
Paul 56, 74, 81, 86, 130, 131, 132, 136, 275

EDEN
Anthony 427

EDWARDS
Carl W. 117, 455

EISENHOWER
Dwight D. 40, 51, 121, 161, 172, 189, 192, 223, 274, 324, 389, 390

ENT
Uzal G. 88, 389

EUBANK
Eugene L. 16

FANNING
Leo F. 455

FELDMANN
Carl L. 88, 189, 246, 265, 457

FELLERS
Bonner 75, 77, 133, 197, 199, 213, 231, 276

FENNELL
Max R. 165, 257, 308, 340, 372, 389, 390

FIELDS
William F. 458

FILIPPI
Anthony 457

FISHER
Albert S. 459

FITZSIMMONS
Paul W. 117, 458

FLETCHER
Jack 245, 258

FLYING TIGERS
7, 31, 34, 44, 45, 78, 79, 93, 94, 101, 122, 125, 185, 199, 309

FOSTER
Irwin 386

FOUNTAIN
Willard 'Chick' 324, 325

GANDHI
Mahatma 85, 188, 248, 363

GARDNER
Edgar W. 432, 455

GAVIN
Ed 220

GEORGE
Harold L. 79, 93, 103, 129, 132, 140

GEROW
Leonard T. 9, 11, 41, 75, 77

GERRY
Clark H. 316

GHORMLEY
Robert L. 296, 376

GOTT
William H. E. 343

GRADY
Henry Francis 363

GRANDEGERALD
Charles 101

HADLOW
Gordon H. 117, 456

HALPRO
.... *See* AAF: Task Force: Halverson Project

HALSEY
William F. 376, 413

HALVERSON
Harry A. 76, 79, 80, 88, 89, 90, 93, 94, 99, 102, 122, 126, 132, 164, 169, 184, 186, 192, 194, 196, 201, 242, 259, 260, 261, 265, 266, 268, 269, 270, 271, 273, 276, 292, 297, 309, 315, 327, 341, 342, 413, 419, 424, 444, 445, 455

HAMILTON
Maxwell 39

HAMMOND
Chester 201

HANNAH
William 27, 33, 120, 319

HARMON
Millard F. 168, 170, 181, 182, 185, 234, 254, 259, 286

HARRIMAN
W. Averell 141, 344, 348

HARRIS
Sir Arthur 32

HARSHBARGER
Clair E. 459

HART
Thomas C. 16, 17, 40, 41, 56, 85

HAYNES
Caleb 165, 198, 372, 413, 414

HAZLETT
Charles L. 456

HEBERT
Richard C. 460

HELFRICH
Conrad E. L. 85

HELMS
Robert W. 457

HIATT
Walker H. 457

HINES
Harry L. 459

HITCH
J. D. 133

HITLER
Adolf 34, 77, 97, 134, 167, 192, 222, 268, 279, 297, 314, 322, 327, 328, 350, 351, 357, 372, 374, 429

HMS
Cornwall 187
Dorsetshire 187
Hermes 187
Prince of Wales 18, 32, 34, 51
Repulse 18, 32, 34, 51

HOBBS
Louis 389

HOLLIS
Sir Leslie 51

HOLLOWAY
Harry 162, 180, 196, 294, 303, 464

HOPKINS
Allen V. 455
Harry 43, 44, 51, 52, 98, 104, 123, 135, 166, 217, 320, 321, 323, 328, 344, 364, 374

HRUBES
Vitus 117, 458

HULL
Cordell 6, 7, 13, 16, 36, 120, 187, 193, 237

HUMPHREYS
Robert P. 429, 444, 455

HUNTER
Charles W. 459

HURD
Charles 175

HURLEY
Patrick J. 391

IZZO
Alphonse 458

JACKSON
Harold J. 456

JAMES
Edwin 46

JN
Akagi 187
Hiryu 187
Shoho 218
Shokaku 219
Soryu 187

JOHNSON
Louis A. 187, 188

JOYNER
William R. 445, 455

KALBERER
   Alfred F. 'Kal'. 88, 242, 246, 248, 265, 270, 309, 445, 446, 455

KAMINSKA
   John E. 458

KANE
   John 186

KENDRICKS
   Edward J. 92, 273, 341, 441, 456

KEPNER
   William E. 88

KESSLER
   Robert 459

KHRUSHCHEV
   Nikita 339

KIDD
   John 398
   John W. 458

KING
   Ernest J. 1, 81, 131, 134, 140, 156, 158, 163, 171, 217, 292, 293, 294, 296, 319, 320, 322, 327, 370, 414

KING FAROUK
   .... 124, 351

KINNANE
   John W. Jr. 460

KIRK
   Alexander C. 41, 120, 126

KIRKALDY
   Robert B. 90, 458

KISER
   James W. 458

KNOX
   Grover L. 457
   William Franklin "Frank" 6, 7, 217, 414

KRAMER
   Harold W. 459
   Robert R. 459

KROCK
   Arthur 47

KUSILAVGE
   Enoch G. 456

KUTER
   Laurence S. 16, 193

KYSAR
   Herbert A. 313, 457

LA LONDE
   Arthur J. 325

LAMPSON
   Sir Miles 124

LANEY
   Francis J. 265, 457

LARUE
   Jesse R. 458

LAVIN
John N. 162, 196, 294

LAYTON
Sir Geoffrey 35, 77

LEAHY
William D. 349

LEAMAN
James H. 459

LIPPENCOTT
Garland A. 431, 456

LIPPMANN
Walter 82, 171, 278

LITTLE
Donald C. 398, 399

LITVINOFF
Maxim 47

LONG
Paul J. 198

LORD GORT 389

LORD HALIFAX
Edward Frederick Lindley Wood (Lord Halifax) 17, 125

LORD LOTHIAN
Philip Henry Kerr (Lord Lothian) 5

LORD MOUNTBATTEN
.... 187, 197

LOVETT
Robert A. 36, 40, 216

LOW
Francis 81

LUNSFORD
Maryon R. 460

LUTON
Baxter C.   458

MACARTHUR
Douglas 9, 10, 11, 12, 13, 14, 15, 16, 17, 18, 35, 38, 39, 40, 41, 42, 44, 51, 53, 54, 85, 139, 164, 166, 167, 168, 172, 175, 184, 185, 214, 257, 262, 294, 327

MAGRUDER
John 31, 56, 93, 101, 102

MAHBOUB
Frank W. 460

MAISKY
Ivan 372

MARSHALL
George C. 7, 8, 9, 10, 11, 13, 16, 17, 35, 38, 40, 47, 49, 51, 52, 54, 75, 82, 119, 131, 134, 140, 156, 157, 163, 168, 169, 175, 185, 187, 189, 191, 193, 194, 201, 214, 215, 217, 239, 258, 259, 260, 271, 272, 273, 274, 275, 277, 278, 291, 292, 293, 294, 296, 314, 315, 318, 319, 320, 322, 324, 329, 339, 348, 350, 352, 362, 374, 375, 391, 414

MARTIN
Frederick L. 17

MAXWELL
    Russell L....... 260, 267, 270, 272, 273, 275, 276, 277, 291, 296, 346, 375, 382, 385

MAYFIELD
    Gilmer E.. 398, 399

MAYHEW
    Wilbur....25, 123, 125, 135, 137, 160, 164, 198, 257, 308, 340, 371, 389

MCCOMB
    Robert H. 460

MCCONNELL
    Jessy C. 117, 457

MCCREA
    John 199

MCGUIRE
    George F. 88, 194, 242, 243, 246, 264, 330, 342, 399, 456

MCJUNKINS
    Keith.. 257

MCKEE
    Richard J.. 457

MCNAIR
    Lesley J. 163

MCNARNEY
    Joseph T. 192, 274

MCNEAL
    Wilbur C... 456

MCNELLY
    George D.. 117, 459

MEDFORD
    John C. 272, 457

MEEK
    Noel W... 457

MENGEL
    Herbert O... 325

METZLER
    Martin E... 458

MILLER
    Henry J. F. 132
    Richard G.... 245, 262, 263, 276, 292, 331, 398, 399, 425, 427, 445, 449, 457

MILLIREN
    James R.. 246, 457

MITCHELL
    Billy 88

MOLOTOV
    Vyacheslav 247, 259, 379, 400

MONTGOMERY
    Bernard................................ 339, 345, 381, 399

MOORE
    Andrew M... 265, 459

MOOTY
    Mark T... 422, 460

MORAN
Frederic S.. 459

MORGENTHAU
Henry...5, 6, 7

MUNSELL
Alexander E... 315

MURRAY
Wallace 125

MUSE
Howard C... 458

MUSSOLINI
Benito... 279, 351

NAPPI
John 457

NAYLOR
Jack R. 455

NERO
Ulysses S............................88, 92, 123, 126, 455

NESBITT
Frederick W. Jr......246, 401, 425, 429, 431, 432, 456

NESTOR
Francis 244, 295, 297, 459

NIMITZ
Chester 202, 213, 292, 293, 294, 314

NORSTAD
Lauris 216, 220, 234, 237, 416

O'CONNOR
John E.. 456

OGLESBY
Sam R... 458

OPERATION
ACROBAT 82
AQUILLA 184, 198, 413, 415
BOLERO........................... 189, 234, 274, 319
Case Blue 295, 308, 314
Crusader.............................. 33, 35, 77, 86, 89, 93, 113, 121, 276, 294, 306, 307, 309, 311, 312, 315, 318, 319, 321, 322, 324, 326, 328, 329, 350, 359, 374, 376, 377, 382
GYMNAST ...74, 75, 84, 85, 87, 158, 274, 318, 319, 322, 328
MAGNET 85, 158
MARS... 390
RING.. 390
ROUNDUP................................189, 318, 320
RUTTER....................................197, 220, 347
SATURN 390, 401
SLEDGEHAMMER.................189, 320, 324
SUPERCHARGE.......381, 382, 383, 384, 385, 387, 388, 399, 400
SUPER-GYMNAST 84, 156
TORCH 218, 305, 322, 328, 339, 343, 345, 346, 349, 357, 364, 370, 373, 374, 383, 387, 389, 400
URANUS... 387, 388
VELVET..... 234, 235, 236, 274, 307, 321, 328, 329, 346, 348, 350, 352, 365, 370, 371, 373, 374, 375, 376, 377, 378, 382, 387, 388, 391, 397, 401, 405
WATCHTOWER.......314, 321, 324, 326, 331, 341, 342, 343, 346, 347, 349, 358, 376
WINTERSTORM..................................... 400

ORRIS
    Harry F. 458

OSGOOD
    Harold D. 456

OSTERHAUS
    Albert F. 460

OWEN
    James C. 117, 459

OWENS
    Thomas G. 456

PARKER
    Lin Jr. 458

PARR
    Earl C. 460

PATRICK
    Anderson T. 455
    Augustus 258

PATTERSON
    Cecil E. Jr. 459

PATTON
    George S. Jr. 384

PAULLIN
    Robert I. 458

PAYNE
    Coy B. 460
    John H. 264, 360, 421, 445, 459

PEARCE
    Conrad R. 460
    Glen H. 460

PEARSON
    Drew 269
    Frank B. 456

PEEK
    Charles O. 458

PERCIVAL
    Arthur E. 131

PERRONE
    Frank 434, 459

PETERSON
    James R. 455

PHILLIPS
    Marshall L. 459
    Thomas "Tom Thumb" 18, 32, 35

PICKERT
    Eldon B. 458

PLANNING
    ABC-1 .... 8, 9, 56, 83
    ABC-2 .... 8, 345
    ORANGE 10, 13, 15, 78
    RAINBOW 8, 9, 13, 15, 16, 25, 83, 84

PLYLER
    William C. 460

PORCH
    Douglas 278

PORTAL
 Sir Charles.............. 47, 50, 74, 84, 87, 168, 247

PORTL
 Gus D.. 459

POUND
 Sir Dudley... 47

POWELL
 Hugh S... 457

PRCHAL
 Louis A... 460

PUNDZAK
 Stephen T.. 117, 460

QUEZÓN
 Manuel L.. 127

RANG
 Francis B. 89, 264, 265, 419, 420, 429, 445, 446, 447, 455

RENDELL
 Robert F... 458

RHOADES
 Richard L. 265, 455

RICHARDS
 William G. 459

RICHARDSON
 George S. 265, 266, 267, 268, 271, 273, 318, 327, 341, 342, 351, 385, 388, 398, 459

RITCHIE
 Neil....................................... 233, 263, 267, 292

ROMMEL
 Erwin 31, 43, 73, 77, 91, 96, 119, 120, 121, 125, 126, 127, 129, 134, 156, 213, 231, 240, 246, 261, 267, 276, 277, 279, 291, 292, 298, 305, 306, 307, 308, 310, 319, 323, 329, 346, 351, 357, 363, 371, 377, 378, 381, 383, 384, 386, 397, 406, 427

ROOSEVELT
 Franklin D. 1, 5, 6, 7, 9, 10, 11, 14, 15, 17, 21, 31, 34, 35, 36, 37, 39, 40, 41, 43, 44, 45, 46, 47, 49, 50, 51, 52, 54, 71, 74, 75, 79, 82, 83, 84, 85, 95, 96, 98, 99, 100, 101, 102, 121, 125, 127, 128, 131, 133, 135, 138, 139, 141, 153, 155, 157, 158, 161, 164, 166, 168, 169, 175, 184, 185, 187, 188, 191, 192, 193, 195, 197, 199, 201, 213, 214, 215, 216, 217, 218, 220, 233, 234, 235, 236, 237, 247, 248, 257, 259, 274, 275, 278, 279, 293, 295, 296, 305, 311, 312, 320, 321, 322, 323, 327, 328, 329, 339, 342, 344, 345, 346, 348, 350, 352, 355, 357, 364, 368, 369, 370, 371, 372, 373, 374, 375, 376, 379, 391, 397, 401, 404, 405, 414, 415, 416, 463, 464, 465, 466, 467, 468, 471

ROSANSKI
 Stanley B.. 459

ROSE
 Joseph.. 257

ROYCE
 Ralph S. "Scotty" ... 96, 132, 220, 297, 312, 325, 460

RUPPERT
 Charles C.. 460

SAIA
    Joe C. 455

SALMON
    Charles E. Jr. 459

SANDERS
    Richard 189, 242, 460

SAYRE
    Francis 127

SCANLON
    Martin F. 133

SCHMIDT
    Ferdinand R. 313, 318, 455

SCHWANABECK
    Alfred L. 460

SECOND FRONT
    70, 115, 135, 152, 153, 161, 185, 191, 193, 209, 275, 321, 337, 345, 364, 469

SEIDEL
    Karl A. 455

SHARP
    Fred D. 82

SHAW
    Charles A. 459

SHEA
    Walter L. 419, 420, 444, 445, 447, 459

SHUMAKER
    Floyd N. 272, 455

    Thomas A. 458

SIBERT
    James 238, 240, 244, 262, 264, 360, 361, 424, 445, 448, 457

SMITH
    Francis H. 90, 460
    Lyman S. 456
    Ray 221
    Thomas A. 457
    Wilfred J. 272, 456

SOLOMON
    James S. 457

SOMERVELL
    Brehon B. 163

SOONG
    T. V. 5, 6, 7, 37, 141, 199, 216, 234, 235, 236, 258, 259, 264, 269, 322

SOUKUP
    Reynold 120

SPAATZ
    Carl A. 16, 345, 349

SPARKS
    Edwin R. 458

SPAULDING
    Perley W. 457

SPRUANCE
    Raymond 242, 258

STALIN
    Joseph 47, 97, 135, 161, 234, 321, 323, 328, 330, 339, 345, 346, 350, 357, 358, 364, 369, 370, 371, 372, 373, 375, 376, 378, 382, 383, 397, 401, 405

STANDLEY
    William H. 372, 375

STARK
    Harold R. 7, 8, 10, 40, 41, 50, 157, 163, 176, 414

STEWART
    William 257, 307

STILWELL
    Joseph W. 75, 93, 157, 169, 171, 172, 187, 190, 192, 194, 195, 197, 198, 199, 200, 201, 213, 215, 216, 223, 233, 238, 239, 273, 274, 275, 291, 293, 295, 296, 322, 344

STIMSON
    Henry L. 6, 7, 12, 14, 32, 34, 35, 36, 39, 40, 41, 45, 51, 52, 163, 201, 217, 274, 275, 278, 326, 329

STORY
    Albert H. 27, 194, 238, 241, 243, 248, 400, 427, 460

STRATEMEYER
    George E. 349

STURKIE
    Howard 330, 336, 339, 340

SWARNER
    Walter C. 455

TAULBEE
    Joseph 257

TAYLOR
    Ula H. 443, 444, 455, 456, 458, 459

TEDDER
    Sir Arthur. 49, 74, 197, 261, 271, 296, 358, 375, 389

THOMAS
    Rowan 27, 42, 50, 52, 56, 175, 191, 193, 233, 307

THOMPSON
    John M. 455
    Robert E. 456

TIMBERLAKE
    Edward J. 373
    Patrick W. 358, 359, 389

TOWERS
    John 245

TROYANOWSKI
    Joseph 459

TUNNO
    David A. 455

TURNER
    Richmond K. 10

UHRICH
    George 242, 243, 246, 313, 455

UMSTEAD
    Arnold M. 459

USS
- *Hornet* 123, 175, 183, 196, 258, 372, 414
- *Kittyhawk* 240
- Langley 137
- *Lexington* 135, 163, 202, 215, 219
- *Neosho* 215, 218
- *Saratoga* 52
- *Sims* 218
- *Tippecanoe* 215
- *Yorktown* 131, 163, 202, 215, 219

VANDENBURG
- Hoyt S. 220

VANNESS
- Harold C. 455

VASQUEZ
- Harold R. 57, 460

VENEGAS
- Raul R. 455

VON BOCK
- Fedor 314

VON MANNSTEIN
- Dieter Wilhelm 240

VON RICHTHOFEN
- Wolfram 349

WALKER
- Howard E. 429, 456
- John R. 458

WALSH
- Martin 238, 460
- Robert L. 244

WALTERS
- Louis L. 455

WARD
- Donald 455

WATERS
- Dillon W. 458

WATTS
- Charles H. 457

WAVELL
- Sir Archibald P. 54, 128, 133, 138, 173, 187, 188, 239, 275

WEDEMEYER
- Albert 163

WEINGART
- Edward F. 455

WELFARE
- Douglas S. 457

WELLES
- Sumner 37, 125

WEST
- Wilbur C. 431, 432, 442, 443, 456, 466

WHANG
- Ping Han 247, 265, 267, 269, 272, 274, 276, 455

WHITLEY
- Lacey A. 455

WHITLOCK
- George B. 272, 387, 388, 459

WICKLUND
　Harold A. 243, 434, 437, 438, 460

WIGGLESWORTH
　H. E. P. 358

WILCOX
　John R. 386, 423, 445, 448, 459

WILHITE
　Irving J. 398

WILKIE
　Wendell 364, 442

WILKINSON
　John W. 373, 419, 420, 421, 422, 423, 424, 425, 445, 447, 448, 459

WILLIAMS
　Bill 266, 272
　Dewey J. 457
　Douglas H. 455
　Jesse C. 458

WILLIS
　Sidney J. 117, 460

WINANT
　John G. 141

WINECOFF
　John M. 460

WITHAN
　Elmer E. 460

WOODY
　Roy O. 458

WOODYARD
　James C. 457

WRIGLEY
　George 325

WYSONG
　Robert T. 458

YANG
　Chieh 192

YELVINGTON
　James L. 460

YERYOMENKO
　Andrey 339, 360

ZIESEL
　Eugene L. 437, 440, 460

ZUCKERMAN
　Paul 267

ABOUT THE AUTHOR

Edward Clendenin has had a life long interest in aviation. He has a degree in Aeronautical Engineering and was employed by several airframe manufacturers. He worked on the AV-8 Harrier, the F-4 Phantom, the F-15 Eagle, the F-16 Fighting Falcon, the F-18 Hornet, and the F-35 Joint Strike Fighter.

In 1999, he attended a reunion of the 376th Bomb Group Veterans Association and has attended nearly every reunion since then. He has held several volunteer positions with the group, including historian.

He is the author of the book The Mission History of the 376th Bomb Group.

www.ingramcontent.com/pod-product-compliance
Lightning Source LLC
Chambersburg PA
CBHW081331080526
44588CB00017B/2588